Geotechnology of Waste Management

Geotechnology of Waste Management

Issa S. Oweis PhD, PE
Vice President, Converse Consultants East, Caldwell, New Jersey
Adjunct Professor of Civil Engineering, New Jersey Institute of Technology, Newark, New Jersey, USA

Raj P. Khera PhD, PE
Professor of Civil and Environmental Engineering, New Jersey Institute of Technology, Newark, New Jersey, USA

Butterworths
London   Boston   Singapore   Sydney   Toronto   Wellington

First published 1990

British Library Cataloguing in Publication Data
Oweis, Issa S.
Geotechnology of waste management.
1. Waste materials. Landfill disposal
I. Title II. Khera, Raj
628.4'4564

ISBN 0–408–00969–1

Library of Congress Cataloging-in-Publication Data
Oweis, Issa S.
Geotechnology of waste management / Issa S. Oweis, Raj Khera.
p. cm.
Includes bibliographical references.
ISBN 0–408–00969–1 :
1. Sanitary landfills. 2. Waste disposal in the ground.
3. Engineering geology. I. Khera, Raj P. II. Title.
TD795.7.093 1990
628.4'4564—dc20

Photoset by Latimer Trend & Company Ltd, Plymouth
Printed and bound in Great Britain at the University Press, Cambridge

Preface

Until the late 1960s dumping and incineration were the primary methods of waste disposal in the United States. The environmental awareness of the public and the subsequent legislation and regulation governing waste disposal required upgrading of existing waste dumps to sanitary landfills and protection of ground water from contamination. Increases in population and rate of generation of waste put economic pressure on the utilization of existing waste disposal sites with due regard for environmental concerns.

The geotechnical aspect of waste management, which is the topic of this book, is an outgrowth of a graduate course taught in the Department of Civil and Environmental Engineering at the New Jersey Institute of Technology. The purpose of the course is to show the application of principles of geotechnical engineering to the design and upgrading of the existing and new waste disposal facilities. The students in this course had taken at least an introductory course in soil mechanics.

The primary emphasis is on the municipal waste, though other types such as mineral wastes from mining industry, industrial wastes, hazardous wastes and dredging wastes are also included. In developing this text we were guided by our own experience and the experience of others that helped shape the contents of this book. Mathematical treatments are presented when necessary and special emphasis is placed on case histories. The book uses both the US customary units of measurements and SI units. Each chapter ends with a list of notations for that chapter and where practical a summary of the chapter. A book on the geotechnology of waste management could be expected to include detailed treatment of solute transport and ground water flow. We have elected to limit the contents to topics that are routinely practiced by geotechnical engineers in the waste management area.

Chapter 1 describes the forms of wastes. The regulations governing the waste disposal are given in Chapter 2. As the regulations change the classification of the waste and the methods governing its disposal will change. Since the design principles governing the waste disposal are rather new, rapid changes are likely to occur in these principles as the technology develops. The major issues in the selection of sites are economic and political and not geotechnical, as described in Chapter 3.

A review of the pertinent geotechnical properties of soils and wastes is provided in Chapters 4 to 6. Leachate and gas generation are discussed in Chapter 7. Physico-chemical aspects of clay structure, its interaction with waste chemicals including the effect on permeability are described in Chapters 8 and 9.

Modification and utilization of old waste disposal sites for construction of facilities is covered in Chapter 10. The final chapter deals with the design of new facilities including liner design, leachate collection systems, gas management, landfill cover, and assessment of health risks at such facilities. Several case histories are included as part of Chapters 10 and 11.

We express our appreciation to the New Jersey Institute of Technology and Converse Consultants East for their support in the preparation of this book.

Finally, we are grateful to Dr. Astrid K. Khera for reading the manuscript and helping in the preparation of the book.

Raj P. Khera and Issa S. Oweis

Conversion of US customary units to SI units

To convert	To	Multiply by
inches (in)	millimetres (mm)	25.40
inches (in)	centimetres (cm)	2.540
inches (in)	metres (m)	0.0254
feet (ft)	metres (m)	0.305
miles (miles)	kilometres (km)	1.61
yards (yd)	metres (m)	0.91
square inches (sq in)	square centimetres (cm^2)	6.45
square feet (sq ft)	square metres (m^2)	0.093
square yards (sq yd)	square metres (m^2)	0.836
acres (acre)	square metres (m^2)	4047
square miles (sq miles)	square kilometres (km^2)	2.59
cubic inches (cu in)	cubic centimetres (cm^3)	16.4
cubic feet (cu ft)	cubic metres (m^3)	0.028
cubic yards (cu yd)	cubic metres (m^3)	0.765
pounds (lb)	kilograms (kg)	0.453
tons (ton)	kilograms (kg)	907.2
one pound force (lbf)	newtons (N)	4.45
one kilogram force (kgf)	newtons (N)	9.81
pounds per square foot (lb/ft^2)	newtons per square metre (N/m^2)	47.9
pounds per square inch (lb/in^2)	kilonewtons per square metre (kN/m^2)	6.9
gallons (gal)	cubic metres (m^3)	0.0038
acre-feet (acre-ft)	cubic metres (m^3)	1233
gallons per minute (gal/min)	cubic metres per minute (m^3/min)	0.0038
newtons per square metre (N/m^2)	pascals (Pa)	1.00
kip	pounds	1000
ton	kip	2.0

Contents

Chapter 1

Forms of waste

1.1 Generation of waste

Waste may be generated in the form of solids, sludges, liquids, gases, and any combination thereof. With increasing industrialization the quantity of waste has increased immensely. Depending on the source of generation, some of these wastes may degrade into harmless products whereas others may be non-degradable and hazardous. Approximately 10–15% of the waste generated in the USA is considered hazardous (Subtitle D Study, 1986). Hazardous wastes pose potential risk to human health and living organisms. Such wastes are not degradable and may have a cumulative detrimental effect.

The Resource Conservation and Recovery Act (RCRA) non-hazardous waste quantities and their primary disposal techniques are given in Table 1.1. Land disposal in the form of landfills, surface impoundment, land application, deep well injection, etc., is the most common form of waste management. About 90% of the off-site disposal of hazardous waste in the United States is on land (US Congress, 1983).

Table 1.1 Annual RCRA non-hazardous waste quantities

Waste type	*Quantity* (× 10^6 ton)	*Disposal method* [a]							
		1	*2*	*3*	*4*	*5*	*6*	*7*	*8*
Municipal	133	×	×	×	×				
Household hazardous	0.1–0.001	×				×	×		
Sewage sludge (dry wt basis)	8.4	×	×	×				×	×
Water treatment sludge	0.21–0.83	×							
MSW combustion ash	2.3	×				×	×		
Industrial non-hazardous	430	×						×	×
Small quantity generator hazardous (< 1000 kg/month) [b]	0.66	×							
Construction and demolition	31.5	×							
Mining (excludes coal mining waste)	1400							×	
Agricultural, gallons/day	1000							×	
Oil and gas (production, processing, etc.), gallons/day	6250							×	

[a] 1, land disposal; 2, ocean disposal; 3, incineration; 4, resource recovery; 5, sewer system; 6, septic tanks; 7, lagoon/surface impoundment; 8, land application.
[b] On-site disposal (sewer, recycling, treatment), off-site disposal (recycling, solid waste facility, or Subtitle C).
Source: Subtitle D Study, 1986.

1.2 Municipal waste

Municipal landfills are heterogeneous mixtures of wastes which are primarily of residential and commercial origin. The composition of the fill materials will depend on the type of commerce and industry. Typically, a municipal landfill consists of food and garden wastes, paper products, plastics and rubber, textiles, wood, ashes and the soils used as cover material. More than 50% of the municipal solid waste (MSW), by weight, is paper and yard waste. Most of the waste is landfilled, and the amount of production of MSW is about 3 pounds per capita per day. Larger objects such as tree stumps, refrigerators, automobile bodies and demolition waste may also be present in MSW. The proportions of these materials may vary from one region to another. The household hazardous waste is mostly landfilled or discharged into sewer or septic systems. Such waste may include cleaners, automotive products, paints and garden products. The constituents of the landfill may vary considerably when wastes from different countries are considered.

The refuse composition is determined, if necessary, by manually sorting large quantities of refuse (0.5–1 ton) over several days or weeks. In order for this method to be meaningful several determinations are needed for various locations of an existing landfill or waste stream. Four or five experienced persons could separate and weigh 1 ton of refuse during a day's work. There is no significant statistical difference in the reliability of the composition determinations for samples above 200 lb (Klee and Carruth, 1970). The other technique is visually to observe the refuse in test pits or waste stream and adjust percentages of national or local averages. A comparison of municipal waste from the USA and countries on four other continents is given in Table 1.2.

Developing countries like India show a lower proportion of salvageable material because it is more economical to recycle these materials. It is interesting to note that as a country becomes more prosperous the proportion of salvageable materials in the refuse increases. This trend may, however, reverse as the available land for disposal of waste becomes more scarce and mandatory recycling legislation increases.

Table 1.2 Typical municipal waste composition percentage by weight

Component	*Country or city*[a]																	
	1	2	3	4	5	6	7	8	9	10	11	12	13	14	15	16	17	18
Metals		1	1	9	9	3	3		3		5	7		2	8	10	7	5
Paper, and paper board	37[b]	25	5	45	42	33	3	38	10	14[b]	22	32	14[b]	8	30	37	54	31
Plastics			1	2	2			8				3				7	2	
Rubber, leather, wood		7	1		3[c]	7	2	12	4		3[c]	7		3	1	6	2	4
Textile		3		1	1	10						4		4	2	2	2	2
Food and yard waste	45[d]	44	45	25	34	15	60	18	74	56[d]	20	36	50[d]	25	16	26	23	16
Glass		1	1	11	8	10	2		7		6	4		3	8	10	5	13
Non-food inorganic	18[e]	19	46	7	1	22	30	24	2	30[e]	43	7	21[e]	55	35	2	5	29

[a] 1, Australia; 2, Bangkok; 3, Beijing; 4, Berkeley, California; 5, Cincinatti, Ohio; 6, Hong Kong; 7, Jakarta, Indonesia; 8, Japan; 9, Korea; 10, Madras, India; 11, New York City; 12, Singapore; 13, Spain; 14, Taiwan; 15, UK; 16, USA; 17, Wayne, NJ; 18, West Germany.

[b] Metal is included under paper and paper board.

[c] Wood only.

[d] Includes wood, bones, etc.

[e] Includes glass, coconut, shells, fibres, etc.

Sources: Sargunan *et al.*, 1986; Aziz, 1986; Hillenbrand, 1986; Subtitle D Study, 1986; Waste Age, 1986.

Thermal destruction and resource recovery is gaining favour as a means of reducing the volume of municipal waste. Dry ash is roughly 20–25% by weight of the unburned waste. Both bottom ash and fly ash are produced from burning of municipal waste. These ashes may contain trace amounts of polychlorinated biphenyls (PCBs), metals, dioxins, polyvinyl chloride (PVC) and other toxic compounds. Because of the sorption characteristics of fly ash, many of these compounds are absorbed by the fly ash. Contaminants leaching from landfills containing such fly ashes may reach groundwater (Clapp *et al.*, 1987).

1.3 Mineral waste

Mineral wastes are generated during the extraction of materials such as metals, fuels, fertilizers, chemicals and clays. Extensive amounts of wastes are produced and stored near mining or extraction operations, which can become a source of soil, water and air pollution if not properly managed. With increasing demand on natural resources, ores of poorer quality are being mined and processed, generating ever larger amounts of waste. Mineral wastes are generally not considered hazardous.

In the mining operation the country or host rock containing the ore is crushed and ground in bar mills or rod mills to release the concentrated mineral being mined. The waste or gangue contains both sharp angular sand-size particles (coarse-grained materials) and smaller clay and colloidal-size particles (fine-grained materials). Before disposal, the fine-grained materials are separated from the coarse-grained fraction. The sandy materials are called tailings and the finer materials when suspended in water are called slimes. These wastes are disposed of by impoundment behind tailing dams, injection into deep wells, and backfilling into used mines. The coarse-grained fraction can be backfilled into the mines, but not the fine-grained fraction as it may cause flooding of the mine.

The coarse fraction of the waste or the tailings, if not hazardous, can be used as construction material. The major difficulty is the disposal of fine-grained material suspended in water. The volumes of these slimes are large and they may have little shear strength. The major challenge to the geotechnical engineer is the reduction of these enormous volumes and the stabilization of the weak materials. In addition, the tailings contain processing chemicals, dissolved salts, detergents and heavy metals, which may pose short and long term threat to the quality of surface and ground water.

1.3.1 Coal and ashes waste

With increasing demand for coal and improved mining technology less productive seams are being mined, resulting in larger amounts of coal and related wastes consisting of sandstone, siltstone, shale, mudstone, clay, etc., in variable proportions. Coarse coal waste is generated during mining and separation and consists of particles larger than 0.02 in (0.5 mm), whereas rock particles larger than 3 in (76 mm) are known as mine rock. The particle shapes are elongated, and break under pressure to form smaller particles. The larger surface area of such particles makes them more susceptible to weathering which is caused by oxidation of pyrite in shale and carbonaceous rocks.

Fine components of coal refuse are in the form of slurries produced during washing of coal and are disposed of behind impoundment dikes constructed out of coarse

Table 1.3 Composition of ash

Constituent	*Fly ash* (%)	*Bottom ash* (%)
Silica (SiO_2)	34–58	21–60
Alumina (Al_2O_3)	20–40	10–37
Iron oxide (Fe_2O_3)	4–24	5–37
Calcium (CaO)	2–27	0–22
Magnesium (MgO)	1–6	0–4
Sulphates (SO_3)	0.5–4	
Water-soluble alkalis	2–6	0–7

Source: Digioia *et al.*, 1977; 1977; Gray and Lin, 1972; Bowders *et al.*, 1987; Edil *et al.*, 1987)

refuse. The relative density of fine waste is lower than that of the coarse waste. If coarse and fine refuse are mixed they form an unstable material and require disposal like slurries. In strip mining additional waste is generated from the removal of overburden before the recovery of coal. This waste may be disposed of separately from the coal waste itself.

In the generation of electricity the use of coal is increasing. About 12–15% of coal ends up as fly ash and bottom ash. Fly ash produced by burning sub-bituminous coal or lignite is called class C. This type of fly ash contains lime and other compounds which impart to it self-cementing properties. This group of fly ash is more common in the western USA. Fly ash formed by burning bituminous coal or anthracite is known as class F. This category of fly ash is primarily generated in the eastern USA. Though it is pozzolanic it shows little self-cementing properties.

In the past fly ash was generally disposed of by a wet method in which it is allowed to settle in ponds. Because of environmental concern the emphasis is now shifting to dry disposal where moisture content of the fly ash is controlled (known as conditioning) and it is compacted like a soil. Typical composition of fly ash is given in Table 1.3. The engineering properties of fly ash are influenced most by free lime and unburnt carbons.

The other major by-product at power plants is flue gas desulphurization sludge waste generated from scrubbers. Some of these materials can be recycled and others require proper disposal.

Certain extraction processes require a considerable amount of energy. Coal may be used to derive the needed energy. At these sites coal, ash and other coal-related wastes are mixed with processing waste.

1.3.2 Phosphatic waste

Phosphate rocks are mined to produce chemical fertilizers. About two-thirds of the phosphate ore, or 'matrix' as it is commonly known, consists of about equal amounts of sand and clay. The sand-size tailings are separated and readily stabilized, but a very large volume of clay suspension is still left for disposal. These clays primarily contain montmorillonite, attapulgite, illite and kaolinite. The large surface area of these clays keeps them in suspension, creating a difficult problem of dewatering and stabilization. Dikes 3–15 m in height are built covering large areas to store clay suspensions with initial solid contents of 2–6%. Sedimentation or free falling of clay particles takes a few days to a few weeks when the solid content is about 10%. Subsequent settlement under self weight of the clay requires 10–100 years for any consequential volume reduction (McVay *et al.*, 1986).

1.4 Industrial waste

Industrial waste may be generated by any of several industries, such as chemical manufacturing and processing, food processing, petroleum refining, plastics and resins, paper and pulp, lumber and wood, and pharmaceuticals. Agricultural waste may include animal waste, irrigation waste, collected field run-off, crop production waste and fertilizers. Some of these wastes may contain various levels of organic and inorganic chemicals and heavy metals.

About 15% of the industrial waste is classified as hazardous. With our increasing understanding of the impact of these wastes on the environment it is likely that more types of waste will be classified as hazardous.

Until 1978 about 80% of industrial waste was disposed of in unlined landfills or in surface impoundments (Brown and Anderson, 1983) as aqueous solutions of organics and inorganics, as organic liquids with organic solutes, and as sludges.

Since November 1981 RCRA prohibits the land disposal of drums containing waste liquids. Even with an adequate liner and waste collection system, bulk liquids may not be disposed of in landfills.

1.4.1 Pulp and paper mill waste

Paper mill sludge is composed of wood bark particles, fibres and fibre particles, talc, silt and clay. It has a consistency of fibrous organic material, with a high initial water content and very unstable structure, and shows large settlements under applied loads.

1.4.2 Flue gas waste

Current regulations require that particulate matter and sulphur dioxide be removed from flue gases. Granular particle sizes which are greater than 100 μm settle out under gravity. Dust particles range in size from 1 to 100 μm. They consist primarily of metal oxides such as rock, ash with the relative density, G_s, between 2 and 3, smaller proportions of organics ($G_s \approx 1$) and heavy metals and their oxides ($G_s > 5$). Particles smaller than 1 μm are caused by dispersion of solid or liquid particles and are called fumes. The range of sizes for various industrial particles is given in Table 1.4 (Mantell, 1975).

Sources for particulate matter are the processing of iron ore, the manufacturing of cement, lead smelting plants, electric furnaces, and fly ash from the combustion of coal, etc. The removal of particulate matter may be done by mechanical separators, fabric filter collectors, electrostatic precipitators or wet scrubbers. The type of equipment used is determined by the nature of the process. Mechanical separators are more efficient for high gas flow, fabric filters are found to be effective in removing zinc oxide fumes from lead smelting, electrostatic precipitators have performed well in collecting fly ash, and wet scrubbers are commonly selected where both gas and particulate matter are present.

The removal of sulphur oxides from flue gases or flue gas desulphurization (FGD) may be through contact with lime or limestone slurries which absorb sulphur oxides.

Table 1.4 Sizes of particles in industrial waste

Type of materials	*Particle size* (μm)
Carbon black	0.01–0.2
Zinc oxide smoke	0.01–0.3
Metallurgical fumes	0.01–2.5
Oil smoke	0.03–1.0
Metallurgical dust	0.7–100
Pulverized coal fly ash	1–100
Cement dust	6–200
Pulverized coal	10–200
Ground limestone	20–800

Source: Mantell, 1975.

This process of wet scrubbing yields waste sludges consisting of calcium sulphate (gypsum), calcium sulphite and often unreacted lime or limestone requiring disposal. Sulphite sludges contain a huge amount of water requiring large areas for land disposal, and their thixotropic nature makes them difficult to dispose of. Sulphate-rich sludges can be mixed with fly ash and placed in compacted fills. Alternatively, double-alkali methods using water-soluble absorbers such as sodium hydroxide, sodium carbonate or sodium sulphite are used. Double-alkali waste products are easier to dewater than the FGD wastes produced using lime or limestone.

Other chemicals that may be present include arsenic, selenium, mercury and heavy metals, further increasing the potential danger of water pollution.

1.5 Dredging waste

Dredged materials are bottom sediments or materials that have been dredged or excavated from navigable waters. Dredged materials comprise natural sediments such as rocks, gravel, sands, silts and clays. These materials may be contaminated by various types of waste or runoff from land. The dredging generates a large amount of waste. A large proportion of this waste is disposed of in the ocean.

The land disposal of dredge waste is a recent phenomenon. It is the result of environmental concerns and regulations. Components which are polluted or cannot be released economically in open waters without causing unacceptable turbidity require land disposal. About 20% of dredged materials is placed in landfills. These are filled hydraulically, placing slurry in diked containment areas. The water content of these slurries ranges from 200 to 300%. They consist primarily of organic silts and clays and may be contaminated with heavy metals and other pollutants.

The materials within a dredged landfill may vary from fairly well distributed to rather inhomogeneous and segregated (Bromwell, 1978). The grain size characteristic of a dredge deposit can be controlled by the discharge velocity, the water level in the deposition pond and the ratio of clay solid to water. If these conditions are properly controlled, depending on the desired results the deposit can be made more stable, have larger storage volume or provide a source of good backfill material. In placing dredged materials the primary effort is toward reducing its volume and not controlling its quality for producing a fill of desirable engineering properties. For detailed procedures used in the placement and control of hydraulic fills see Turnbull and Mensur (1972), and for its engineering characteristic and classification see Whitman (1970).

The dredge fines are clays and silts of high compressibility. Because of their high compressibility and low strength they are the least desirable materials from a geotechnical engineering viewpoint. The soil deposited from the lowest water–clay ratio, because of its higher clay content, will have the smallest bearing capacity (Bromwell, 1978).

1.6 Hazardous waste

About 90% of industrial hazardous wastes in the USA is produced as liquids; 60% of this waste is organic liquids and 40% inorganic liquids. Four classes of hazardous waste have been classified (US Environmental Protection Agency, 1980):

Type 1 Aqueous inorganic
Type 2 Aqueous organic
Type 3 Organic
Type 4 Hazardous sludges, slurries and solids

Water is the solvent in type 1. Solutes are mostly inorganics such as those produced by the electroplating industry like cadmium, cyanides and metals. Organics (carbon compounds) are the solutes for type 2 and are produced by the pesticide industry. In type 3, both the solutes and the solvents are organics such as oil-based paints and motor oil. Type 4 includes sludges generated by various dewatering, filtration or treatment processes. Examples are sludges from the petroleum refining industry and treatment plants. Industrial wastes are the major source of hazardous waste (US Environmental Protection Agency, 1980). Table 1.5 shows the type of hazardous substances produced by various industries.

Table 1.5 Representative hazardous substances within the industrial waste stream

Industry	*As*	*Cd*	*CH*[a]	*Cr*	*Cu*	*CN*	*Pb*	*Hg*	*MO*[b]	*Se*	*Zn*
Battery		×		×	×						×
Chemical manufacturing			×	×	×			×	×		
Electrical and electronic			×		×	×	×	×		×	
Electroplating and metal finishing		×		×	×	×					×
Explosives	×				×		×	×	×		
Leather				×					×		
Mining and metallurgy	×	×		×	×	×	×	×		×	×
Paint and dye		×		×	×	×	×	×	×	×	
Pesticide	×		×			×	×	×	×		×
Petroleum and coal	×		×				×				
Pharmaceutical	×							×	×		
Printing and duplicating	×			×	×		×		×	×	
Pulp and paper								×	×		
Textile				×	×				×		

[a] Chlorinated hydrocarbons and polychlorinated biphenyls.
[b] Miscellaneous organics such as acrolein, chloropicrin, dimethyl sulphate, dinitrobenzene, dinitrophenol, nitroaniline and pentachlorophenol.
Source: Matrecon Inc. (1980).

Summary

Waste is generated in solid, sludge, liquid and gaseous forms. The amount of waste generated and the types of hazardous materials present in the waste stream increase with increasing industrialization of the country. As the new regulations governing waste disposal take effect it is becoming difficult to find suitable sites for waste disposal. Future emphasis must be on waste reduction and recycling rather than waste disposal.

References

Aziz, M. A. (1986) Surface disposal of refuse: Geotechnical considerations, technologies and environmental impacts, *Int. Symp. on Env. Geot.*, Vol. 1. Ed. Fang, H. Y., pp. 81–90

Bowders, J. J., Usmen, M. A. and Gidley, R. A. (1987) Stabilized fly ash for use as low-permeability barriers, *Geotechnical Practice for Waste Disposal '87*, University of Michigan, Ann Arbor, Michigan, June 15–17, Geot. Special Publ. No. 13, pp. 320–333

Bromwell, L. G. (1978) Properties, behavior and treatment of waste fills, *ASCE, Met. Section, Seminar: Improving Poor Soil Conditions*, Oct. New York

Brown, K. W. and Anderson, D. C. (1983) *Effect of Organic Solvents on the Permeability of Clay Soils*, EPA-600/S2-83-061

Clapp, T. L., Kosson, D. S. and Ahlert, R. C. (1987) Leaching characteristics of residual ashes from incineration of municipal solid waste, *Proc. Second International Conference on New Frontiers for Hazardous Waste Management*, EPA/600/9-87/018F, pp. 19–27

DiGioia, A. M., Meyers, J. F. and Niece, J. E. (1977) Design and construction of bituminous fly ash disposal sites, *Proc. Geot. Practice for Disposal of Solid Waste Materials*, ASCE, University of Michigan, Ann Arbor, pp. 267–284

Edil, T. B., Berthous, P. M. and Vesperman, K. D. (1987) Fly ash as a potential waste liner, *Geotechnical Practice for Waste Disposal '87*, Proc. Spec. Conf. Geot. Special Publ. No. 13, Ed. Woods, R. D.

Gray, D. H. and Lin, Y. K. (1972) Engineering properties of compacted fly ash, *J. Soil Mech. and Found. Div.*, **98,** No. SM4, 361–380

Hillenbrand, S. (1986) One billion people can't be wrong, *Waste Age*, Oct.

Klee, A. J. and Carruth, D. (1970) Sample weights in solid waste composition, *J. Sanitary Eng. Div., Proc. ASCE*, **SA4,** August, 945–953

Mantell, C. L. (1975) *Solid Wastes: Origin, Collection, Processing and Disposal*, John Wiley & Sons, New York

Matrecon Inc. (1980) *Lining of Waste Impoundment and Disposal Facilities*, EPA 530/SW-870C

McVay, M., Townsend, F. and Bloomquist, D. (1986) Quiescent consolidation of phosphatic waste clays, *J. Geot. Eng., ASCE,* **112,** No. 11, 1033–1052

Sargunan, A., Mallikarjun, N. and Ranapratap, K. (1986) Geotechnical properties of refuse fills of Madras, India, *Int. Symp. on Env. Geot.*, Vol. 1, Ed. Fang, H. Y., pp. 197–204

Subtitle D Study (1986) Phase 1 Report, EPA/530-SW-86-054, USEPA

Turnbull, W. J. and Mansur, C. I. (1973) Compaction of hydraulically placed fills, *J. Soil Mech. and Found. Div., ASCE,* **99**, No. SM11, 939–956.

US Congress (1983) *Technologies and Management Strategies for Hazardous Waste Control*, Office of Technology Assessment, Washington DC

US Environmental Protection Agency (1980), *Classifying Solid Waste Disposal Facilities*, USEPA, Cincinnati, Ohio, March, SW-828

Waste Age, An introduction, Jan. 1986, p. 124

Whitman, R. V. (1970) Hydraulic fills to support structural loads, *J. Soil Mech. and Found. Div., ASCE,* **96,** No. SM1, 23

Chapter 2

Regulations governing solid waste disposal

2.1 Introduction

In the last two decades many of the industrialized countries have been regulating generation, disposal and management of waste. In the USA the primary emphasis has been on waste disposal rather than waste reduction. In Western European countries Governments have played a much larger role in pollution management. The primary emphasis has been on waste reduction through reuse, recycling, clean technology development, etc. (Piasecki and Davis, 1987). In recent years many of the States in the USA are finding that they have no place to safely dispose of the enormous amounts of waste being generated and they are enacting regulations requiring recycling, waste reduction, and alternative technology development as well.

In this chapter, primarily the geotechnical aspects of the federal regulations governing the existing and new waste disposal facilities are presented. As far as the individual states are concerned some relevant regulations are included for the state of New Jersey.

2.2 Federal standards

The Federal regulations governing waste are given in the Resource Conservation and Recovery Act of 1976 as amended, and Code 40 of the Federal Regulations. The objective of the act is '. . . to promote the protection of health and environment and to conserve valuable materials and energy resources . . .' The Act is administered by the Office of Solid Waste of the Environmental Protection Agency (EPA). In order to comply with RCRA, the regulations by EPA and the State regulations should be carefully studied. The Federal EPA regulations are explained in Title 40 of the Code of Federal Regulations (40 CFR). The regulations are often updated; amendments and updates are published in the Daily Federal Register. A collection is published annually in a revision edition of CFR.

RCRA contains specific definitions of terms that are mentioned in the regulations for solid waste management. The reader should note that the term 'solid waste' under RCRA is broader than just garbage. It includes liquids and contained gases. In order for material to be termed 'hazardous waste', it should first be 'solid waste'. EPA criteria for classifying a waste as *hazardous* are based on several factors such as flamability, corrosion, toxicity, persistence or degradability in nature, its potential of accumulation in living tissue, substantial or potential hazard to human health or the environment when improperly treated, stored, transported, disposed or otherwise managed, etc.

Solid waste under RCRA covers all materials (liquids and contained gas included). Exceptions are domestic sewage, *in situ* mining waste, special nuclear waste and non-point source discharge industrial waste. The term 'solid waste management' means the systematic administration of activities that provide the storage, source, separation, collection, transportation, transfer, processing, treatment and disposal of solid waste (RCRA Section 1004 (28)).

2.3 Federal guidelines

Subtitle C of the RCRA establishes a detailed 'cradle to grave' provision for hazardous waste management. The technical requirements for EPA regulations are explained in Title 40 of the CFR, parts 260–270. If the solid waste is ruled non-hazardous, the State programs developed under Subtitle D, RCRA apply. Subtitle D, as it currently exists (1988), provides Federal guidelines to be implemented by the states. The objective is to 'assist in developing and encouraging methods for the disposal of solid waste which are environmentally sound . . .'. Under the authority of RCRA, the EPA in 1979 issued the Criteria of Classification of Solid Waste Disposal Facilities and Practices (40 CFR 257). The Criteria contain Federal guidelines to be enforced by the states, local governments, or citizen suits. Under the Criteria, a state would agree to develop a plan for prohibiting new open dumps and upgrading or closing existing open dumps. The Criteria provide minimum environmental performance standards for sanitary landfills. Facilities violating the Criteria are termed open dumps. The Hazardous and Solid Waste Amendments (HSWA) to RCRA in 1984 required EPA to conduct a Subtitle D study to determine if the current criteria are adequate to protect human health and the environment from groundwater contamination.

Subtitle D wastes are all solid wastes under RCRA that are not subject to hazardous waste regulations under Subtitle C. Some hazardous materials are excluded from Subtitle C and come under Subtitle D. These are household hazardous waste (HHW) and small quantity generator solid waste (SQG); that is, those generating waste between 100 and 1000 kg/month. Before the amendment of 1984, generators of less than 1000 kg/month were exempt from RCRA regulations. HHW includes automotive oil and fuel additives, drain openers, oven cleaners, wood and metal polishes and cleaners, grease and rust solvents, carburetor and fuel injection cleaners, air conditioning refrigerants, starter fluids, paint thinners, paint removers, adhesives, herbicides, pesticides, fungicides, and wood preservatives. SQG mostly includes used lead–acid batteries (62%) and spent solvents (18%). In addition to HHW and SQG, the following wastes are excluded from the regulations as being hazardous (40 CFR 261.4):

1. Agricultural waste; animal waste; fertilizers, crop produce, etc.
2. Fly ash, bottom ash, sludge waste, flue gas emission, control waste from combustion of coal or other fossil fuels.
3. Oil and gas waste; brine and drilling muds.
4. Mining waste including coal and phosphate rock; products of crushing; screening, washing and flotation activities; mine overburden (including uranium mines) returned to the mine site.

5. Construction and demolition debris, lumber, roofing and sheeting scraps, broken concrete, asphalt, brick, stone, wallboard, glass, and other materials.
6. Sludge from water and wastewater treatment facilities.
7. Cement kiln dust.

PCBs (polychlorinated biphenyls) are excluded from RCRA hazardous waste regulations although RCRA requires testing for them. However, they are regulated under the Toxic Substance Control Act (40 CFR 761).

2.4 Subtitle C regulations

2.4.1 Identification

A solid waste is termed hazardous if:

1. The waste is listed as hazardous waste.
2. The waste meets specific criteria on ignitability, corrosivity, reactivity and toxicity.

2.4.2 Listed wastes

The basis for listing wastes as hazardous are the organic and inorganic constituents present that are listed in Appendices VIII of 40 CFR 261. The listed hazardous wastes include generic wastes from non-specific sources (28 sources as of 1986, including common solvents and metal treatment wastes, etc.; 40 CFR 261.31), wastes from specific sources (85 sources as of 1986, including wood preservatives industry, organic chemical industry, etc.; 40 CFR 261.32), and 'discarded commercial chemical products, off-specification species, container residues, and spill residues thereof'. The last group of wastes (663 products as of 1986) are considered acutely hazardous (197 products) or toxic wastes (466 products) only if they are discarded. They are not considered hazardous if they are stored or used in manufacturing a product (40 CFR 261.33). The group includes products such as naphthalene, trichloroethylene, phenol, acetone and benzene.

To test for these constituents, EPA has specified test methods for organic and inorganic chemicals.

2.4.3 Specific criteria

A solid waste is ruled hazardous if it is ignitable, corrosive, reactive or 'EP toxic' (40 CFR 261.21–263.24).

2.4.4 Ignitability

If a representative sample of a solid waste is liquid, it is ruled ignitable if it has a flash point of $<60°C$ (140°F). If the waste is not liquid, it is ruled ignitable if it is capable, under standard temperature and pressure, of causing fire through friction, adsorption of moisture, or spontaneous chemical changes and, when ignited, burns so vigorously and persistently that it creates a hazard.

2.4.5 Corrosivity

A solid waste is ruled corrosive if its representative sample is aqueous and has a solid waste 2⩾ pH ⩾12.5. A solid waste is also ruled corrosive if its representative sample is liquid and it corrodes steel at a rate greater than 6.35 mm (0.25 in) per year at a test temperature of 55°C (130°F). The corrosivity criteria do not cover wastes that do not contain liquids.

2.4.6 Reactivity

A solid waste is ruled reactive if:

1. It is normally unstable and readily undergoes violent change without detonating; reacts violently with water; or when mixed with water forms potentially explosive mixtures, or generates toxic gases, vapours or fumes in a quantity sufficient to present a danger to human health or the environment.
2. It is readily capable of detonation or explosive decomposition or reaction at standard temperature and pressure.
3. It is capable of detonation or explosive reaction when subjected to a strong initiating source or when heated under confinement.
4. It is a cyanide or a sulphide-bearing waste which, when exposed to pH conditions between 2 and 2.5, can generate toxic gases, vapours or fumes in a quantity sufficient to present a danger to human health or the environment.

2.4.7 EP toxicity

A solid waste is ruled hazardous if any of the contaminants exceeds the maximum concentration given in Table 2.1, which is 100 times the limit for safe drinking water.

The determination of concentrations is based on the 'extraction procedure' (EP) toxicity test. In broad terms, the procedure requires separation of a minimum 100 g sample into solid and liquid phases. The solid portion is then placed in an extractor with 16 times its weight of deionized water, and 0.5N acetic acid is added, if necessary, to maintain a pH of 5 ± 0.2. Extraction is performed by agitating the mixture for 24 h.

Table 2.1 Maximum concentration of contaminants for characteristics of EP toxicity

Contaminant	*Maximum concentration* (mg/l)
Arsenic	5.0
Barium	100.0
Cadmium	1.0
Chromium	5.0
Lead	5.0
Mercury	0.2
Selenium	1.0
Silver	5.0
Endrin (1,2,3,4,10,10-hexachloro-1,7-epoxy-1,4,4a,5,6,7,8,8a-octahydro-1,4-endo,endo-5,8-dimethanonaphthalene)	0.02
Lindane (1,2,3,4,5,6-hexachlorocyclohexane, γ-isomer)	0.4
Methoxychlor (1,1,1-trichloro-2,2-bis-(*p*-methoxyphenylethane)	10.0
Toxaphene ($C_{10}H_{10}Cl_8$, technical chlorinate, camphene, 67–69% chlorine)	0.5
2,4-D (2,4-dichlorophenoxyacetic acid)	10.0
2,4,5-TP-Silver (2,4,5-trichlorophenoxypropionic acid)	1.0

Deionized water is then added and the material in the extractor is separated into its liquid and solid phases using a specified filtration procedure. The extract is then analysed for the contaminants in Table 2.1.

2.5 Operation and closure

Landfills, waste piles and surface impoundments are the three common facilities for disposal of solid wastes. RCRA requirements distinguish between facilities that were in existence as of 19 November 1980 and new facilities. A hazardous waste facility in existence on 19 November 1980 qualifies for 'interim status'.

2.5.1 Interim status facilities

The regulations, in broad terms, require monitoring of the groundwater during the operation of the facility, a final cover for closure and maintenance of the final cover for post-closure (40 CFR 265).

2.5.2 Groundwater monitoring

The regulations require a minimum of one hydraulically up-gradient monitoring well and a minimum of three hydraulically down-gradient monitoring wells. The objective is to determine the facility's impact on the groundwater in the uppermost aquifer. The purpose of the up-gradient wells is to determine the background groundwater quality since such wells are not affected by the facility. The purpose of the down-gradient wells is to detect any 'statistically significant amounts of hazardous waste or hazardous waste constituents that migrate from the waste management area to the uppermost aquifer'. Testing is required for the parameters in Table 2.2 to characterize the suitability as drinking water supply.

Parameters establishing groundwater quality are chloride, iron, manganese, phenol, sodium and sulphate.

Indicators of groundwater contamination are pH, specific conductance, total organic carbon and total organic halogen.

Tests for the quality parameters are to be conducted at least annually while contamination indicators are tested at least semi-annually. Initial background concentrations of all the parameters for all the wells are determined quarterly for 1 year after the effective date of the regulations. Groundwater monitoring must continue through the active life of the facility and for 30 years after closure. The owner or operator can obtain a partial waiver from some or all the requirements, if he can demonstrate that there is a 'low potential for migrating hazardous waste or hazardous waste constituent from the facility via the uppermost aquifer to water supply wells (domestic, industrial or agricultural) or to surface water'. The groundwater monitoring requirements may also be waived for surface impoundments if such impoundments are used to neutralize wastes that are hazardous, or listed as hazardous, solely because of their corrosivity. If hazardous constituents are ruled to have entered the groundwater, the regulations require quarterly testing for groundwater quality until final closure. No remedial action is required by the regulation as a result of the monitoring program.

Table 2.2 EPA interim primary drinking water standards

Parameter	*Maximum level*
Arsenic	0.05 mg/l
Barium	1.0 mg/l
Cadmium	0.01 mg/l
Chromium	0.05 mg/l
Fluoride	1.4–2.4 mg/l
Lead	0.05 mg/l
Mercury	0.002 mg/l
Nitrate (as N)	10.00 mg/l
Selenium	0.01 mg/l
Silver	0.05 mg/l
Endrin	0.0002 mg/l
Lindane	0.004 mg/l
Methoxychlor	0.1 mg/l
Toxaphene	0.005 mg/l
2,4-D	0.1 mg/l
2,4,5-TP-Silver	0.01 mg/l
Radium	5 pCi/l
Gross α-radiation	15 pCi/l
Gross β-radiation	4 mrem/year
Turbidity[a]	1 /TU
Coliform bacteria	1 /100 ml

[a] Applicable only to surface water supplies.

2.5.3 Closure

After wastes are no longer disposed, the regulations require the owner or the operator to close the facility to minimize post-closure escaping of hazardous waste into the ground, subsurface water, surface water, or the atmosphere. For landfills the regulations, as a minimum, require the installation of a final cover and a run-off and run-on control (40 CFR 265.310).

The regulations require 'containment' for surface impoundments and waste piles. Containment earthen dikes are required for surface impoundments. When leachate is hazardous an impermeable base is required together with run-off and run-on control systems for a waste pile. The closure requirements for a surface impoundment provide the option for removal of standing liquids, waste and waste residues, liners, and underlying and surrounding contaminated soils. If these requirements cannot be fulfilled, then the rules for landfills apply, including provisions of a final cover.

The closure for waste pile requires removal or decontamination of waste residues, liner, and contaminated subsoils. If full removal or decontamination is impractical, the closure requirements for landfills apply.

2.6 Standards for new hazardous waste facilities

New hazardous waste facilities may not legally operate without an RCRA permit. The conditions for a permit are based on the standards explained in 40 CFR 264. The standards pertain to many parameters, including the location of the facility, its design, and monitoring factors.

2.6.1 Location

The regulations prohibit locating a hazardous waste facility within 61 m (200 ft) of a fault which has had displacements in Holocene time. A hazardous waste facility located in a 100-year flood plain must be designed, constructed, operated and maintained to prevent a washout of any hazardous waste by a 100-year flood. Exceptions to the rule can be made if the owner can demonstrate that waste can be removed safely before flood waters can reach the facility or that no adverse environmental and health effects result if washout occurs (40 CFR 264.18).

2.6.2 Groundwater protection standards

Groundwater is ruled contaminated if it contains hazardous constituents in concentrations exceeding those listed in Table 2.3 (40 CFR 257.4) if the background levels are lower.

The groundwater can also be ruled contaminated if the concentrations of certain organic and inorganic chemicals are significantly above the up-gradient groundwaters which have not been affected by the facility. In this case 'corrective action' may be necessary.

The regulations require 'sufficient' number of wells to represent the groundwater quality up gradient and down gradient. As a first step the owner has to have a plan to monitor for 'indicator parameters' (e.g. pH, specific conductance, total organic carbon or total organic halogen) of waste constituents or reaction products to provide a reliable indication of the presence of hazardous constituents. The owner or operator is required to determine annually the rate and direction of groundwater flow in the uppermost aquifer. If a 'statistically significant' increase in hazardous waste constituents is detected down gradient and is ruled to be affecting the groundwater, compliance monitoring would be required. In this case all monitoring wells must be tested for the hazardous waste constituents listed in Appendix VIII, 40 CFR 261. If the standards are violated based on compliance monitoring, then the owner or operator has to initiate a corrective action program 'that prevents hazardous

Table 2.3 Groundwater quality standards for new hazardous waste facilities

Constituent	*Maximum limit* (mg/l)
Arsenic	0.05
Barium	1.0
Cadmium	0.01
Chromium	0.05
Lead	0.05
Mercury	0.002
Selenium	0.01
Silver	0.05
Endrin	0.0002
Lindane	0.004
Methoxychlor	0.1
Toxaphene	0.005
2,4-D	0.1
2,4,5-TP-Silver	0.01

constituents from exceeding their respective concentration limits' by removing them or treating them in place. The acceptable limits of hazardous waste constituents are specified by EPA as a part of the compliance monitoring program.

2.6.3 Design and closure

Surface impoundments (40 CFR 264, Sub-part K) used for disposal of hazardous waste must have two or more liners and a leachate collection system between such liners. The top liner must be constructed of a material that the waste constituents do not migrate into, and the lower liner must be designed to prevent migration of the constituents through the liner during the active life of the facility including post-closure. The lower liner may be of natural earthen materials with a thickness not less than 3 ft and permeability not more than 10^{-7} cm/s. The permeability for primary and secondary leachate collection systems is 10^{-2} cm/s. Containment dikes are required for all surface impoundments to prevent failure in the event of a leak. The regulations exempt double-lined surface impoundments from groundwater monitoring and response requirements. The impoundment (and underlying liners) must be located entirely above the seasonal high water table.

Closure of a surface impoundment facility must be accomplished by either removal of all contaminated materials (including the liner and subsoils) or stabilization of the waste to support safely a final cover having a permeability less than or equal to the permeability of any bottom liner system or natural subsoils present.

Waste piles (40 CFR 264, Sub-part L) are temporary disposal facilities and must have clay liners during their active life, including the closure period. Other requirements (leachate collection, run-off control, double liner, exemption, etc.) are similar to those of landfills.

The requirements for landfills (40 CFR 264, Sub-part N) are similar to those for surface impoundment facilities: a run-off management system to collect and control the water volume resulting from a 24-h 25-year storm, and run-on control to prevent flow on to the active portion of the landfill from at least 25-year storm. The leachate collection system must be designed to ensure that the leachate depth over the liner does not exceed 30 cm (1 ft). The final cover must have a permeability less than or equal to the permeability of any bottom liner or natural subsoil.

For both landfills and surface impoundments there are exceptions to the double-liner requirements. In any case, at least a single synthetic liner is required.

2.7 Subtitle D program

Under RCRA Subtitle D the regulations governing non-hazardous waste are developed and implemented by State and local authorities with regulatory directions and technical assistance from the Federal Government. The criteria are contained under 40 CFR 257 Criteria for Classification of Solid Waste Disposal Facilities and Practices. Major elements of the criteria that are of geotechnical interest are for surface and groundwater, explosive gases, and flood plains.

2.7.1 Surface water

A facility is prohibited from discharging pollutants into the waters of the United States and is in violation of the requirements of the National Pollutant Discharge

Elimination System (NPDES) under Section 402 of the Clean Water Act, as amended. 'Pollutant' means dredge spoil, solid waste, garbage, rock, sand, sewage sludge, etc. The waters of the United States include 'wetlands', which means those areas that are inundated or saturated at such frequency and duration to support vegetation typically adapted for life in saturated soil conditions. If the facility is not operating in compliance with or has not applied for a 402 NPDES permit, then the surface water criterion is violated. The point-source discharge is prohibited if it violates the legal requirements of state-wide quality management plans under Section 208 of the Clean Water Act, as amended. Under the 208 plan, some control and practices are required to reduce water pollution, which may include run-off control, treatments, etc.

The surface water criteria require that the facility shall not cause a discharge of dredged material or fill material in the water that is in violation of the requirements under Section 404 of the Clean Water Act, as amended. To avoid violations, the facility must be in compliance with or have applied for a 404 (dredge or fill) permit.

2.7.2 Groundwater

The groundwater criteria require that the facility or practice shall not contaminate an underground drinking water source beyond the solid waste boundary or other boundary defined if several conditions are met. Groundwater is ruled contaminated if the concentrations of some organic and inorganic chemicals (Table 2.3) exceed those given in 40 CFR 257.4. If the down-gradient value of any of the criteria contaminants significantly exceeds the up-gradient values, the facility may also be in violation of the groundwater criteria.

2.7.3 Explosive gases

The criteria are violated if the concentration of methane (CH_4) exceeds 25% of the lower explosive limit (LEL) in facility structures (buildings, sheds), utility lines, or at the 'property boundary'. A methane concentration of 5% represents the LEL.

2.7.4 Municipal landfills

The EPA has recently proposed new rules that regulate the way municipal solid waste landfills are sited, designed, built and operated (Federal Register, 30 August 1988). The new rules are under new part 40 CFR 258. Unsuitable sites include those located in a 100-year flood plain, wetlands, seismic impact zones and geologically unstable areas (e.g. sites within sinkholes). New landfills are prohibited within 61 m (200 ft) of a fault that has had displacement in Holocene time.

Under the new rules the landfills would have to be monitored for explosive gases in facility structures (action level in 25% of LEL) and the facility property boundary (action level in LEL). The design of new landfills would require the use of liners, leachate collection system and final covers. The details are based on state-established goals for groundwater protection. The factors to be assessed are the site hydrogeological and climatic characteristics, volume and quality of leachate, groundwater quality, and proximity of groundwater use. Existing landfills would have to be covered to 'prevent' infiltration.

Under the proposed rules groundwater monitoring is required and the states are asked to establish an action level for some 230 parameters for corrective action. Semiannual sampling is required for 15 indicator parameters (e.g. chemical oxygen demand, chloride, volatile organics, heavy metals). If 'statistically significant' variance from background levels for two indicator parameters, or one or more heavy metals or volatile organic contents, is indicated, quarterly sampling is required for all the parameters.

2.8 State regulations

Implementation scope and standards for the Subtitle D program among the states vary. The New Jersey standards deal in detail with various aspects of geotechnical construction and quality control. The regulations require a minimum scope for a geotechnical investigation for new solid waste facilities which are classified as classes I, II and III. Class III receives only inert non-putrifiable waste, class I receives all types of non-hazardous waste. The design requirements vary and the least stringent is for class III. The liner requirements depend on the hydraulic conductivity of the foundation geological materials. A 3 ft (0.9 m) clay liner is required where the hydraulic conductivity is 10^{-6} cm/s or less. For geological materials with hydraulic conductivity greater than 10^{-5} cm/s, a dual composite liner system is required. Each tier consists of a synthetic liner in contact with earthen or admixture liner (clay, asphalt concrete, Portland cement, soil bentonite). The two liners are separated by a leachate detection system. Various requirements of the New Jersey and other state regulations are referenced in subsequent chapters.

References

Piasecki, B. W. and Davis, G. A. (1987) *America's Future in Toxic Waste Management: Lessons from Europe*, Quorum Books, New York

Chapter 3

Site selection

3.1 Introduction

There are several issues that have impact for site selection for a solid waste land disposal facility. In broad terms, the three major issues are environmental, economic and political. The geotechnical and hydrogeological parameters fall within the environmental category. The political factor is heavily impacted by public attitude. For a siting study to achieve public acceptance, citizen groups should participate in identifying the siting criteria and their relative importance. The ultimate goal is to select a site where the greatest protection to the environment is provided in the event that the technology, presumably affording protection, fails. In this regard some states are identifying areas where waste facilities could be located safely.

3.2 The siting process

The siting process involves several stages. The purpose of each stage is to narrow the list of possible sites. This ultimately leads to one or more sites for detailed investigation and analysis.

Table 3.1 Siting criteria

Economic	*Socioeconomic*	*Environmental*
Development land	Archaeological and historical sites	Wetlands
Slope	Dedicated land	Threatened or endangered species
Utilities	Land use	Slope
Site access	Noise impact	Air quality
Flexibility	Sensitive receptors	Odours
Capacity	Social impact	Aesthetic impact
Development, operation and maintenance cost		Flood plains
Compatibility with existing solid waste management systems		Soils
		Geology
		Groundwater
		Monitoring
		Surface and groundwater hydrology
		Topography

The factors that may be considered in assessing the suitability of a site are shown in Table 3.1. As exemplified in Table 3.1, the outcome of a siting study depends on the relative importance of different criteria.

3.2.1 Sussex County

The siting methodology for Sussex County (Converse TenEch, 1981) is illustrated in Table 3.2. The numbers in parentheses indicate the importance or weight given to that category. These numbers were established with the participation of citizen groups and not entirely by professionals. As a result, slopes were given more weight (41) than the permeability of the subsoils (29.6). The siting criteria adopted by the New Jersey Department of Environmental Protection (DEP) are consistent with Table 3.2.

Table 3.2 Siting methodology in Sussex County

Level			
1	Initial screening: apply broad exclusionary criteria		
2	Screening: apply geotechnical/hydrogeological exclusionary criteria; rank sites based on geotechnical/hydrogeological criteria	Soils (74)[a]	Permeability (29.6), pH (7.4), cation exchange capacity (14.8), surficial soils (22.2)
		Geology (74)	Bedrock type (20.6), bedrock continuity (16.4), faults (37)
		Groundwater (99)	Aquifer yield (26.4), aquifer use (26.4), groundwater quality (9.9), seasonal water table (23.1), groundwater flow system (13.2)
		Monitoring aspects (28), cover material (34), slope (41)	
3	Rank sites based on further hydrogeological and environmental criteria	Land use (57), aesthetic impact (54), site access (43), sensitive areas (59), utilities (45), soils (83), hazards (40), surface and groundwater hydrology (94), geology (85), topography (70) (slope, erodibility run-on/ run-off), flora and fauna (42), air quality (59), noise (54), odours (65)	
4	Rank further based on eight additional economic and socioeconomic criteria; select a site or sites for detailed specific investigation and testing		

[a] Number in parentheses indicates importance attached to the category – see text.

In level 1, exclusionary criteria (fatal flaws) are applied. The exclusionary criteria normally encompass regulatory exclusions (if any) and may include inland or coastal wetlands, distance from water supply wells, agricultural preservation areas (APA), flood plains, archaeological and historical sites, dedicated lands, habitats for threatened or endangered species, and developed land. In this stage, four exclusionary criteria (fatal flaws) were applied. The screening stage narrowed the possible sites to 91 and eliminated approximately 50% of the county.

Level 2 screening involves application of further exclusionary criteria and ranking the remaining sites based on the geotechnical/hydrogeological criteria shown in Table 3.2. Exclusions could include areas with high seasonal water tables, areas with very steep slopes, critical recharge areas, heavily fractured rocks, rocks with solution channels, active faults, etc.

The significance of the criteria with respect to environmental performance is summarized in Table 3.3. A site that is considered good with respect to a specific criterion could be given a suitability rating of 3, a moderate site a rating of 2, and a fair site a rating of 1. Other options for numerical ranking could be formulated. The words ‘good’, ‘moderate’, ‘fair’, etc., are somewhat subjective but must have a logical basis. A subsoil permeability of less than 10^{-7} cm/s is considered good, since this value is specified for a natural liner by many regulatory agencies.

Sites with a high cation exchange capacity (CEC) can be considered good. A CEC >25 mg/100 g (mEq/100 g) could have a rating of 3; a CEC <15 a rating of 1. Soils with a pH >6.5 are positively effective in the removal of metal cations. Thus, subsoils with a pH >6.5 are considered good for siting whereas those with values between 5 and 6.5 are moderate and those with a pH <5 are fair. Sites where the groundwater quality is poor are considered good sites. ‘Poor’ quality may be defined in terms of total dissolved solids (TDS) as $>10\,000$ mg/l. ‘Fair’ sites may be defined as those where the TDS is <500 mg/l. In seismically active areas, the distance to a causative fault with the potential for seismic instability could be used for rating.

After the number of possible sites is reduced in the level 2 process, further ranking is performed at level 3 based on hydrogeological and other criteria. After this level of screening, the geotechnical/hydrogeological suitabilities of the top three or four sites are usually within a few score points of each other. Level 4 screening is usually based on economic and socioeconomic issues only. The advantage of this process is its ability to advance sites with the most hydrogeological attributes for consideration.

3.2.2 Hunterdon County

The Hunterdon County siting study (Rogers *et al.*, 1985) rated the site based on the criteria in Table 3.4. Suitability numbers ranked from 5 (most suitable) to 1 (least suitable). The ranking was performed after applying 14 exclusionary criteria. The criteria covered distances to airports of 5000–10 000 ft, sites with wetlands and head streams within 500 ft, sites with glacial outwash, limestone and pre-Cambrian material, etc. In the ranking system the rock type formed the basis for rating the suitability for groundwater protection.

After the suitability numbers, S, and weight, W, are assigned for each category, the ranking score, C, is computed as:

$$C=\sum_{i=1}^{n} W_i \cdot S$$

where i is the criterion number and n the number of criteria (e.g. $n=6$ for level 2 Sussex screening in Table 3.2, and $n=21$ for Hunterdon ranking in Table 3.4).

Table 3.3 Definitions and significance of siting criteria

Criteria	*Sub-criteria*	*Definition*	*Significance*
Soils	Permeability	Soil property that governs the rate at which water moves through it	Subsoil permeability impacts release of pollutants to groundwater; lower subsoil permeability is preferable for siting
	pH	Indication of acidity and alkalinity (pH 7 = neutral)	Characterizes tendency of soil to absorb heavy metals; greater pH is preferable for siting
	Cation exchange capacity	Capacity of soil to exchange cations expressed as a sum for all exchangeable cations	Indicates the ability of soil to attenuate some contaminants, particularly heavy metals; higher CEC is preferable for siting
	Surficial soils	Unconsolidated materials at the Earth's surface	Affects degree of attenuation and the need for liners; surficial soils with lower permeability are preferable for siting
Geology	Bedrock and outcropping		Carbonate rocks are susceptible to solution; fractured rock facilitates pollution migration; sites with more overburden are preferable
	Continuity and mass permeability	Related to open discontinuities, solution channels	Controls the potential for migration of contaminants
	Faults	Mapped planes or zones of rock fracture along which displacement has occurred	Impacts the stability of a facility and potential release of pollutants
Groundwater	Aquifer/well yield	Relates to a geological formation or group of formations capable of yielding usable quantities of groundwater to wells or springs	Sites with high aquifer capabilities may be off-limits for some facilities
	Aquifer use	Use of aquifer could be within a specified distance from facility, potential, or sole source (solely or principally supply drinking water to a large percentage of a populated area)	Impacts the water supply; aquifers with low actual or potential use are preferable for siting purposes; sole-source aquifers are considered very significant even if yields are low
	Groundwater quality	The natural quality of groundwater as measured against drinking water standard	Areas with poor natural groundwater quality represent more suitable locations, all else being equal
	Groundwater flow system	Refers to the occurrence and movement of groundwater with regard to direction and velocity	Sites where direction of groundwater flow is away from use, or where flow is upward, or where water is deep, are preferable all else being equal

Table 3.3 (*cont'd*)

Criteria	*Sub-criteria*	*Definition*	*Significance*
	Seasonal high groundwater level	The maximum level to which groundwater is expected to rise	Unsaturated zones act as a barrier (no direct mixing) between base of facility and groundwater. Most regulations specify a minimum of 1.5 m (5 ft)
Monitoring aspects	Monitoring aspects	Refers to RCRA requirements for groundwater monitoring	Sites that are easier to monitor (e.g. presence of a layer of sand and gravel) or a known discharge body (e.g. lake)
Cover	Cover material	Refers to earth material (available on site) used for daily waste sealing	Sites with abundance of workable and relatively impervious soils are preferable, all else being equal
Slope	Slope	Deviation of the land surface from the horizontal measured as the average topographic relief for the site	Impacts release of contaminants, site development and operation. Slopes greater than 15% or 22% are considered too steep
Surface and groundwater hydrology	Proximity to streams/lake	Refers to overland proximity and protected uses of the nearest lake or stream	Impacts opportunity for runoff and contaminants polluting lakes/streams
	Proximity to wells/aquifer	See Aquifer under Geology above	Impacts groundwater resources; sites closer than 800 m (2500 ft) to a high yield well (70 g/m) may be excluded
	Proximity to flood-prone areas	Land areas inundated by flood of specified frequency (usually 100 years)	Impacts transport of hazardous waste
	Proximity to recharge areas	Refers to lands draining to existing or planned storage reservoirs	Impacts drinking water supply
Topography	Slope erodibility	Migration of soil particles by surface water or other natural phenomena	The potential of soil erosion impacts facility construction and operation
	Run-on and run-off	Run-off refers to rainwater or leachate that drains overland away from the facility; run-on refers to drainage overland on to any part of the facility	Sites with little need for control of run-on from upland and slow run-off are preferable. Run-on is usually controlled by berms, stream diversion, etc. Run-off control is impacted by velocity of water traversing the site

Table 3.4 Siting criteria and weights (Hunterton County Siting Study)

Criteria	*Weighting*
Groundwater protection	100
100-year flood plain	95
Streams, headwaters, wetlands	95
Farmland (Agricultural development area, ADA)	83
Developed land	78
Wooded buffer	69
Office, residences	66
Farms outside ADA	65
Loss of woodland	64
Downwind residences	61
Truck traffic – residences	59
Sensitive public use	58
Truck traffic	48
Public open space	45
Trout	45
Relocation of residences	41
Historic sites	39
Disposal costs	39
Rock type	36
Transportation costs	30
Pedestrian right-of-way	25

Data for the siting studies are usually found in the files or libraries of state and federal agencies, and in the libraries of private engineering and environmental firms.

3.2.3 Data sources in the United States

US Geological Survey (USGS) data available include geological index maps, professional papers, water supply papers, topographic maps and other data.

US Department of Agriculture (USDA) Soil Conservation Services (SCS) data include surveys of surface soils described in agricultural terms. Data such as pH, CEC, seasonal water levels, etc., are provided for the counties surveyed.

State Geological Survey publications provide excellent detailed geological maps covering local areas.

Aerial photography is available in 9-in forms with overlap for stereoscopic viewing. Scales could range from 1:12 000 to 1:80 000 photos used for topographic–geological mapping. Drainage patterns are available from USGS or SCS.

Imagery obtained by satellite with scales varying from 1:1 000 000 to 1:250 000 are available from Earth Research Observation System Date Center (EROS). Skylab imagery from an orbit 770 miles above the Earth and aerial photography produced by National Aeronautics and Space Agency (NASA) are also available from EROS.

Summary

Major issues in site selection are environmental, economic and political, with political factors being impacted heavily by public opinion. Criteria considered important by the public may not necessarily be the soundest environmentally, but are the most crucial for getting acceptance or rejection.

References

Converse/TenEch (1981) *Identifying Potential Sanitary Landfill Disposal Sites/Systems For Sussex County, New Jersey*, Prepared for the Sussex County Board of Chosen Freeholders

Rogers, Golden and Halpern (1985) *Hunterdon County Landfill Evaluation and Siting Study*, prepared for County of Hunterdon, Flemington, New Jersey

Chapter 4

Index properties

4.1 Soils

Soils are multi-phase systems consisting of mineral particles of different sizes and shapes and voids filled with gases and/or liquids. Several relationships are used to express the state of a soil. They provide useful information about the state of a soil. A typical soil element, its phase diagram, and the corresponding notations are shown in Figure 4.1. The basic relationships are:

Porosity	$n = V_v/V$
Void ratio	$e = V_v/V_s$
Degree of saturation	$S = 100V_w/V_v$ (%)
Water content	$w = 100W_w/W_s$ (%)
Unit weight of water	$\gamma_w = W_w/V_w$
Unit weight of solids	$\gamma_s = W_s/V_s$
Specific gravity of solids	$G_s = \gamma_s/\gamma_w$
Soil mass	
Total unit weight	$\gamma = W/V$
	$= \gamma_w(G_s + Se)/(1 + e)$
Dry unit weight	$\gamma_d = W_s/V$
	$= \gamma_w G_s/(1 + e)$
	$= \gamma/(1 + w)$
Buoyant unit weight	$\gamma_b = \gamma - \gamma_w$

A useful relationship is: $Gw = Se$.

The specific gravity (or relative density) for most inorganic soils varies over a narrow range (2.6–2.8) and is commonly assumed to be 2.65 for sands and 2.7 for clays. Some of the soil characteristics such as its texture and plastic properties are useful indices for certain engineering properties. The index properties are usually assessed as part of a site investigation process.

Some typical values of index properties are provided in Table 4.1. Some of these values are estimated in the field while others are determined in the laboratory. The values for cohesionless soils vary between a narrow range whereas for cohesive soils there is a large variation.

Coarse-grained soils consist of particles ranging from boulders to fine sands. The distribution of the particle sizes in a soil mass have certain influence on its engineering properties. Grain size distribution curves for some soils are shown in Figure 4.2. The nominal particle size corresponding to 10% passing by weight is known as the

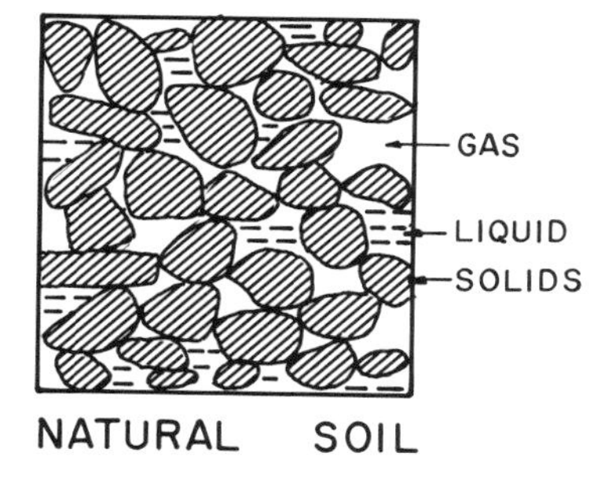

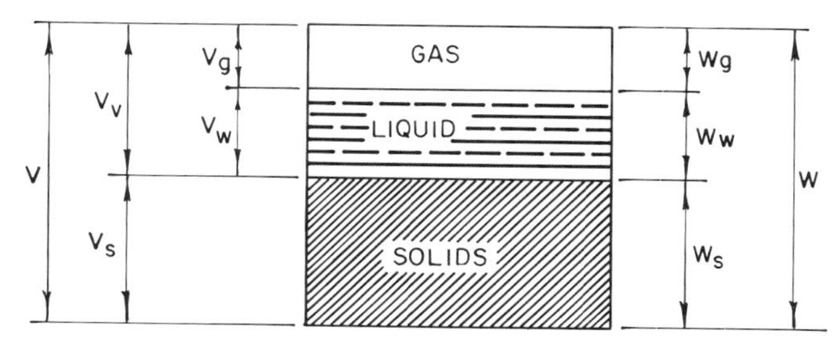

Figure 4.1 Phase diagram of a soil showing its weight–volume relationships

Table 4.1 Typical values of soil index properties

Soil description	*Void ratio*		*Porosity (%)*		*Dry unit weight (lb/ft³)*		*Saturated unit weight (lb/ft³)*	
	Max.	*Min.*	*Max.*	*Min.*	*Min.*	*Max.*	*Min.*	*Max.*
Uniform sand	1.0	0.4	50	29	83	118	84	136
Silty sand	0.9	0.3	47	23	87	127	88	142
Clean well graded sand	0.95	0.2	49	17	85	138	86	148
Silty sand and gravel	0.85	0.14	46	12	89	146	90	155
Sandy or silty clay	1.8	0.25	64	20	60	135	100	147
Well graded gravel, sand, silt and clay mixture	0.7	0.13	41	11	100	148	125	156
Inorganic clay	2.4	0.5	71	33	50	112	94	133
Colloidal clay (50% < 2 μm)	12	0.6	92	37	13	106	71	128

Source: NAVFAC DM 7.1, 1982.

effective particle diameter and is represented by D_{10}. Other particle characteristics that affect the soil behaviour are particle shape, roundness or angularity, hardness, etc. (Lambe and Whitman, 1969).

As shown in Figure 4.3, cohesive soils may exist in a liquid, plastic, semi-solid or solid state. The limiting water contents for these states are arbitrarily defined and are known as liquid limit w_l, plastic limit w_p, and shrinkage limit w_s (the Atterberg limits).

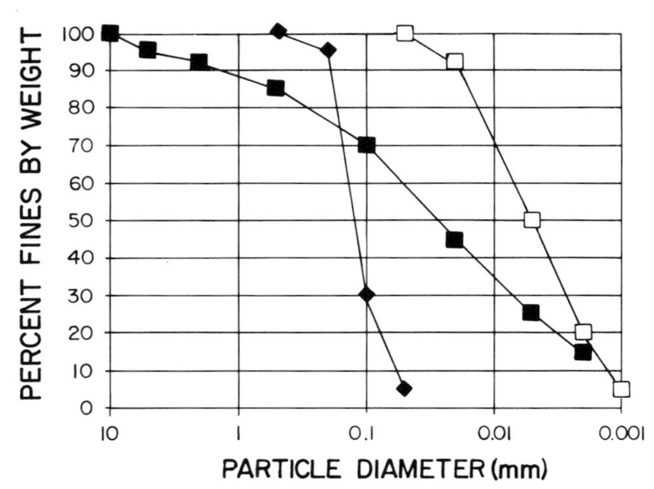

Figure 4.2 Particle distribution curves for: ■, a glacial till; □, silt and clay; ◆, fine sand

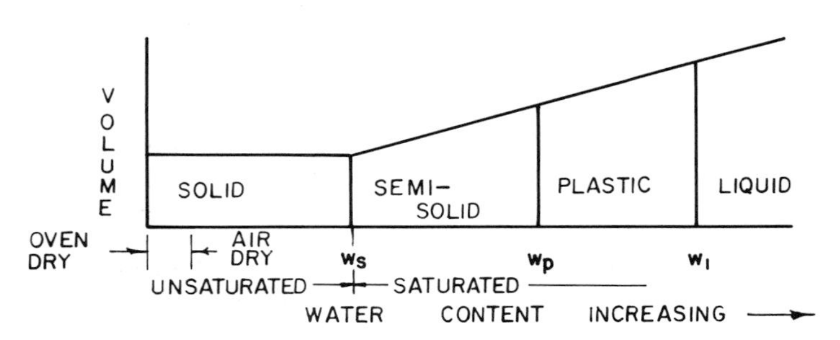

Figure 4.3 Soil consistency

The plasticity index is given by:

$$I_p = w_l - w_p \quad (4.1)$$

I_p is used in classification of soils and has been correlated with strength and other properties.

Liquidity index is given by:

$$I_L = (w_n - w_p)/I_p \quad (4.2)$$

where w_n is the natural water content.

If the liquidity index of an undisturbed soil is >1, the soil is susceptible to strength loss upon disturbance. For I_L between 0 and 1 the soil behaves plasticly.

For organic soils the specific gravity is less than 2.6. The moisture content is usually high (Table 4.2) and the unit weight is lower than that of inorganic soils.

Standard laboratory procedures for index properties tests are given in Appendix A4.1.

Table 4.2 Index properties of organic materials

Soil type	*Natural water content* (%)	*Total unit weight* (lb/ft³)	*Specific gravity*	*Liquid limit* (%)	*Plasticity index* (%)
Fibrous peat	500–1200	60–70	1.2–1.8	—	—
Fine-grained peat	400–800	60–70	1.2–1.8	400–900	200–500
Silty peat	250–500	65–90	1.8–2.3	250–500	150–350
Sandy peat	100–400	70–100	1.8–2.4	100–400	50–150
Organic clay	50–200	70–100	2.3–2.6	65–150	50–150
Organic sand or silt	30–125	90–110	2.4–2.6	30–100	NP–40

Source: NAVFAC DM 7.1, 1982.

4.2 Soil classification

The most commonly used classification method is the Classification of Soils for Engineering Purposes (ASTM D 2487–85). This may be used in conjunction with the Description and Identification of Soils (ASTM D 2488–84). In this system the soils are placed in three major categories:

1. *Highly organic soils.* A soil is classified as Peat (PT) if it is composed of decomposing vegetable tissues, has fibrous to amorphous texture, a spongy consistency, is dark brown to black in colour and has organic odour.
2. *Fine-grained soils.* If 50% or more by dry weight of the test specimen passes a No. 200 sieve, the soil is classified as a fine-grained soil.
3. *Coarse-grained soils.* If more than 50% by dry weight of the test specimen is retained on a No. 200 sieve, the soil is classified as a coarse-grained soil.

Coarse and fine-grained soils are further subdivided into 14 groups (Table 4.3).

The notations for soil types and their states are shown in Table 4.4. The textural divisions of coarse-grained soils are given in Table 4.5.

4.3 Fine-grained soils

In the classification of fine-grained soils Atterberg limits are determined for the materials passing No. 40 sieve and plotted on the plasticity chart shown in Figure 4.4. Clays plot on or above the A line and have a plasticity index $\geqslant 4$. Silts fall below the A line and have a plasticity index < 4. The U line in the plasticity chart defines the upper boundary for natural soils.

If soil classification is at the border line of two groups, use both symbols with a separating slash (for example, CL/CH if w_l is close to 50).

Group names of the fine-grained soils are shown in Table 4.6. The presence of gravel and sand modifies the group names as follows.

If $< 30\%$ but $\geqslant 15\%$ of the specimen is retained on No. 200, add 'with sand' if sand is greater than gravel (for example, lean clay with sand), or add 'with gravel' if gravel greater than sand (for example, lean clay with gravel).

Table 4.3 Soil classification chart

Criteria for assigning group symbols and group names using laboratory tests[a]				*Soil classification*	
				Group symbol	*Group name*[b]
Coarse-grained soils (> 50% retained on No. 200 sieve)	Gravels (> 50% of coarse fraction retained on No. 4 sieve)	Clean gravels (< 5% fines[c])	$C_u \geqslant 4$ and $1 \leqslant C_c \leqslant 3$[e]	GW	Well graded gravel[f]
			$C_u < 4$ and/or $1 > C_c > 3$[e]	GP	Poorly graded gravel[f]
		Gravels with fines (> 12% fines[c])	Fines classify as ML or MH	GM	Silty gravel[f,g,h]
			Fines classify as CL or CH	GC	Clayey gravel[f,g,h]
	Sands (≥ 50% of coarse fraction passes No. 4 sieve)	Clean sands (< 5% fines[d])	$C_u \geqslant 6$ and $1 \leqslant C_c \leqslant 3$[e]	SW	Well graded sand[i]
			$C_u < 6$ and/or $1 > C_c > 3$[e]	SP	Poorly graded sand[i]
		Sands with fines (> 12% fines[d])	Fines classify as ML or MH	SM	Silty sand[g,h,j]
			Fines classify as CL or CH	SC	Clayey sand[g,h,j]
Fine-grained soils (≥ 50% passes the No. 200 sieve)	Silts and clays (liquid limit < 50)	Inorganic	PI > 7 and plots on or above A line	CL	Lean clay[k,l,m]
			PI < 4 or plots below A line	ML	Silt[k,l,m]
		Organic	$\frac{\text{Liquid limit – oven dried}}{\text{Liquid limit – not dried}} < 0.75$	OL	Organic clay[k,l,m,n] Organic silt[k,l,m,o]
	Silts and clays (liquid limit ≥ 50)	Inorganic	PI plots on or above A line	CH	Fat clay[k,l,m]
			PI plots below A line	MH	Elastic silt[k,l,m]
		Organic	$\frac{\text{Liquid limit – oven dried}}{\text{Liquid limit – not dried}} < 0.75$	OH	Organic clay[k,l,m,p] Organic silt[k,l,m,q]
Highly organic soils		Primarily organic matter, dark in colour, organic odour		PT	Peat

[a] Based on the material passing the 3 in (75 mm) sieve.
[b] If field sample contained cobbles or boulders, or both, add 'with cobbles or boulders, or both' to group name
[c] Gravels with 5–12% fines require dual symbols: GW–GM well graded gravel with silt, GW–GC well graded gravel with clay, GP–GM poorly graded gravel with silt, GP–GC poorly graded gravel with clay.
[d] Sands with 5–12% fines require dual symbols: SW–SM well graded sand with silt, SW–SC well graded sand with clay, SP–SM poorly graded sand with silt, SP–SC poorly graded sand with clay.
[e] $C_u = D_{60}/D_{10} \quad C_c = \frac{(D_{30})^2}{D_{10} \times D_{60}}$
[f] If soil contains ≥ 15% sand,add 'with sand' to group name.
[g] If fines classify as CL–ML, use dual symbol GC–GM, or SC–SM.
[h] If fines are organic, add 'with organic fines' to group name.
[i] If soil contains ≥ 15% gravel, add 'with gravel' to group name.
[j] If Atterberg limits plot in hatched area, soil is a CL–ML, silty clay.
[k] If soil contains 15–29% plus No. 200, add 'with sand' or 'with gravel', whichever is predominant
[l] If soil contains ≥ 30% plus No. 200, predominantly sand, add 'sandy' to group name.
[m] If soil contains ≥ 30% plus No. 200, predominantly gravel, add 'gravelly' to group name.
[n] PI ≥ 4 and plots on or above A line.
[o] PI < 4 or plots below A line.
[p] PI plots on or above A line.
[q] PI plots below A line.

Table 4.4 Notations for soil type and state

Soil type or state	*Symbol*	*Remarks*
Gravel	G	
Sand	S	
Well graded	W	$C_u \geq 4$ and $1 \leq C_c \leq 3$ for gravel $C_u \geq 6$ and $1 \leq C_c \leq 3$ for sand
Poorly graded	P	$C_u < 4$ and/or $1 > C_c > 3$ for gravel $C_u < 6$ and/or $1 > C_c > 3$ for sand
Silt	M	
Clay	C	
Liquid limit < 50	L	
Liquid limit > 50	H	
Organic	O	
Peat	PT	

Table 4.5 Textural division of soils

Soil name	*Particle sizes* (mm (in))	*Sieve size* (in) *or US sieve No.*
Boulders	> 300 (12)	
Cobbles	300–75 (12–3)	
Gravel		
Coarse	75–19 (3–$\frac{3}{4}$)	3 to $\frac{3}{4}$
Fine	19–4.75 ($\frac{3}{4}$–0.187)	$\frac{3}{4}$ to No. 4
Sand		
Coarse	4.75–2.00	No. 4 to No. 10
Medium	2.00–0.425	No. 10 to No. 40
Fine	0.425–0.075	No. 40 to No. 200
Clays and silts	0.075	

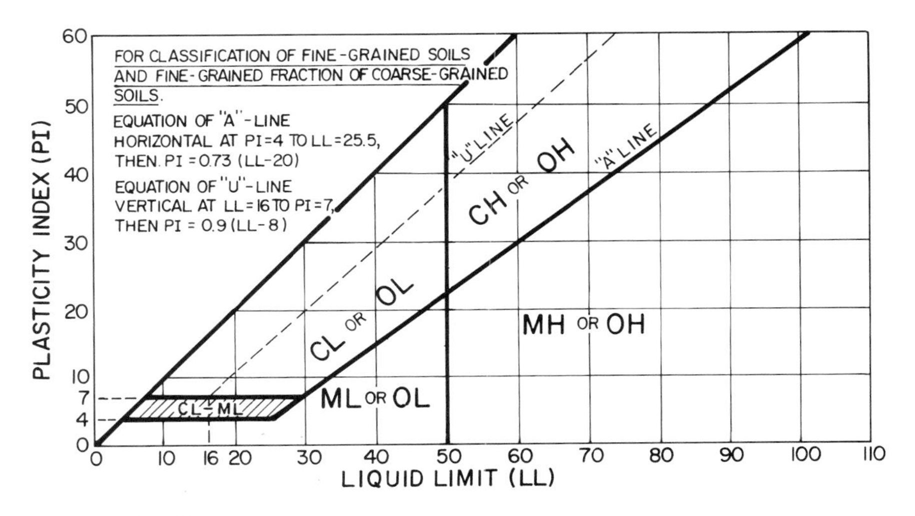

Figure 4.4 Plasticity chart

Table 4.6 Group names of fine-grained soils

Group name	*Symbol*	W_l	W_l *oven dry*	*PI*	*Remarks*
Lean clay	CL	<50		>7	Plots on or above the A line
Fat clay	CH	$\geqslant 50$		>7	Plots on or above the A line
Silty clay	CL–ML			4–7	Plots on or above the A line
Silt	ML	<50		<4	Plots below the A line
Elastic silt	MH	$\geqslant 50$			Plots below the A line
Organic silt	OL	<50	$<0.75w_l$	<4	Or plots below the A line
Organic clay	OL	<50	$<0.75w_l$	>4	Or plots above the A line
Organic silt	OH	>50	$<0.75w_l$		Plots below the A line
Organic clay	OH	>50	$<0.75w_l$		Plots on or above the A line

If $<50\%$ but $\geqslant 30\%$ of the specimen is plus No. 200, add 'sandy' if gravel is $<15\%$ (for example, sandy silty clay) or add 'gravelly' if sand is $<15\%$ (for example, gravelly silty clay). If sand is greater than gravel and gravel is equal to or greater than 15%, add 'with gravel' at the end (for example, sandy fat clay with gravel). With sand and gravel components reversed this soil description will be gravelly fat clay with sand.

Soils for which the liquid limit after oven drying drops to $<75\%$ of its value before oven drying are classified as organic soils.

4.4 Coarse-grained soils

Coarse-grained soils are those in which $>50\%$ of the sample is retained on a No. 200 (75 μm) sieve. If $>50\%$ of the coarse fraction is plus No. 4, it is classified as gravel, otherwise sand. If fines are present (materials pass a No. 200 sieve) the proportion of their weight is determined and grouped as $<5\%$, 5–12% and $>12\%$. A particle size distribution curve is plotted for the soils. The coefficient of curvature,

$$C_c = D_{30}^2 / D_{10} \times D_{60}$$

and the coefficient of uniformity,

$$C_u = D_{60} / D_{10}$$

are computed. D_{10}, D_{30} and D_{60} are the nominal particle diameters for 10, 30 and 60% fines by weight on the particle size distribution curve. C_c and C_u are used for characterizing soil gradation (well graded, W; poorly graded, P).

4.4.1 Less than 5% fines

If the proportion of gravel is greater than that of sand, $C_u \geqslant 4$ and $1 \leqslant C_c \leqslant 3$, it is a well graded gravel (GW), otherwise it is a poorly graded gravel (GP). If sand is also present add 'with sand' for sand $\geqslant 15\%$ (for example, well graded gravel with sand, GW).

If the proportion of sand is equal to or greater than that of gravel, $C_u \geqslant 6$ and $1 \leqslant C_c \leqslant 3$, class it as well graded sand (SW), and otherwise as a poorly graded sand

(SP). If gravel is also present add 'with gravel' for gravel ⩾15% (for example, poorly graded sand with gravel, SP).

4.4.2 Between 5 and 12% fines

Determine C_u, C_c and type of fines, that is, silt (M), clay (C), or silty clay (CL–ML). Use dual symbols, such as GW–GM, if the proportion of gravel is greater than that of sand, or SP–SC if the proportion of sand is equal to or greater than that of gravel. To the group name, add 'with silt' for silt fines (for example, well graded gravel with silt), 'with clay' for clay fines, and 'with silty clay' for silty clay fines. If the proportion of gravel is greater than that of sand, and sand is ⩾15%, add 'and sand' at the end (for example, well graded gravel with silt and sand, GW–GM). Similarly, if the proportion of sand is equal to or greater than that of gravel, and gravel is ⩾15% add 'and gravel' (for example, poorly graded sand with silty clay and gravel, SP–SC).

4.4.3 Greater than 12% fines

For fines >12%, determine if they are silt (M), clay (C), or silty clay (CL–ML) fines.

If <15% of the coarse fraction is sand, classify the soil as silty, clayey, or silty clayey gravel (for example, silty, clayey gravel, GC–GM). Add 'with sand' if sand is ⩾15% (for example, silty, clayey gravel with sand, GC–GM).

If <15% of the coarse fraction is gravel name the soil as silty, clayey, or silty clayey sand (for example, clayey sand, SC). Add 'with gravel', if gravel is ⩾15% (for example, clayey sand with gravel, SC).

4.5 Municipal waste

The unit weight of refuse varies widely because of the large variations in the waste constituents, state of decomposition, degree of control during placement such as thickness or absence of daily cover, amount of compaction, total depth of landfill, the depth from which a sample is taken, etc. Household refuse (garbage) usually weighs more than industrial waste. Table 4.7 shows the range of uncompacted and compacted densities for MSW components. As illustrated in Table 4.7, the waste components play a major role in the unit weight. The unit weight of the refuse, as placed, may be estimated if the components of the waste stream are known. Considering, for example, the US national average (EPA, 1986) in Table 4.8, the estimated in-place unit weight for well compacted refuse (excluding soil cover) would be 30.3 lb/ft^3 (4.76 kN/m^3). For an average of 6 in of soil cover at 120 lb/ft^3 (18.85 kN/m^3), the average in-place unit weight for the well compacted landfill would be 35 lb/ft^3 (5.5 kN/m^3) and 28 lb/ft^3 (4.4 kN/m^3) for normal compaction.

Figure 4.5 (Shoemaker, 1972) exemplifies the effect of the layer thickness and the number of passes. For well operated landfills, the layer thickness would usually be around 2 ft, yielding a typical unit weight of about 35 lb/ft^3 (5.4 kN/m^3). For larger thickness the unit weight drops substantially. The effect of the number of passes depends on the composition of refuse. For a waste layer composed mainly of old rubber tires, increasing the number of passes rarely makes a significant difference. The usual assumption is that four to five passes is the limit beyond which further compaction is generally insignificant. In well operated landfills such control is achieved. An experienced operator can judge the optimum number of passes.

Table 4.7 Unit weight of uncompacted and compacted refuse components

Waste component	*Uncompacted unit weight* (lb/ft^3)	*Water content* (*% of dry weight*)	*Ratio of compacted to uncompacted weight*	
			Normal compaction	*Well compacted*
Food waste	8–30	50–80	2.9	3.0
Paper and paper board	2–8	4–10	4.5	6.2
Plastics	2–8	1–4	6.7	10
Textiles	2–6	6–15	5.6	6.7
Rubber and leather	6–16	1–12	3.3	3.3
Yard waste	4–14	30–80	4.0	5.0
Wood	8–20	15–40	3.3	3.3
Glass	10–30	1–4	1.7	2.5
Metals	3–70	2–6	4.3	5.3
Ash, brick, dirt	20–60	6–12	1.2	1.3

After Tchobanoglous *et al.*, 1977.

Table 4.8 Estimating unit weights of refuse for US average

Waste component	*% of total waste*	*Assumed uncompacted weight* (*lb/ft^3*)	*Volume per ton of refuse* (ft^3)		
			Uncompacted	*Normal compaction*	*Well compacted*
Food waste	8	18	9	3.1	3.0
Paper and paper board	37	4	185	41.1	30.3
Plastics	7	4	35	5.2	3.5
Textiles	2	4	10	1.8	1.5
Rubber and leather	2	9	4.4	1.3	1.3
Yard waste	18	7	51.4	12.9	10.3
Wood	4	15	5.3	1.6	1.6
Glass	10	12	16.7	9.8	6.7
Metals	10	12	16.7	9.8	6.7
Ash, brick, dirt	2	30	1.3	1.1	1.0
Total volume			334.8	87.7	65.9
Average unit weight (lb/ft^3) = 2000/334.8			6	22.8	30.3

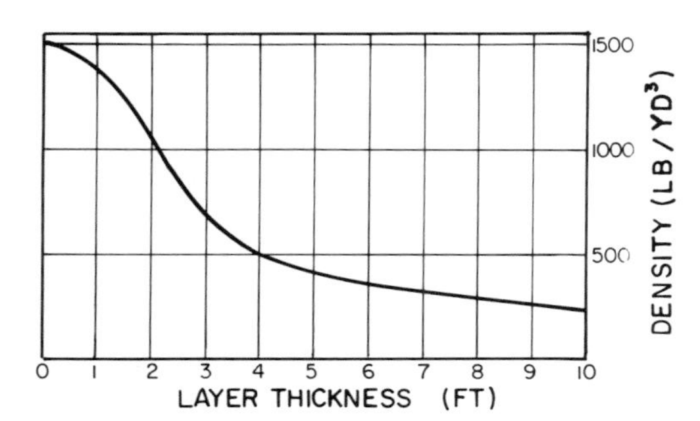

Figure 4.5 Effect of layer thickness on density (Reproduced from Shoemaker, 1972)

Landfills with large percentages of rubble (bricks, broken concrete, etc.) have high unit weights. For typical mixed domestic and industrial sources the use of 40 lb/ft^3 (6.28 kN/m^3) is reasonable. The refuse also has the capability to adsorb water in addition to its original moisture content. This additional water has been estimated at 1.8 in/ft (Fenn *et al.*, 1975) or about 9 lb/ft^3 (1.41 kN/m^3). Thus for the waste depicted in Table 4.8 the wet unit weight may range from 37 to 43 lb/ft^3 (5.8–6.8 kN/m^3) depending on the degree of compaction. In some old landfills underlain by clayey soils, the rate of leachate generation would be larger than the rate at which leachate percolates into the foundation soils. In such cases, a leachate mound develops inside the landfill (see Figure 4.6). Beneath the leachate mound a unit weight of refuse of 65–70 lb/ft^3 (10.2–11 kN/m^3) is usually used. Table 4.9 summarizes published values for unit weights of various types of refuse.

The average bulk unit weight of a landfill can be indirectly estimated if the foundation soils are normally consolidated or lightly overconsolidated clays for which effective consolidation pressure and excess pore pressure can be estimated (Figure 4.6). The average unit weight of the landfill γ_l is then given by:

$$\gamma_l = (\sigma'_{vc} - \sigma'_o + \delta u)/h_l \tag{4.3}$$

where σ'_o is the effective overburden pressure or consolidation pressure before placement of the landfill; σ'_{vc} is the effective consolidation pressure due to landfill load (from the laboratory consolidation test); δu is the excess pore pressure in the foundation soil at the point sampled; and h_l is the height (depth) of landfill.

This method was used for estimating the average unit weight of a landfill in the Hackensack Meadows, New Jersey, receiving domestic and industrial waste. The average unit weight of 42 lb/ft^3 (6.6 kN/m^3) was determined for the active portion of the landfill where the depth of refuse was about 100 ft. In the older portion of the landfill, the average unit weight was calculated as 62 lb/ft^3 (9.74 kN/m^3).

4.5.1 Test pit

Test pits are typically dug 10 ft deep and the weight and volume of the waste removed are determined. The volume of the test pit is determined either by approximate measurement or by lining and filling it with water. Only the materials near the surface of the landfill can be investigated. A large number of tests may be needed because of the variation in landfill composition. The results usually show a large scatter.

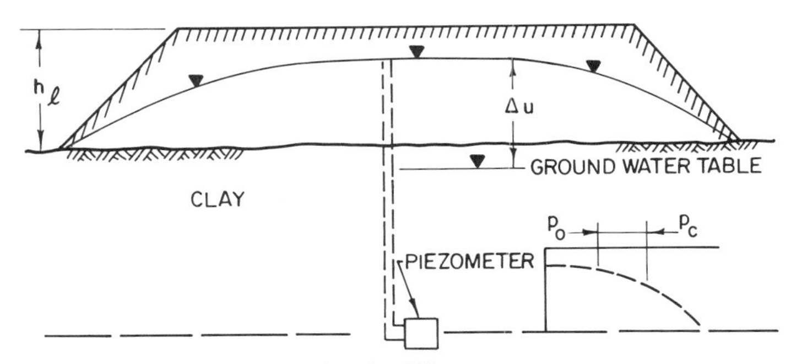

Figure 4.6 A leachate mound in a landfill

Table 4.9 Unit weights for landfill materials

Description and state	*Total unit weight*	
	lb/ft³	kN/m³
Municipal waste		
Poor compaction	18–20	2.8–3.1
Moderate to good compaction	30–40	4.7–6.3
Good to excellent compaction	55–60	8.6–9.4
Baled waste	37–67	5.5–10.5
Shredded and compacted	41–67	6.4–10.5
In situ density	35–44	5.5–6.9
Active landfill with leachate mound	42	6.6
North-east US active landfill	30–40	4.6–6.3
Incinerator residue		
Poorly burnt	46	7.2
Intermediate burnt	75	11.8
Well burnt	81	12.6
Ashes	41–52	6.4–8.2
Hazardous waste landfill site		
75 ft deep waste with soil cover	101	15.9
40–50 ft deep dry dust and soil	30–110	4.6–17.3
62 ft deep waste average		
Kiln dust, sludge tar, creosote and soil	73	11.5
Dust	46	7.2
Tars	104	16.3
Contaminated soils	69	10.8
75 ft deep chemical solutions and scrap metals mixed with contaminated soil	63–74	9.9–11.6
30–40 ft deep landfill with 90–95% waste in metal drums	90	14.1

Sources: Bromwell, 1978; Collins, 1977; Oweis and Khera, 1986; Peirce *et al.*, 1986; Sargunan *et al.*, 1986; Shoemaker, 1972; Sowers, 1973; Tchobanoglous *et al.*, 1977.

4.5.2 Field test sections

This technique requires an area about 300 ft long and 150 ft wide to be filled with refuse using usual daily field parameters (layer thickness, compaction equipment and passes). The weight of refuse is determined from the number of trucks delivering the refuse. The volume is determined by optical surveys of dimensions.

This technique is reasonably reliable for assessing the unit weight before advanced decomposition. The results are typically 35–45 lb/ft³ (5.5–7.1 kN/m³) for average compaction.

4.5.3 Refuse inventories

For a landfill where good records are kept, the weight of refuse dumped can be reasonably assessed. The volume is determined from optical surveys and approximate estimates of the settlement of the foundation where it comprises compressible materials.

4.5.4 Sampling

For refuse in an advanced state of decomposition, tube samples could be obtained in a manner similar to that used for soils, and the unit weight is determined using standard procedures.

4.5.5 Drilling

Under suitable conditions large-diameter holes may be drilled using coring buckets. The unit weight is determined by weighing the refuse removed and estimating the volume of the hole. Figures 4.7 and 4.8 show wet and dry unit weights determined using this method for a site in southern California. The in-place unit weights were established using a 12 in (304 mm) diameter coring bucket. Immediately after removal from the hole the materials were weighed. Volumes were estimated assuming a slightly enlarged diameter to account for imperfections in the diameter of the hole and the height of the bulk sample. The indicated densities may be slightly higher than those actually existing. The conclusions derived from the data are:

1. The unit weight of the refuse material increased with increasing depth.
2. The wet weight of the newer fill was slightly higher than the wet unit weight of the older fill, largely because of higher moisture content of the newer fill.
3. The dry unit weight of the newer fill and the older fill were approximately equal.

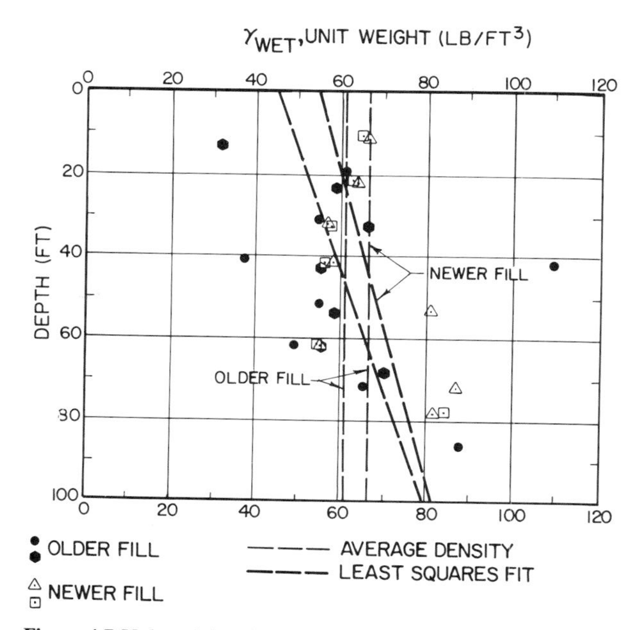

Figure 4.7 Unit weight of landfill materials from field tests

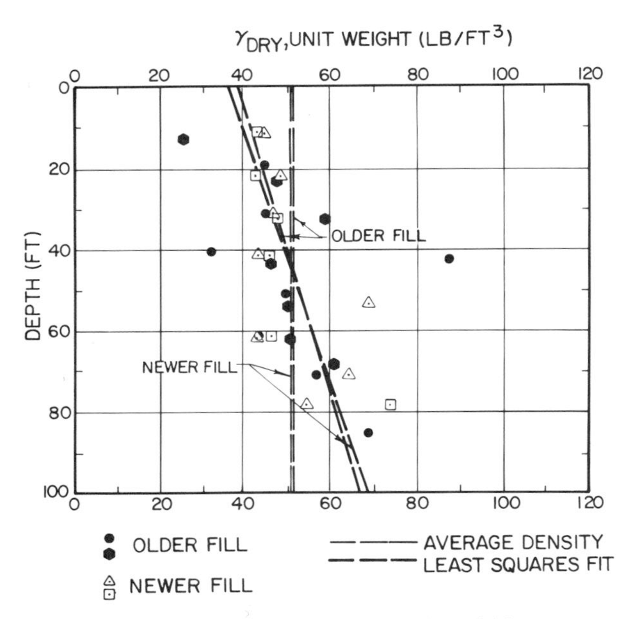

Figure 4.8 Unit weight of landfill materials from field tests

4.5.6 Moisture content

The range of moisture contents for municipal waste is 15–40% of dry weight. The moisture content is usually determined by heating the sample at 40–60°C for an hour or more to avoid burning of waste. Typical ranges for refuse components are provided in Table 4.7.

4.6 Mineral waste

The index properties of mineral and industrial wastes are affected by, among other things, the nature of the ore, the chemical process used, the age of the fill, groundwater conditions, organic materials present and their nature, chemical and biological degradation, leaching, and climatic conditions.

The properties of fly ash vary from plant to plant, and even in the same plant they may change from time to time. Several factors such as the type of coal, type of furnace, burning temperature, etc., contribute to these variations. Index properties of some of these materials are given in Table 4.10 and Figure 4.9. Note that the incinerator residue has particle sizes similar to that of coarse coal waste and the phosphatic clay waste has the smallest size particles with the highest values of plasticity index. Only the average grain size curve is shown for each of the materials. However, the actual grain size curves may vary considerably from those shown and even for a given material they may vary from source to source. For example, for class F and class C fly ashes, some statistical data for grain size are shown in Table 4.11.

Table 4.10 Index properties of mineral waste

Description	W_l	I_p	G_s	*pH*
Combined coal waste			2.1 – 2.6	
Coarse coal waste	2 – 20	NP	1.75 – 2.7	
Fine coal waste	20 – 65	NP	1.4 – 2.22	
Fly ash		NP	2.1 – 2.65	5.2 – 12.3
Bottom ash		NP	2.30 – 2.80	
FGD				
Sulphite-rich	24 – 65	8 – 20	2.46 – 2.72	>12
Sulphate-rich	20 – 22	NP	2.2 – 2.5	>10
Alumina red mud				
Fines	41 – 46	6 – 9	2.84 – 3.27	
Sand			3.16 – 3.27	
Phosphatic clay	100 – 200	90 – 145	2.5 – 2.8	7 – 8
Paper mill waste	70 – 320		1.9 – 2.3	

NP, non-plastic.
Sources: Bromwell and Oxford, 1977; Chae and Gurdziel, 1976; Charlie, 1977; Hagerty *et al.*, 1977; Holubec, 1976; Krizek *et al.*, 1987; Mabes *et al.*, 1977; Maynard, 1977; Moulton *et al.*, 1976; Somogyi and Gray, 1977; Ullrich and Hagerty, 1987.

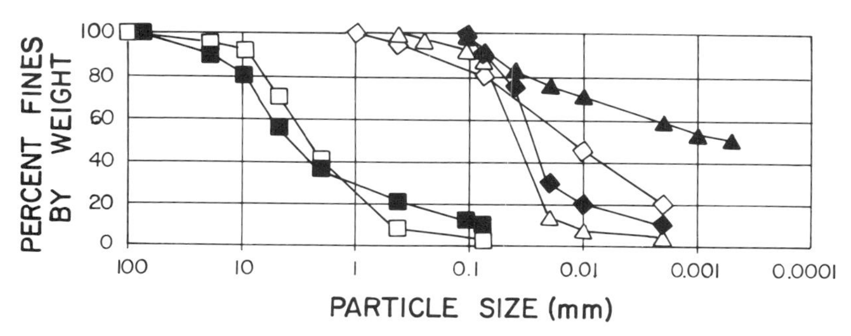

Figure 4.9 Grain size distribution curves for mineral waste: □, coarse coal; ◇, fine coal; △, fly ash (New Jersey); ■, incinerator residue; ◆, FGD sludge; ▲, phosphate clay

Table 4.11 Statistical grain size data for fly ash

	Class F fly ash			*Class C fly ash*		
Particle size (mm)	*Mean value*	*Standard deviation*	*Coefficient of variation* (%)	*Mean value*	*Standard deviation*	*Coefficient of variation* (%)
D_{85}	0.079	0.063	80.0	0.063	0.020	31.7
D_{50}	0.023	0.015	66.9	0.022	0.011	50.0
D_{15}	0.0057	0.0048	64.8	0.0084	0.0082	97.6

Source: McLaren and DiGioia, 1987, Table 1, p.689, reproduced by permission.

Organic contents of paper mill waste may range from 40 to 90% with water content ranging between 45 and 740%. The liquidity index from field sludge samples has been reported to range between 1 and 1000 (Charlie, 1977).

4.7 Dredging waste

Overall about 85% of the dredge materials consist of particles smaller than sand (Boyd *et al.*, 1972) and may contain a large proportion of organic silts and clays (Krizek and Salem, 1977a). In the vicinity of disposal pipes sands dominate, with little or no clay-size particles. The effective size (D_{10}) decreases as the distance from the inlet pipes increases. Index properties of these materials are shown in Table 4.12.

The data from the Atterberg limit tests by Bromwell (1978) showed that all the points plot close to the A line, except for a brackish water environment for which the liquid limit was somewhat higher. Murdock and Zeman (1975) reported a considerable decrease in the plastic properties when a soil was either air dried, oven dried or freeze dried. The collapse of swelling minerals, the decomposition of organics, and oxidation of iron was believed to be responsible for the reduced plasticity. The drop in the plastic properties has been reported for natural organic soils as well (Terzaghi and Peck, 1967).

The dry unit weight of the dredge materials ranges between 51 and 64 lb/ft^3 (8 and 10 kN/m^3). This unit weight was reported to increase by about 4% per year (Krizek and Salem, 1977b).

Table 4.12 Index properties of dredge materials

Description	*Sand* (%)	*Silt* (%)	*Clay* (%)	*Organic* (%)	w_l	I_p	*G*
Toledo, Ohio	14–19	46–50	31–40	4–8	55–95	12–57	
Mobile, Alabama	7	18	75	5	100	35	2.72
Average USA	16	50	33		80–120	15–55	2.65
West Germany	0–5	35–60	10–40				
Seawater	5–95	5–70	0–40		45–212	25–65	

Sources: Bromwell, 1978; Carrier *et al.*, 1983; Haliburton, 1977; Krizek and Salem, 1977a; Lacasse *et al.*, 1977; Long and Demars, 1987; Rizkallah, 1987.

Summary

Weight–volume relationships and index properties are defined. A unified classification system for soils is described. Methods of determining unit weight of municipal wastes are presented. These unit weights may vary from 18 to 90 lb/ft^3 depending on the amount of compaction and the constituents of the waste.

Index properties of mineral, industrial and dredge waste are tabulated. Considerable variations in these properties is found as there are a large number of factors that affect these properties. Methods of determining some of the index properties are given in Appendix A4.

Notation

C_c coefficient of curvature
C_u coefficient of uniformity
d effective particle diameter
D_{10} particle diameter for 10% fines by weight
D_{30} particle diameter for 30% fines by weight
D_{60} particle diameter for 60% fines by weight
e void ratio
h_l height (depth) of landfill
I_p plasticity index
I_L liquidity index
n porosity
S degree of saturation
V total volume of soil
V_s volume of solids
V_v volume of voids
V_w volume of water
w water content (%)
W total weight of soil
w_l liquid limit
w_n natural water content
w_p plastic limit
w_s shrinkage limit
W_s weight of solids
W_w weight of water
γ total unit weight of soil
γ_b buoyant unit weight of soil
γ_d dry unit weight of soil
γ_s unit weight of solids
γ_w unit weight of water
σ'_o effective overburden pressure before landfill
σ'_{vc} effective consolidation pressure due to landfill
Δu excess pore pressure in the foundation soil

References

Boyd, M. B., Saucier, J. W., Keeley, R. O., Montgomery Brown, R. D., Mathis, D. B. and Guice, C. J. (1972) *Disposal of Dredge Spoil*, Dredge Material Research Program Technical Report H-72-8, US Army Engineer Waterways Experiment Station, Vicksburg, Miss

Bromwell, L. G. (1978) Properties, behavior and treatment of waste fills, *ASCE, Met. Section, Seminar—Improving Poor Soil Conditions*, October 1978, New York

Bromwell, L. G. and Oxford, T. P. (1977) Waste clay dewatering and disposal, *Proc. Conf. on Geotechnical Practice for Disposal of Solid Waste Materials*, ASCE, Ann Arbor, Michigan, June 1977, pp. 541–548

Carrier, III, W. D., Bromwell, L. G. and Somogyi, R. (1983) Design capacity of slurried mineral waste ponds, *J. Geot. Eng., ASCE*, **109**, No. 5, 699–718

Chae, Y. S. and Gurdziel, T. J. (1976) New Jersey fly ash as structural fill. In *New Horizons in Construction Materials*, Vol. 1, Ed. Fang, H. Y., Envo Publishing Co. Inc., Bethlehem, Pa, pp. 1–13

Charlie, W. A. (1977) Pulp and papermill solid waste disposal: a review, *Proc. Conf. on Geotechnical Practice for Disposal of Solid Waste Materials*, ASCE, Ann Arbor, Michigan, June 1977, pp. 71–86

Collins, R. J. (1977) Highway construction of incinerator residue, *Proc. Conf. on Geotechnical Practice for Disposal of Solid Waste Materials*, ASCE, Ann Arbor, Michigan, June 1977, pp. 246–266

Fenn, D. G., Hanley, K. J. and Deseare, T. U. (1975) *Use of the Water Balance for Predicting Leachate Concentration from Solid Waste Disposal Sites*, EPA-530/SW-168, USEPA, Cincinnati, Ohio

Haliburton, T. A. (1977) Development of alternatives for dewatering dredged materials, *Proc. Conf. on Geotechnical Practice for Disposal of Solid Waste Materials*, ASCE, Ann Arbor, Michigan, June 1977, pp. 615–631

Hagerty, D. J., Ullrich, C. R. and Thacker, B. K. (1977) Engineering properties of FGD sludges, *Proc. Conf. on Geotechnical Practice for Disposal of Solid Waste Materials*, ASCE, Ann Arbor, Michigan, June 1977, pp. 23–40

Holubec, I. (1976) Geotechnical aspects of coal waste embankments, *Can. Geot. J.*, **13,** No. 1, 27–39

Krizek, R. J. and Salem, A. M. (1977a) Field performance of a dredge disposal area, *Proc. Conf. on Geotechnical Practice for Disposal of Solid Waste Materials*, ASCE, Ann Arbor, Michigan, June 1977, pp. 358–383

Krizek, R. J. and Salem, A. M. (1977b) Time-dependent development of strength in dredgings, *Journal of the Geotechnical Engineering Divison, ASCE*, **103,** No. GT3, 168–184

Krizek, R. J., Chu, S. C. and Atmatzidis, D. K. (1987) Geotechnical properties and landfill disposal of FGD sludge. In *Geotechnical Practice for Waste Disposal '87*, Proc. Special Conf. Geot., Special Publ. No. 13, Ed. Woods, R. D., pp. 625–639

Lacasse, S. M., Lambe, T. W., Maar, W. A. and Neff, T. L. (1977) Void ratio of dredged material, *Proc. Conf. on Geotechnical Practice for Disposal of Solid Waste Materials*, ASCE, Ann Arbor, Michigan, June 1977, pp. 153–168

Lambe, T. W. and Whitman, R. V. (1969) *Soil Mechanics*, John Wiley & Sons, New York

Long, R. P. and Demars, K. R. (1987) Shallow ocean disposal of contaminated dredge material, *Geotechnical Practice for Waste Disposal '87*, Proc. Special Conf. Geot., Special Publ. No. 13, Ed. Woods, R. D., pp. 655–667

Mabes, D. L., Hardcastle, J. H. and Williams, R. E. (1977) Physical properties of Pb–Zn mine-process waste, *Proc. Conf. on Geotechnical Practice for Disposal of Solid Waste Materials*, ASCE, Ann Arbor, Michigan, June 1977, pp. 103–117

McLaren, R. J. and DiGioia, A. M. (1987) The typical engineering properties of fly ash, *Geotechnical Practice for Waste Disposal '87*, Proc. Special Conf. Geot., Special Publ. No. 13, Ed. Woods, R. D., pp. 683–697

Maynard, T. R. (1977) Incinerator residue disposal in Chicago, *Proc. Conf. on Geotechnical Practice for Disposal of Solid Waste Materials*, ASCE, Ann Arbor, Michigan, June 1977, pp. 773–792

Moulton, L. K., Rao, S. K. and Seals, R. K. (1976) The use of coal associated wastes in the construction and stabilization of refuse landfills. In *New Horizons in Construction Materials*, Vol. 1, Ed. Fang, H. Y., Envo Publishing Co. Inc., pp. 53–65

Murdock, A. and Zeman, A. J. (1975) Physico chemical properties of dredge spoil, *ASCE, Journal of the Waterways, Harbors and Coastal Engineering Division*, **101,** No. WW2, 201–214

NAVFAC DM 7.1 (1982) *Soil Mechanics Design Manual*, Naval Facilities Engineering Command

Oweis, I. A. and Khera, R. (1986) Criteria for geotechnical construction on sanitary landfills, *Int. Symp. on Env. Geot.*, Vol. 1, Ed. Fang, H. Y., pp. 205–222

Peirce, J., Sallfors, G. and Murray, L. (1986) Overburden pressures exerted on clay liners, *ASCE, Envir. Eng.*, **112,** No. 2, 284

Rizkallah, V. (1987) Geotechnical properties of polluted dredged material, *Geotechnical Practice for Waste Disposal '87*, Proc. Special Conf. Geot. Special Publ. No. 13, Ed. Woods, R. D., pp. 759–771

Sargunan, A., Mallikarjun, N. and Ranapratap, K. (1986) Geotechnical properties of refuse fills of Madras, India, *Int. Symp. on Env. Geot.*, Vol. 1, Ed. Fang, H. Y., pp. 197–204

Shoemaker, N. B. (1972) Construction techniques for sanitary landfill, *Waste Age*, March/April, 24–25, 42–44

Somogyi, F. and Gray, D. H. (1977) Engineering properties affecting disposal of red muds, *Proc. Conf. on Geotechnical Practice for Disposal of Solid Waste Materials*, June 1977, pp. 1–22

Sowers, G. F. (1973) Settlement of waste disposal fills, *8th Int. CSMFE*, Moscow, USSR National Society for Soil Mechanics and Foundation Engineering, pp. 207–210

Tchobanoglous, G., Theisen, H. and Eliassen, R. (1977) *Solid Wastes*, McGraw-Hill, New York

Terzaghi, K. and Peck, R. B. (1967) *Soil Mechanics in Engineering Practice*, 2nd Ed., John Wiley & Sons, New York

Ullrich, C. R. and Hagerty, D. J. (1987) Stabilization of FGC wastes, *Geotechnical Practice for Waste Disposal '87*, Proc. Special Conf. Geot. Special Publ. No. 13, Ed. Woods, R. D., pp. 797–811
US Environmental Protection Agency (EPA) (1986) 'Subtitle D study', Phase IR, EPA/530-SW86-054, October

Appendix A4

A4.1 Index properties tests

Water content (ASTM D 2216)

A representative test specimen in its natural moist state is weighed and then dried in an oven at a temperature of 110° ± 5°C to a constant weight.

Unit weight (AASHTO 233)

This technique is suitable for cohesive soils in their natural or compacted state and for soils that remain intact during sampling. A block of soil coated with a known amount of paraffin is weighed to obtain its net weight. The sample is then immersed in an overflow volumeter for determination of its net volume.

Specific gravity (ASTM D 854)

The procedure is used for soils passing No. 4 (4.75 mm) sieve. The soil may be at its natural water content or oven dried. A pycnometer and distilled water are used and the soil suspension is deaired. The specific gravity values are based on a water temperature of 20°C.

Particle size analysis (ASTM D 422)

A dry sample of soil is passed through a series of sieves and the weight retained on each sieve is recorded. In instances where the soil contains plastic fines which may adhere to sand particles it is necessary to wash the sample through a sieve No. 200 (75 μm). The grain size distribution for fines is determined using a hydrometer graduated to read grams/litre. The method is based on Stokes's law according to which different particle sizes settle at different velocities.

Liquid limit (ASTM 243)

Most natural soils have a water content below their liquid limit. It is generally necessary to add water (distilled) to the soil for the test. The specimen is placed in a standard cup and divided by a grooving tool. The moisture content at which the two soil halves come together at the bottom of the groove under the impact of 25 blows is defined as the liquid limit.

Plastic limit (ASTM 424)

The soil specimen is rolled into $\frac{1}{8}$ inch thread. The moisture content at which this thread breaks or shows cracks is the plastic limit.

Shrinkage limit (Corps of Engineers Method Engineering Manual EM 1110-2-1906)

The shrinkage limit is the moisture content at which the soil is saturated and any loss of moisture will not cause a decrease in volume.

A4.2 *In situ* tests

The above techniques for determinations of index properties are used in the laboratory. Sometimes these tests must be conducted in the field: however, the results are less reliable than the laboratory results.

A4.2.1 Moisture content

Nuclear method (ASTM 3017)

This method measures *in situ* water content by directing neutrons into the soil and measuring the intensity of neutrons reflected back. The equipment is calibrated to determine the water content as weight of water per unit volume of material.

Gas burning method

A moist sample is weighed, placed in a pan and allowed to dry on a stove to a constant weight with occasional stirring.

Speedy meter (AASHTO T-217)

The moisture content is determined by measuring the pressure resulting from the acetylene gas generated due to the reaction between soil moisture and calcium carbide.

A4.2.2 Unit weight

Sand cone method (ASTM D1556)

The test consists of excavating a hole and determining its volume using sand. The material removed is weighed and tested for moisture content.

Nuclear moisture/density gauge, for shallow depth (ASTM 2922)

The wet density is determined by placing a gamma radiation source and a detector on, into, or adjacent to the material to be tested. The intensity of the radiation detected is converted into wet density using a proper calibration.

Chapter 5

Compressibility and strength

5.1 Geostatic stress

The stresses due to the self-weight of a deposit are known as geostatic stresses. The unit weight γ of a natural soil deposit usually increases with depth. Where the ground surface is horizontal the vertical stress σ_v, at a given depth z, may be determined as

$$\sigma_v = \int_0^z \gamma \, dz \qquad (5.1)$$

Alternatively, if there are more than one type of materials present, the stress at any given depth is computed by adding the stress from each of the layers above that point:

$$\sigma_v = \Sigma_1^n \gamma_n z_n \qquad (5.2)$$

If groundwater is present, then neutral pressure u_0, due to static head of water h_w, is computed as

$$u_0 = \gamma_w h_w \qquad (5.3)$$

where γ_w is the unit weight of water.

5.2 Effective stress

When an external load is applied to a soil or refuse mass, the total stresses increase. The stress increases occur in each of the three phases; that is, solid, liquid, and gas. The stresses developed within the pore spaces are known as pore pressures. The total stress minus the pore pressure is called effective stress and is given by

$$\sigma' = \sigma - u \qquad (5.4)$$

where σ is the total stress and u the pore pressure.

Horizontal stress within a soil mass is given by:

$$\sigma_h = K_0 \sigma'_v \qquad (5.5)$$

where K_0 is the coefficient of lateral stress.

The value of K_0 has a large range and is affected by the nature of the material and its stress history.

5.3 Stresses from surface loads

The changes in vertical stress due to surface loads are determined from the solutions based on elasticity theory. Most of these solutions assume the materials under stress to be semi-infinite, elastic and isotropic. The surface loads are considered completely flexible and the shearing stresses between the loaded area and the underlying soils are negligible.

The increase in vertical stress due to surface loads is determined from the influence diagrams which have been prepared for various types of loads and is given by

$$\sigma_v = Ip \tag{5.6}$$

where I is the influence value and p the intensity of surface load.

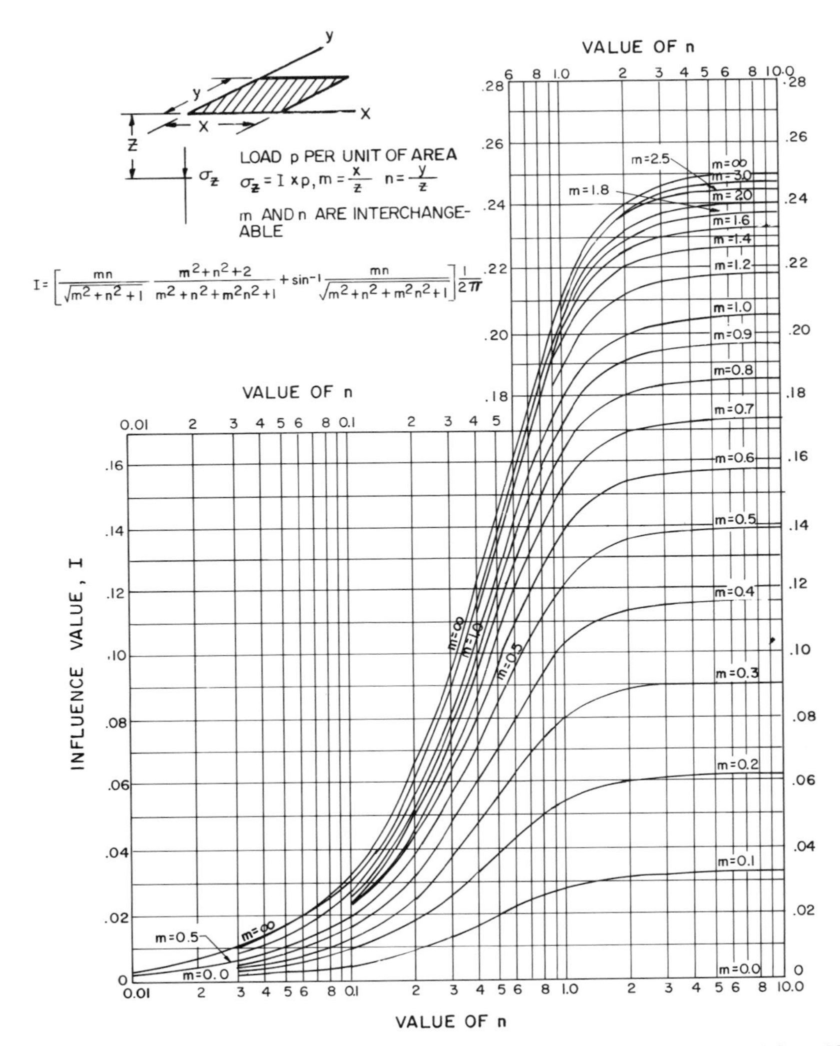

Figure 5.1 Influence values for a rectangular uniform surface load (Reproduced from NAVFAC, 1982)

The influence values for rectangular surface load are shown in Figure 5.1, for an embankment with symmetrical triangular cross-section of infinite length in Figure 5.2, for triangular loads of finite length in Figure 5.3, and for an embankment load of infinite length in Figure 5.4.

Example 5.1

Find the stress increase due to a long embankment shown in Figure 5.5. The unit weight of the embankment is $18\,kN/m^3$.

Solution Extend the slope of the embankment to meet at L. (Refer to Figures 5.5 and 5.2.)

The increase in stress at A from JLN is

$2z/b = 2\times 6/24$ or 0.5
$2x/b = 2\times 3/24$ or 0.25
$\sigma_v = 18\times 8\times 0.62$ or 89.3 kPa

Decrease in stress at A from triangle KLM, which must be placed directly on the soil, is shown as K′L′M′:

$2z/b = 2\times 6/12$ or 1.0
$2x/b = 2\times 3/12$ or 0.5
$\sigma_v = 18\times 4\times 0.39$ or 28.1 kPa

Net increase in stress $= 89.3 - 28.1$ or 61.2 kPa.

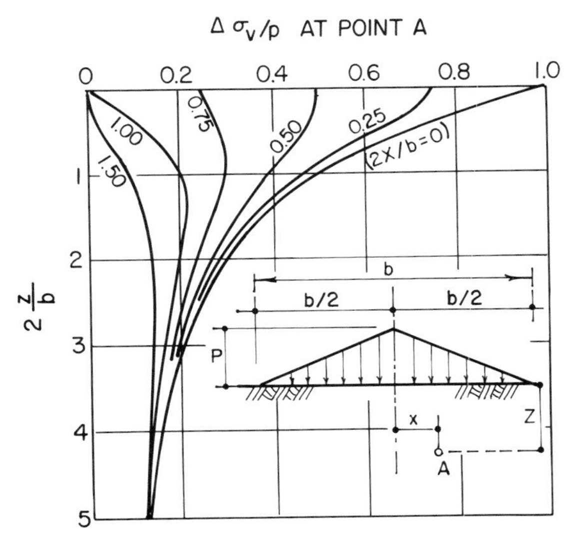

Figure 5.2 Influence values for vertical stress from infinitely long triangular embankment loads (Reproduced from Stamatopoulos and Kotzias, 1985, figure 4.7, p. 89, by permission of John Wiley & Sons Inc.)

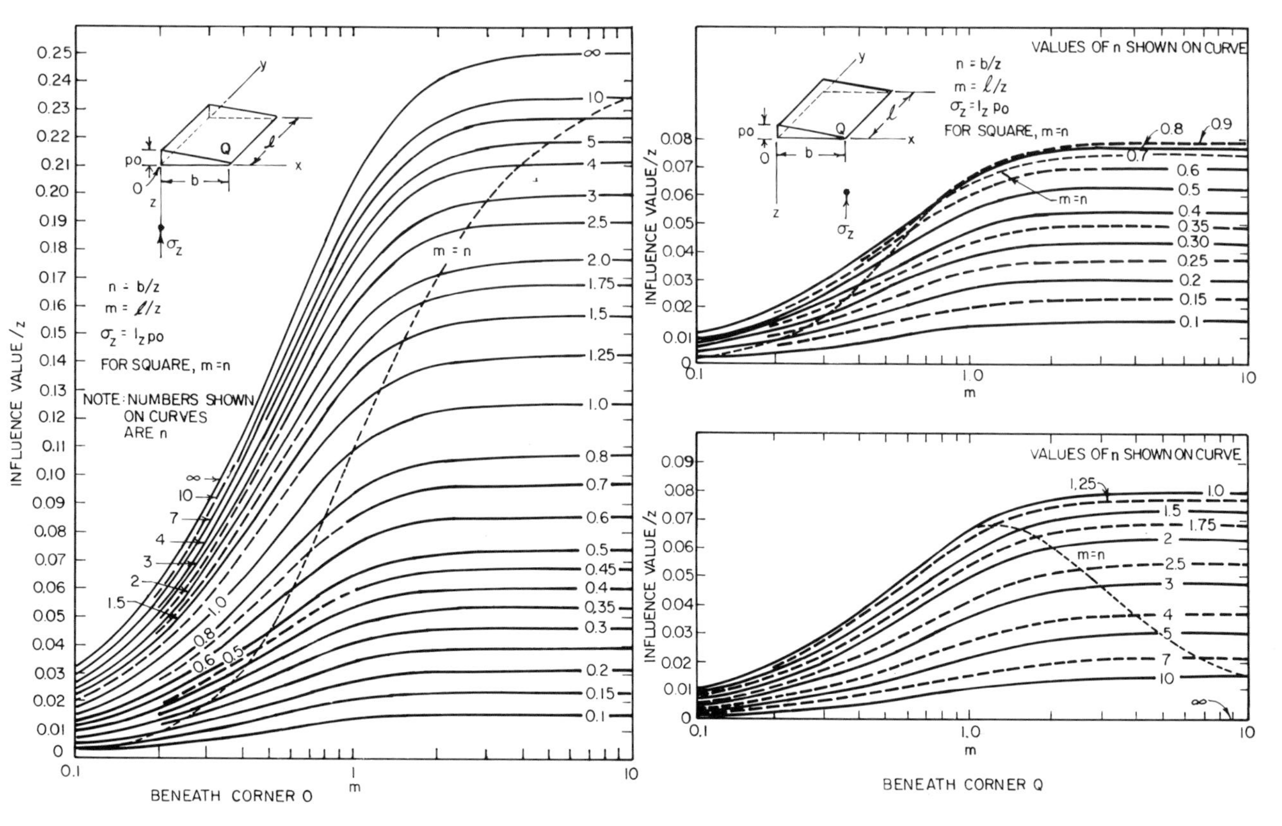

Figure 5.3 Influence values for vertical stress from triangular embankment loads with rectangular plan area (Reproduced from NAVFAC, 1982)

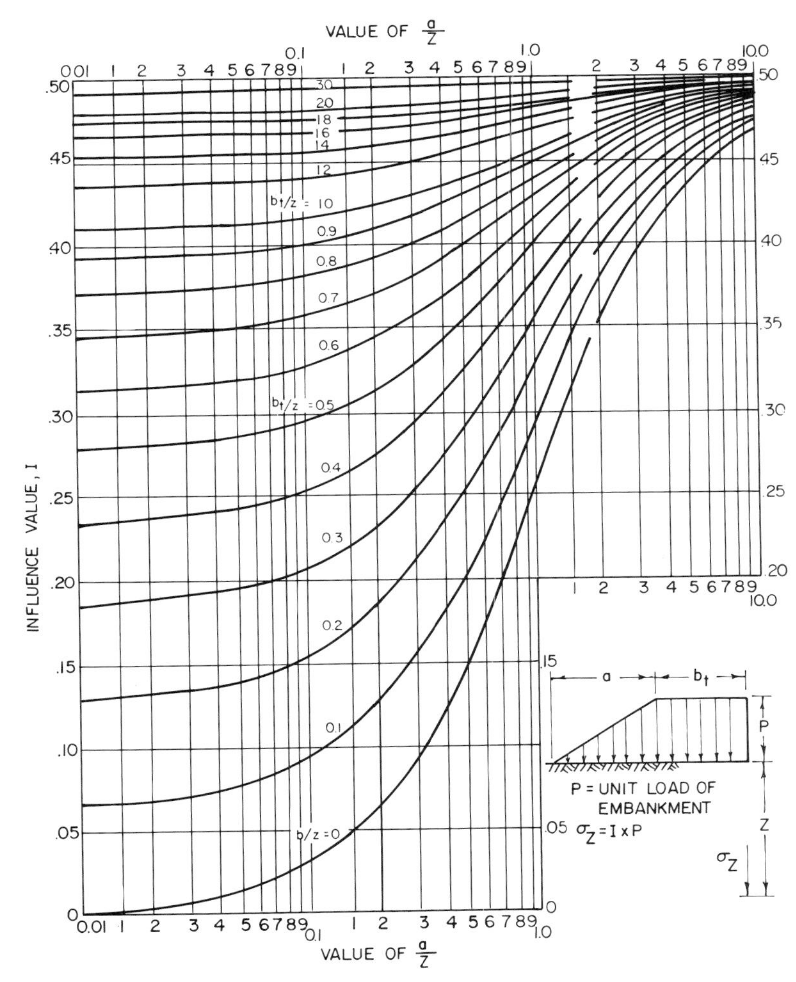

Figure 5.4 Influence values for vertical stress from infinitely long embankment loads (Reproduced from NAVFAC, 1982)

Example 5.2

An embankment with a unit weight of 18 kN/m^3 is shown in Figure 5.6. Increases in stresses are required at 6 m below points A and B.

Solution Stress at A′: assume uniform load on rectangles AEFG and AIHG, and uniformly increasing load on AEKB and AIJB.

For stress contribution from AEFG and AIHG,

$x/z = 12/6$ or 2.0
$y/z = 15/6$ or 2.5

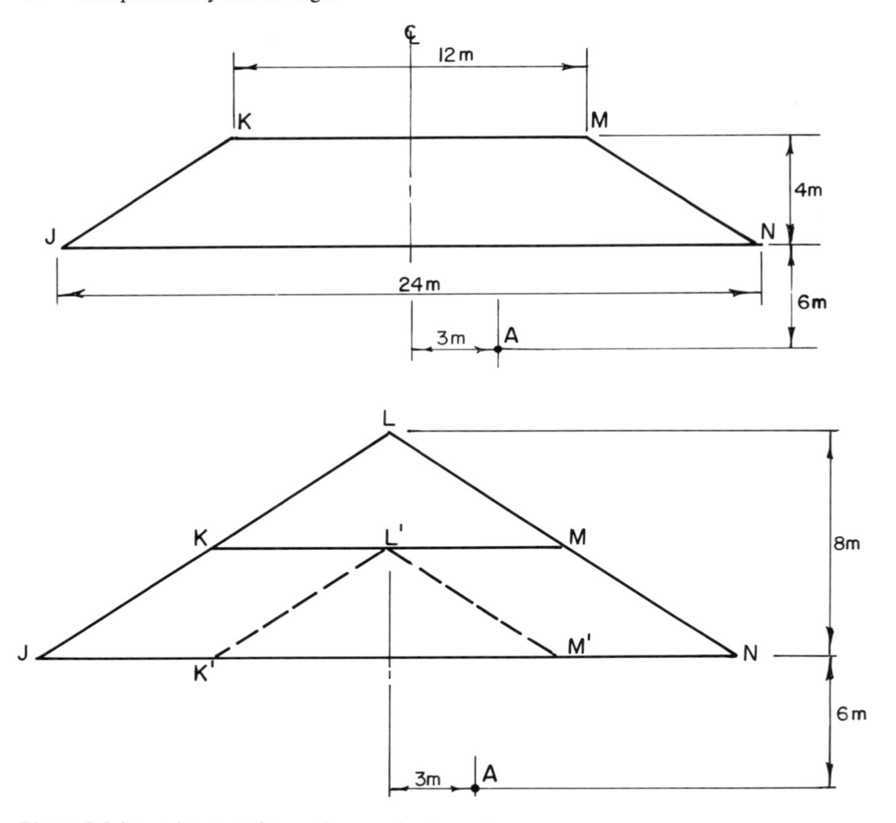

Figure 5.5 Stress increase from a long embankment

From Figure 5.1,

$I = 0.236$

$\sigma_v = 2 \times 18 \times 4 \times 0.236$ or 34.0 kPa

and, for stress contribution from AEKB and AIJB, use Figure 5.3:

$b/z = 6/6$ or 1.0

$l/z = 12/6$ or 2.0

$I = 0.123$, considering contribution from both sides of AB,

$\sigma_v = 2 \times 18 \times 4 \times 0.123$ or 17.7 kPa

Therefore, stress increase at $A' = 30.0 + 17.7$ or 47.7 kPa.

Stress at point B′: assume uniform load on rectangles BKFG and BJHG, and uniformly increasing load on AEKB and AIJB.

For stress contribution from BKFG and BJHG, use Figure 5.1:

$x/z = 12/6$ or 2.0

$y/z = 18/6$ or 3.0

$\sigma_v = 2 \times 18 \times 4 \times 0.238$ or 38.9 kPa

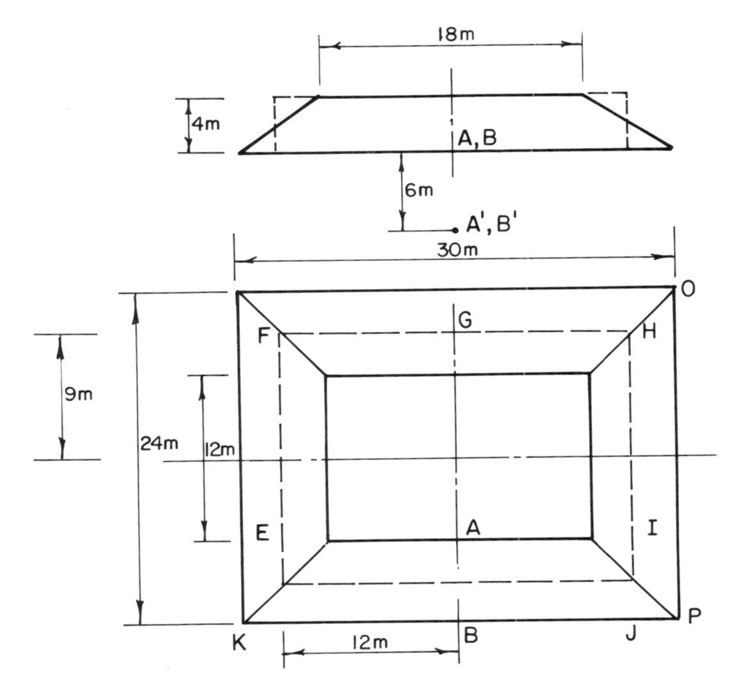

Figure 5.6 Stress increase from a rectangular embankment

and, for stress contribution from AEKB and AIJB, use Figure 5.3:

$b/z = 6/6$ or 1.0
$l/z = 12/6$ or 2.0
$\sigma_v = 2 \times 18 \times 4 \times 0.123$ or 17.7 kPa

Therefore, stress increase at B′ = 38.9 − 17.7 or 21.2 kPa.

5.4 Settlement

The settlement in fine-grained or soft material is assumed to have three components (Figure 5.7); namely, initial settlement, δ_i, settlement due to primary consolidation, δ_c, and settlement due to secondary compression, δ_s.

$$\text{Total settlement, } \delta_t = \delta_i + \delta_c + \delta_s \quad (5.7)$$

In saturated soils the initial or distortional settlement occurs quickly and may be accompanied by a lateral expansion. If the soil is not saturated there would be an immediate settlement due to compression of air or gases in the voids. The settlement due to primary consolidation is determined from consolidation theory. Secondary compression comes from creep and other causes. The total settlement is dependent on the nature of the soil, its stress history, geological conditions, relative thickness of the compressible layer with respect to the width of the loaded area, intensity of the load, and rate of application.

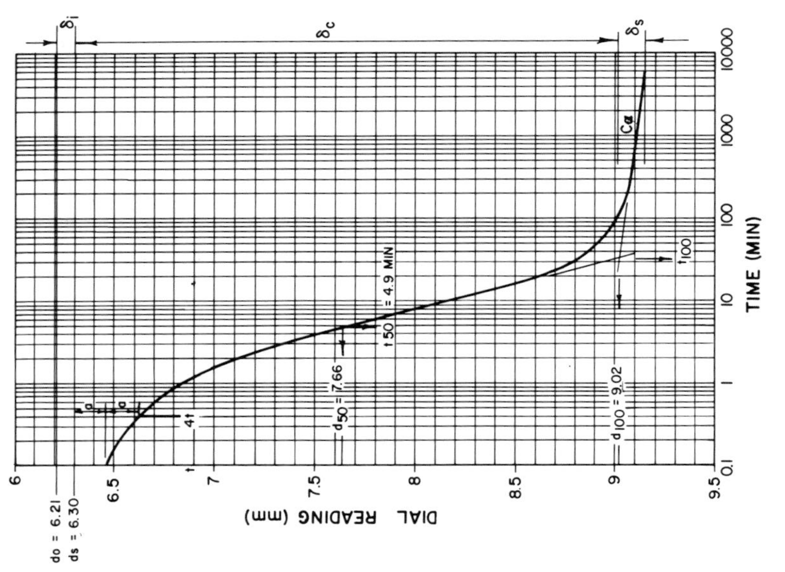

Figure 5.7 (a) Dial reading *versus* log T

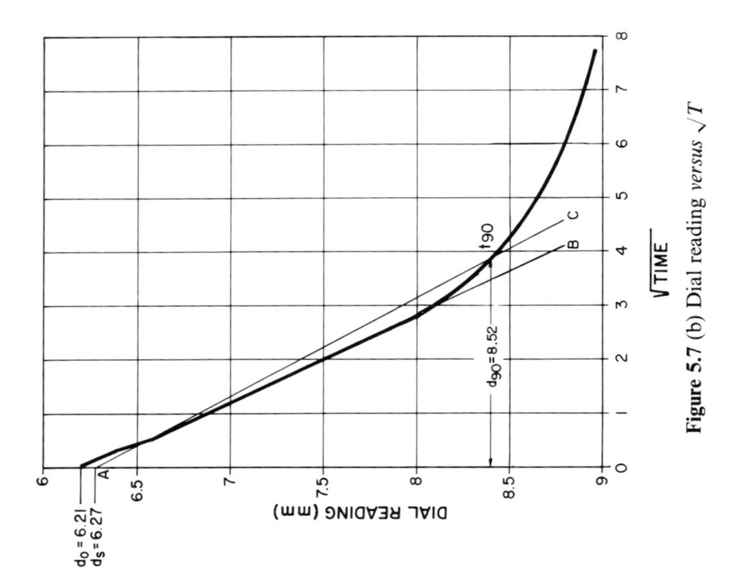

Figure 5.7 (b) Dial reading *versus* $\sqrt{T}$

5.4.1 Initial settlement

The initial settlement is estimated from

$$\delta_i = pbI(1-\nu^2) \div E_u \tag{5.8}$$

where p is the average stress on the soil surface, b the width of the loaded area, I the influence factor (which depends on shape and rigidity of the load), ν Poisson's ratio, and E_u the undrained modulus.

The approximate value of the influence factor for a long rectangular load on a compressible soil of thickness H and Poisson's ratio between 0.3 and 0.5 is given (Stamatopoulos and Kotzias, 1985) by:

$$I = 0.094H/b \qquad \text{for } H/b \leqslant 4 \tag{5.9a}$$

$$I = 0.32 + 0.013H/b \qquad \text{for } H/b > 4 \tag{5.9b}$$

During preloading most materials are stressed close to their load capacity. E_u for soft clays under such conditions is about $100s_u$ to $200s_u$, where s_u is the undrained shear strength. For stresses below yield (e.g. one-third yield stress) E_u may vary from $500s_u$ to $1000s_u$. For an experimental technique to determine E_u see Lee and Valliappan (1974). For waste materials such as those existing at municipal landfills it is generally not possible to determine the magnitude of δ_i.

5.4.2 Consolidation settlement

The settlement due to consolidation is determined from one-dimensional consolidation theory (Terzaghi, 1943), which assumes that the soil is saturated and homogeneous, the flow is one-dimensional, the coefficient of consolidation is constant for a given load increment, the vertical displacement is small, and the load is applied instantaneously.

Based on an arithmetic plot of stress–strain curve, the coefficient of volume compressibility, m_v, is the slope of the curve for a given load increment $d\sigma'_v$ (Figure 5.8a):

$$m_v = d\varepsilon / d\sigma'_v = 1/D \tag{5.10}$$

where D is the constrained modulus, ε the strain (dH/H), and H the thickness of soil.

The settlement for a small load increment for which the stress–strain is linear is given by Equation 5.11a, otherwise the curve is approximated by a series of straight lines and Equation 5.11b is used.

$$\delta_c = Hm_v\, d\sigma'_v \tag{5.11a}$$

$$= H\Sigma(m_v\, d\sigma'_v) \tag{5.11b}$$

$$= H\Sigma(d\sigma'_v/D) \tag{5.11c}$$

A laboratory void ratio (e) or strain (ε) *versus* stress curve on a semi-log plot is shown in Figure 5.8b in which:

σ'_{vo} = effective overburden stress
σ'_{vm} = maximum past-consolidation stress
σ'_{vc} = effective consolidation stress
C_c = compression index
= slope of virgin compression line

$= de/\log(\sigma'_2/\sigma'_1)$

C_r = recompression index

CR = compression ratio

$= d\varepsilon/\log(\sigma'_2/\sigma'_1)$

e_o = initial void ratio

In a normally consolidated soil $\sigma'_{vm} = \sigma'_{vo}$, and the consolidation settlement is obtained using the slope of the virgin compression line:

$$\delta_c = \frac{HC_c}{(1+e_0)}[\log(\sigma'_{vc}/\sigma'_{vo})] \tag{5.11d}$$

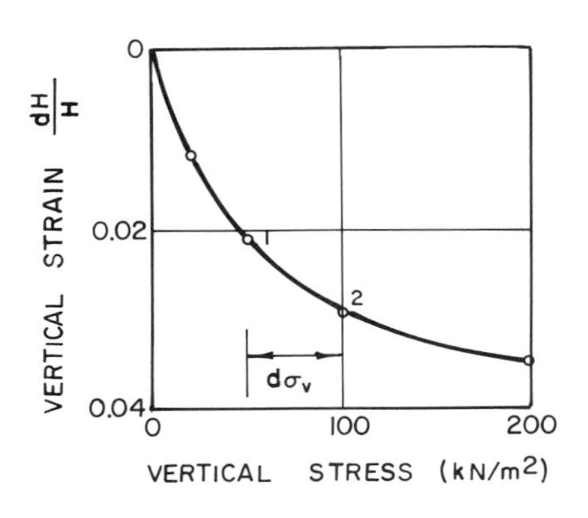

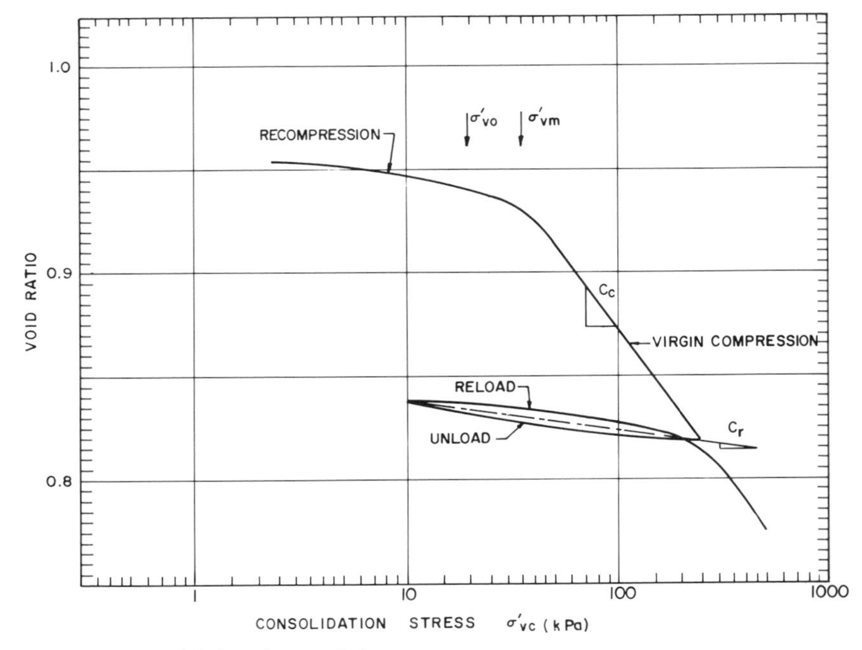

Figure 5.8 Consolidation characteristics

$$= H\,\mathrm{CR}\,[\log(\sigma'_{vc}/\sigma'_{vo})] \tag{5.11e}$$

where CR is the slope of the virgin portion of strain–log pressure curve.

In an overconsolidated soil, $\sigma'_{vm} > \sigma'_{vo}$, and the consolidation settlement is given by:

$$\delta_c = \frac{H}{(1+e_0)}[C \log(\sigma'_{vm}/\sigma'_{vo}) + C_c \log(\sigma'_{vc}/\sigma'_{vm})] \tag{5.11f}$$

$$= H[\mathrm{RR}\log(\sigma'_{vm}/\sigma'_{vo}) + \mathrm{CR}\log(\sigma'_{vc}/\sigma'_{vm})] \tag{5.11g}$$

where RR is the slope of the recompression portion of the strain–log pressure curve and is called the recompression ratio.

In the absence of consolidation test data and for preliminary settlement computations the value of C_c may be estimated. For inorganic soils with sensitivity and liquid limit less than 4 and 100, respectively, the compression index is given by

$$C_c = 0.009(w_l - 10) \tag{5.12}$$

For organic soils and peat,

$$C_c = 0.0115 w_n \tag{5.13}$$

where w_l is the liquid limit and w_n the natural water content.

Several other empirical correlations may be found in Holtz and Kovacs, 1981.

5.4.3 Rate of consolidation

When a saturated soil of low permeability is subjected to an external load an excess pore pressure, u_e, develops. The differential equation governing the dissipation and distribution of the excess pore pressure is

$$\partial u_e/\partial t = c_v(\partial^2 u_e/\partial z^2) \tag{5.14}$$

where t is the time, z the depth of a soil layer measured from the top, c_v the coefficient of consolidation for vertical flow ($= k/\Gamma_w m_v$), Γ_w the unit weight of water, and k_v the hydraulic conductivity in a vertical direction.

The solution to Equation 5.14, assuming initial excess pore pressure to be equal to the applied stress and drainage at top and bottom, is shown in Figure 5.9. The dimensionless parameters Z (depth factor), T_v (time factor), and U_z (consolidation ratio) are:

$$Z = z/H_d \tag{5.15}$$

$$U_z = 1 - (u_e/u_i) \tag{5.16}$$

$$T_v = c_v t/(H_d)^2 \tag{5.17}$$

where H_d is the shortest distance to the drainage boundary, u_i the initial excess pore pressure, and u_e the excess pore pressure at time t.

The average degree of consolidation, U_v, is obtained by summation of the consolidation ratio over the entire depth of the stratum. The solution to the theoretical equation is graphically represented in Figure 5.10 and tabulated in Table 5.1. The actual equation can be represented fairly precisely by the empirical equations $T_v = \pi\,(U_v)^2/4$ for $U_v < 0.6$, and $T_v = 1.781 - 0.9332\log(100 - U_v\%)$ for $U_v > 0.6$.

The coefficient of consolidation is determined using curve-fitting techniques and laboratory data. A semi-log and a square root of time fitting method are illustrated by examples.

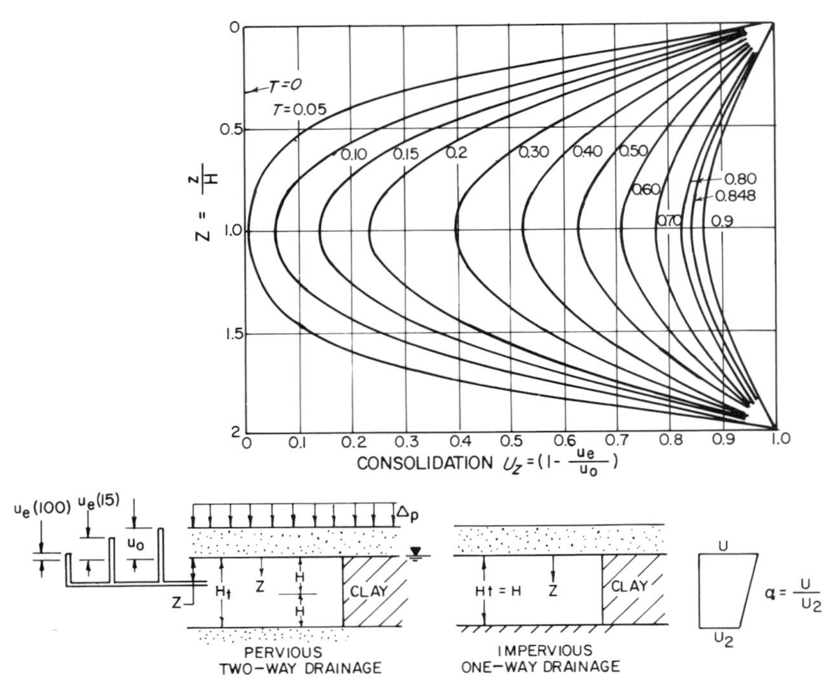

Figure 5.9 Dimensionless parameter for consolidation ratio (Reproduced from NAVFAC, 1982)

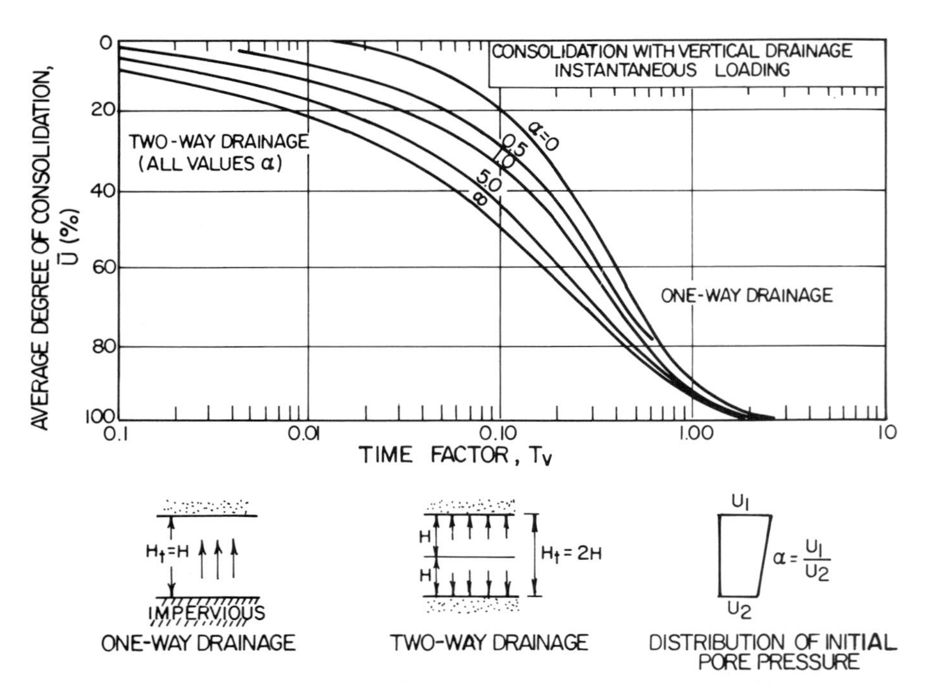

Figure 5.10 Average degree of consolidation (Reproduced from NAVFAC, 1982)

Table 5.1 Relationship between average degree of consolidation and dimensionless time factor

Average degree of consolidation, U_v (%)	*Dimensionless time factor, T_v*
0	0.00
10	0.007
20	0.031
30	0.071
40	0.126
50	0.197
60	0.286
70	0.403
80	0.567
90	0.848
95	1.128
100	∞

Example 5.3

The height of a test specimen before the application of a stress of 200 kPa (2 ton/ft^2) is 21.24 mm. The time settlement curve is represented by Figure 5.7(a). Determine c_v for this load increment.

Solution The distance d_0 to d_s represents the immediate compression.

Average sample thickness at end of primary consolidation

$= 21.24 - (9.02 - 6.30) \div 2$

$= 19.88$ mm

Drainage height, H_d

$= 19.88/2$

$= 9.94$ mm (drainage at both ends)

Based on Figure 5.7a, for the laboratory test specimen for an average degree of consolidation of 50%, $t = 4.9$ min. From Table 5.1 for $U_v = 50\%$, the value of $T_v = 0.197$. Using Equation 5.17, $c_v = 2.04$ m^2/year.

In the square root of time curve the data lie on a straight line up to 60% consolidation, and the abscissa of the theoretical curve at 90% consolidation is 1.15 times the abscissa of the extension of this straight line (Taylor, 1948).

Example 5.4

Do Example 5.3 using a square root of time *versus* dial reading plot.

Solution Line AB (Figure 5.7b) is the best fit for the early data. Line AC has a slope 1.15 times that of AB. The distance d_0 to d_s represents the immediate compression.

Average specimen height at end of primary compression

$= 21.24 - (8.52 - 6.27)\ (10/9) \div 2$

$= 19.99$

Drainage height H_d=
$=19.99/2$
$=9.995$ mm (drainage at both ends)

From Figure 5.7(b), for an average degree of consolidation of 90% (d_{90}), $t=15.20$ min. From Table 5.1, for $U_v=90\%$, $T_v=0.848$. Using Equation 5.17, $c_v=2.92\ \text{m}^2/\text{year}$.

Generally the value of c_v from the square root of time curve is larger than that obtained with the semi-logarithmic plot. From the laboratory values, c_v is frequently low and results in an over-estimation of time for the settlement. A more realistic value of the coefficient of consolidation can be obtained from field piezometer measurements.

Example 5.5

The soil profile at a given site consists of a 20 m thick layer of a compressible material between an upper and a lower layer of sand. The water table is at the top of the compressible layer and the tip of a piezometer is 6.5 m below the top. A surface load causes a stress increase of 100 kPa. After 15 days the pressure head is 14.97 m and after 75 days it is 13.01 m. Determine the value of c_v.

Solution Initial head=[(100/9.8)+6.5] or 16.70 m and the depth factor, $Z=z/H_d=6.5/10$ or 0.65.

Time (days)	*Total head* (m)	*Excess head* (m)	$(1-u_e/u_i)$	*T* (*From Figure 5.9*)
0	16.70	10.5	$1-10.20/10.20=0$	
15	14.97	8.45	$1-8.45/10.20=0.172$	0.11
75	13.01	6.51	$1-6.51/10.20=0.362$	0.21

Therefore,

$$c_v=\frac{(H_d)^2(T_2-T_1)}{t_2-t_1}$$

$$=\frac{(10)^2(0.21-0.11)}{75-15}$$

$$=0.2\ \text{m}^2/\text{day}$$

The value of c_v depends on the magnitude of the stress and the stress increment ratio, i.e. the ratio of the applied stress to the existing stress. Both c_v and permeability decrease as the stress increment ratio climbs. If horizontal drainage is expected to dominate (as may be the case where the thickness of the compressible layer is much larger than the width of the loaded area, the horizontal permeability k_h is greater than the vertical permeability k_v, or vertical drains are used) the coefficient of consolidation for horizontal flow may be estimated as

$$c_h = c_v \cdot k_h / k_v$$

For materials that are extremely variable, only field permeability tests can provide reasonable values of k_h.

If there are more than one compressible layers with different properties or if the load-induced stress varies with depth, there will be more than one value for c_v. An equivalent thickness is computed for each of the layers using c_v values for one of the layers as the basis. With layer i as the reference the equivalent thickness, H_{ne}, of the other n layers is

$$H_{ne} = H_n (c_{vi}/c_{vn})^{0.5} \qquad (5.18)$$

where H_n is the actual thickness of layer n, c_{vi} is the coefficient of consolidation for layer i, and c_{vn} is the coefficient of consolidation for layer n.

The system is treated then as a single compressible layer with the coefficient of consolidation, c_{vi}, and a total equivalent thickness, H_{Te}, given by (NAVFAC, 1982):

$$H_{Te} = \sum_{i=1}^{i=n} H_{ie} \qquad (5.19)$$

The settlement at any time δ_{ct} is given by

$$\delta_{ct} = U_v \, \delta_c \qquad (5.20)$$

5.4.4 Secondary compression

The settlement after the primary consolidation is termed secondary compression. The settlement, δ_s, is given by

$$\delta_s = C_\alpha H_p \log(t/t_p) \qquad (5.21)$$

where $C_\alpha = d\varepsilon/\log(t/t_{100})$ (see Figure 5.7), t is the time at which secondary compression is needed, t_p is the time for completion of primary consolidation, and H_p is the thickness of the compressible layer at the end of primary consolidation.

In normally consolidated soils the coefficient of secondary compression, C_α, may be estimated as:

$$C_\alpha = w_n \times 10^{-4} \qquad (5.22)$$

C_α is much larger in the virgin compression range than in the recompression range. When the stress increment ratio is low, the relative contribution from secondary compression to the total settlement is larger.

5.4.5 Settlement of municipal waste

Settlement of municipal waste is calculated from Equations 5.11 and 5.21. The parameters CR and C_α are based on observations (see Chapter 10).

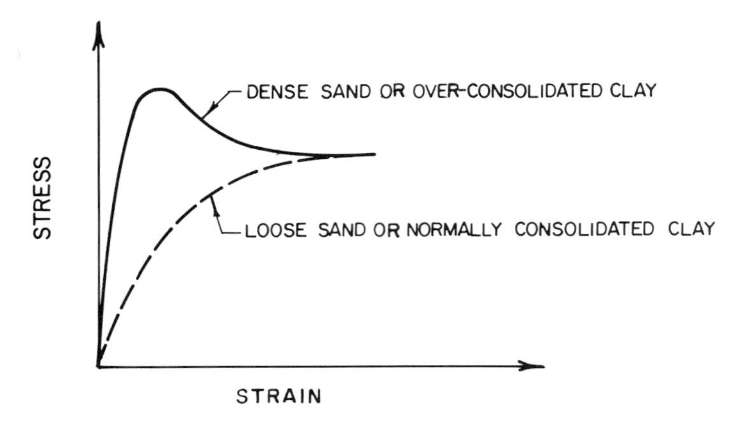

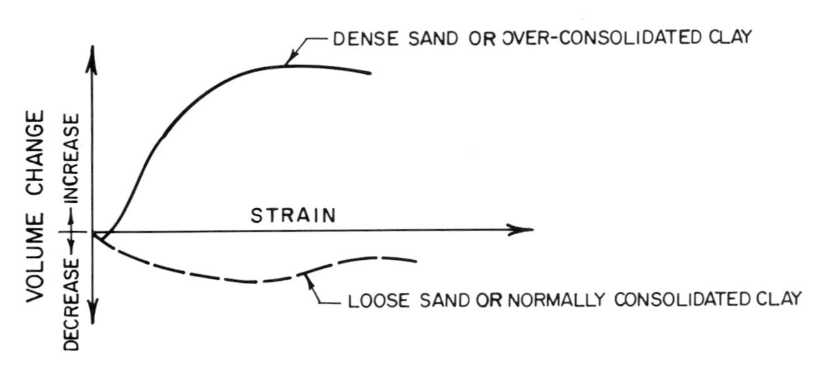

Figure 5.11 Non-linear stress–strain curves of cohesive and granular soils

5.5 Shear strength

In soils and waste materials the relationship between stress and strain is non-linear and time dependent. Volume changes develop from the applied normal and shearing stresses (Figure 5.11).

5.5.1 Failure theories

The most commonly used strength theories are those of Coulomb and Mohr. According to Coulomb the shear strength, τ_f, is expressed in terms of cohesion, c, and angle of friction, φ (Figure 5.12a).

$$\tau_f = c + \sigma \tan \varphi \tag{5.23}$$

Terzaghi introduced the concept of effective stress and showed that effective stress governs the strength:

$$\tau_f = c' + (\sigma - u) \tan \varphi' \tag{5.24}$$

The effective stress strength parameters c' and φ' can be determined from drained tests or tests with pore pressure measurements.

According to the failure theory by Mohr, a critical combination of normal and shearing stresses results in failure on a certain plane. A failure envelope is defined by a curve joining the points representing the failure stresses on these planes. It is often difficult to establish the orientation of the failure plane in a test specimen. The failure envelope is then obtained by a curve tangential to the circles representing stress at failure. Frequently, for stresses within a limited range, the Mohr failure envelope is approximated by a straight line which is then known as the Mohr–Coulomb rupture line (Figure 5.12b).

In laboratory strength tests a test specimen is subjected to a constant rate of deformation while the resulting load changes are measured, and volume changes or pore water pressure are measured. The following are some of the tests in common use.

5.5.2 Direct shear test

In a direct shear test a specimen is mounted in a shear box consisting of a fixed lower half and a movable upper half which is displaced horizontally at a constant rate of deformation. Failure is assumed to occur on the horizontal plane. In this test the state of stress within the specimen cannot be determined accurately nor can the pore water pressure be measured. Proper drainage may be assured only if the soil has a high permeability. This type of test is not considered appropriate for fine-grained materials unless it is of sufficiently long duration. Large shear boxes are often used in the field.

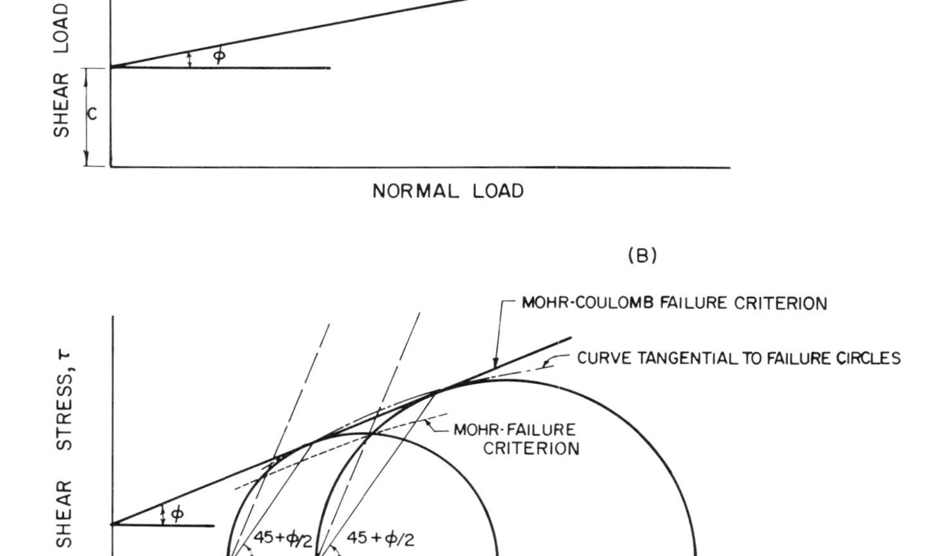

Figure 5.12 (A) Coulomb failure line. (B) Mohr–Coulomb failure criterion

5.5.3 Unconfined compression test

In an unconfined compression test a cylindrical specimen is subjected to an axial compression. Stiff soils show a distinct failure plane, whereas considerable bulging is displayed by soft soils. This test can be performed only on intact cohesive or cemented soils. If the materials to be tested possess any defects such as cracks, fissures, slickensides, seams of sand or silt, etc., this test should not be used. The failure line for typical tests on saturated cohesive soils is shown in Figure 5.13. The test is of short duration; there is no drainage.

5.5.4 Vane shear test

Vane shear equipment consists of a small rod with four vanes at one end and a calibrated spring at the other end. The vane is forced into the soil and rotated mechanically. The resulting torque is measured and from this the shear strength of the soil is determined. A tore vane is similar in principle and can also be used in the field.

5.5.5 Triaxial test

The triaxial test offers the most flexibility. In this test a cylindrical specimen is enclosed in flexible membranes and mounted between a cap and a pedestal with provisions for drainage and application of axial load. This assembly is surrounded by a pressure chamber. The load is applied in two stages. In the first stage a confining pressure is applied and in the second stage a constant rate of axial deformation is applied while the resulting load is measured. The time for completion of the first stage and the rate of deformation in the second stage are selected based on the nature of soil, imposed drainage conditions, sensitivity of the testing equipment, etc. If effective stress strength parameters are desired either the pore water pressure is measured or the specimen is permitted to drain.

5.5.6 Pore water pressure

Where pore water pressure measurements are desired the soil must be saturated. To achieve complete saturation all drainage lines are purged of air. A back pressure of 200–500 kPa is applied to the pore fluid while the chamber pressure is maintained 25–50 kPa higher than the back pressure. After the specimen has reached equilibrium, the chamber pressure is raised by a small increment $d\sigma_c$ (say 25 kPa) and the change in

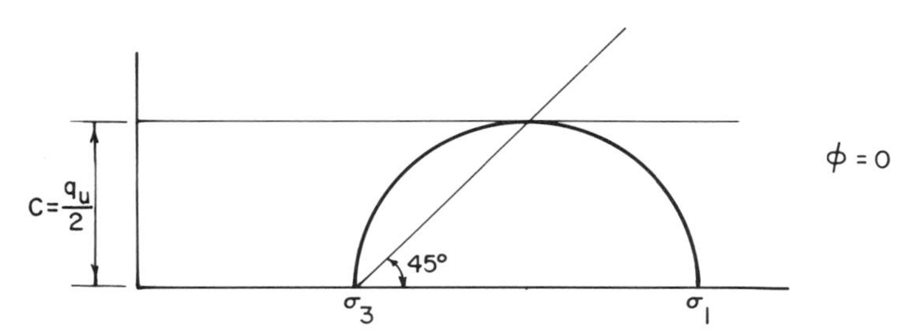

Figure 5.13 Results of the unconfined compression test

pore pressure du is monitored until it reaches equilibrium. Full saturation is assumed once a value greater than 0.95 is obtained for pore pressure parameter B (Skempton, 1954):

$$B = \mathrm{d}u/\mathrm{d}\sigma_c \qquad (5.25)$$

where dσ_c is the applied change in confining pressure and du is the resulting change in pore pressure.

Since volume changes occur during the application of shearing stresses, if drainage is not permitted, pore pressure will show changes. These changes are expressed in terms of pore pressure parameter A (Skempton, 1954):

$$A = \mathrm{d}u/(\sigma_1 - \sigma_3) \qquad (5.26)$$

where σ_1 is the major principal stress and σ_3 the minor principal stress.

For details on back pressure saturation see Black and Lee (1973), and for triaxial testing and interpretation see Bishop and Henkel (1962), and Berre (1981).

5.5.7 Unconsolidated undrained test

In an unconsolidated undrained (UU) test the soil is not allowed to drain during the application of confining pressure or axial load. In the absence of drainage no changes in void ratio occur so the strength remains constant. The diameter of all the stress circles at failure is, therefore, the same. When total stress circles are plotted their position on the normal stress axis is determined by the magnitude of the confining pressure, σ_c. Ideally, for a saturated soil the Mohr–Coulomb line is represented by a horizontal line, that is $\varphi = 0$ (Figure 5.13). When effective stress is used in plotting a single circle is obtained as any change in σ_3 results in corresponding change in the pore pressure.

5.5.8 Consolidated undrained test

During the first stage of testing in the consolidated undrained (CU) test, the test specimen is allowed to drain or consolidate at a confining pressure σ'_c. At the end of this stage the void ratio of the specimen corresponds to the applied σ'_c. In the second stage drainage is not permitted but pore pressure may be measured. The Mohr–Coulomb line shows a positive angle of friction and may also show a cohesion intercept.

For normally consolidated fine-grained soils the rupture line corresponding to the total stresses passes through the origin (Figure 5.14), $c = 0$ and φ is positive. The value of pore pressure parameter A is close to one. The effective stress circles move toward the origin, therefore the angle of friction corresponding to effective stress φ' is greater than φ.

Overconsolidated cohesive soils exhibit a cohesion intercept. Figure 5.15 shows the typical shape of strength envelope where identical test specimens are subjected to increasing σ'_c in the first stage of testing and then sheared. As the soil has a tendency to dilate negative pore pressure develops during the application of principal stress difference. The value of pore pressure parameter A is close to zero or even negative. The effective stress circles move away from the origin for lower σ'_c. As the consolidation stress is increased the specimen approaches normal consolidation and pore pressure becomes positive and the effective stress circles move toward the origin.

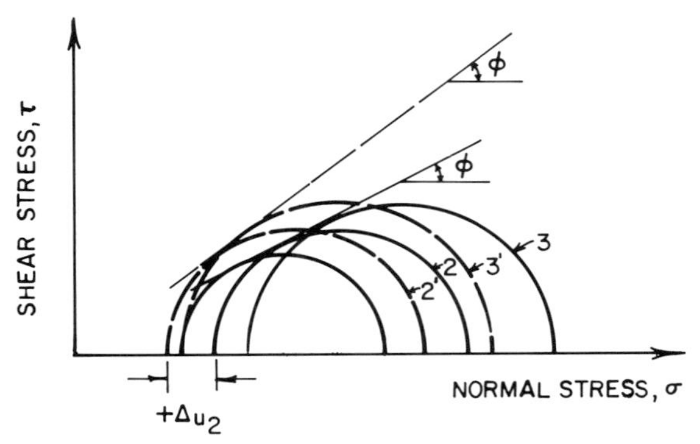

Figure 5.14 Results of consolidated undrained tests on normally consolidated clays

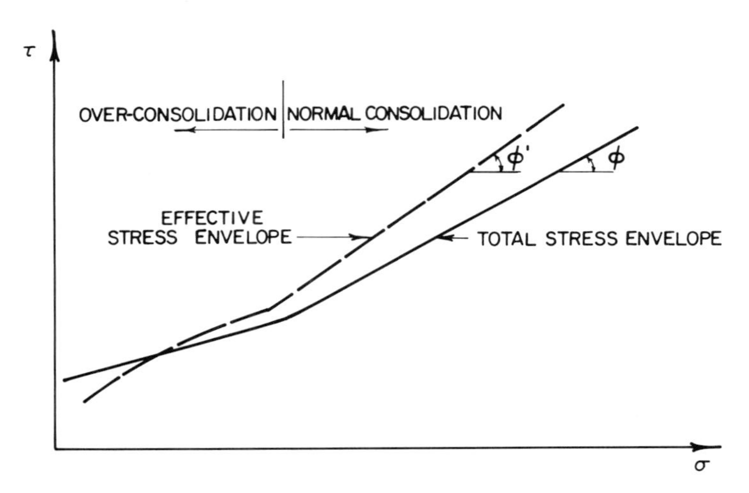

Figure 5.15 Effect of consolidation stress on the rupture curve of over-consolidated soils

5.5.9 Consolidated drained test

In the consolidated drained (CD) test a specimen is permitted to consolidate at each of the two stages of testing. The rate of deformation is such that no appreciable excess pore pressure develops during the second stage of the test. Strength envelopes are essentially those corresponding to effective stress (Figures 5.14 and 5.15). For well draining soils such as clean sand, CD tests are used. For soils having low a coefficient of permeability CD tests are rarely used because of the long time required for completion of the test.

5.6 Selection of strength parameters

Strength parameters are selected based on the materials encountered, drainage conditions, duration of construction, etc. For fine-grained soils and cases where the duration of failure is such that drainage cannot occur within the time span of failure, use total stress strength parameters determined by undrained tests. Total stress analysis is used for short-term stability or conditions immediately after construction.

For foundation on very soft soils progressive failure may occur requiring special considerations (Duncan and Buchignani, 1975).

If boundary conditions are such that effective drainage can take place, use drained or effective stress strength parameters. Such conditions will prevail in most coarse-grained soils except where the loading is rapid, such as during earthquake. In cases where steady-state seepage is occurring in embankments or in excavations in stiff soils long-term stability analysis requires the use of effective stress strength parameters.

5.7 Strength of natural soils

Undisturbed samples are extremely difficult to obtain for cohesionless soils. Strength may be estimated from Figure 5.16 (NAVFAC, 1982).

For normally consolidated plastic soils initial estimates of cohesion may be obtained from:

$$c/\sigma_v' = 0.10 + 0.004 I_p \tag{5.27}$$

where σ_v' is the effective vertical stress and I_p the plasticity index (%).

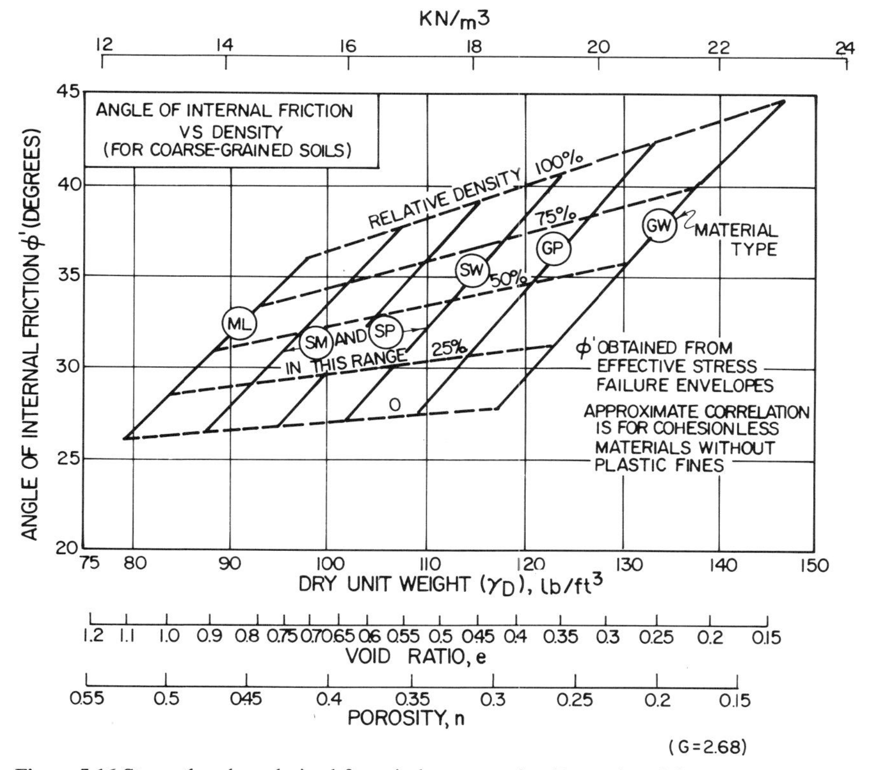

Figure 5.16 Strength values derived from index properties (Reproduced from NAVFAC, 1982)

The angle of friction φ' may be estimated by the following (Holtz and Kovacs, 1981):

$$\varphi' = \begin{cases} 36° - 0.25I_p & \text{for } I_p < 20 \\ 31° - 0.20(I_p - 20) & \text{for } 20 < I_p < 50 \\ 25° - 0.06(I_p - 50) & \text{for } 50 < I_p < 100 \end{cases} \quad (5.28)$$

5.8 Strength of municipal waste

Tests carried out on compacted bales of waste (Fang *et al.*, 1977) show that the friction angle varies from 15° to 25°, while the cohesion remains constant at 1300 lb/ft^2 (65 kPa). As shown in Figure 5.17 no definite failure load could be identified from these tests, therefore the load at 15–20% strain was assumed as failure load.

Figures 5.18 and 5.19 show the plan and section of a test loading conducted on a sanitary landfill in Southern California. Up to 38 ft (11.6 m) of fill with an in-place density of 125 lb/ft^3 (19.6 kN/m^3) were placed at an average rate of 1.6 ft/day (0.3 m/day). The landfill foundation is a massive conglomerate underlain by moderately cemented and thickly bedded siltstone. Figure 5.20 shows the data for the slope inclinometer. A movement of about 17 in (430 mm) is indicated in the upper portion. The movement accelerated with fill placement. When fill placement ceased, the rate of movement decreased markedly. Evaluation of the progress of the horizontal and vertical movements indicated that the lateral movements correlated well with the vertical movements due to compression of refuse. Secondary lateral movements occurred because of spreading of the refuse material below the test fill.

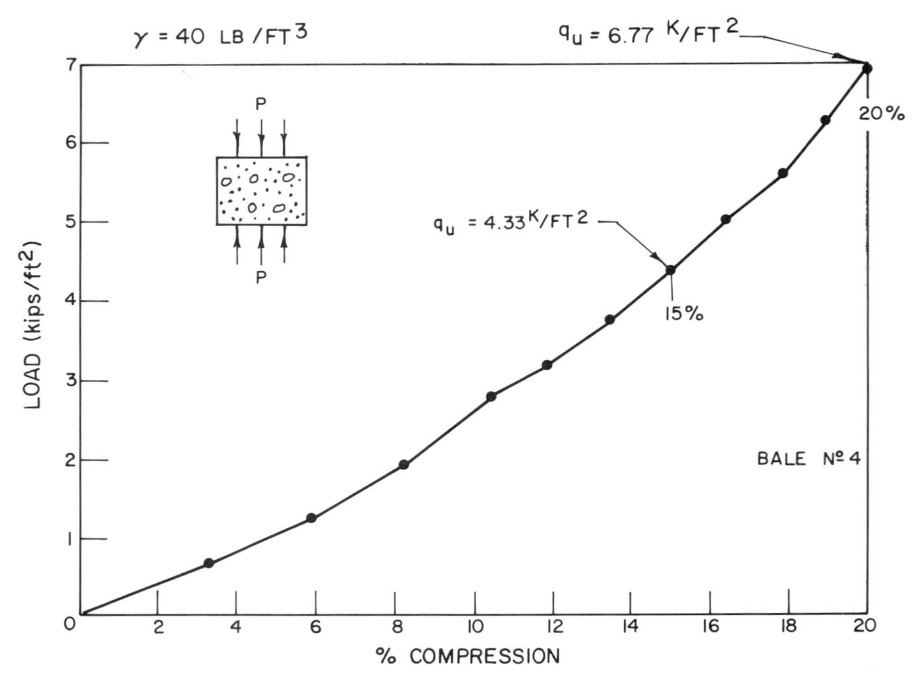

Figure 5.17 Stress–strain behaviour of compacted municipal waste (Reproduced from Fang *et al.*, 1977, by permission)

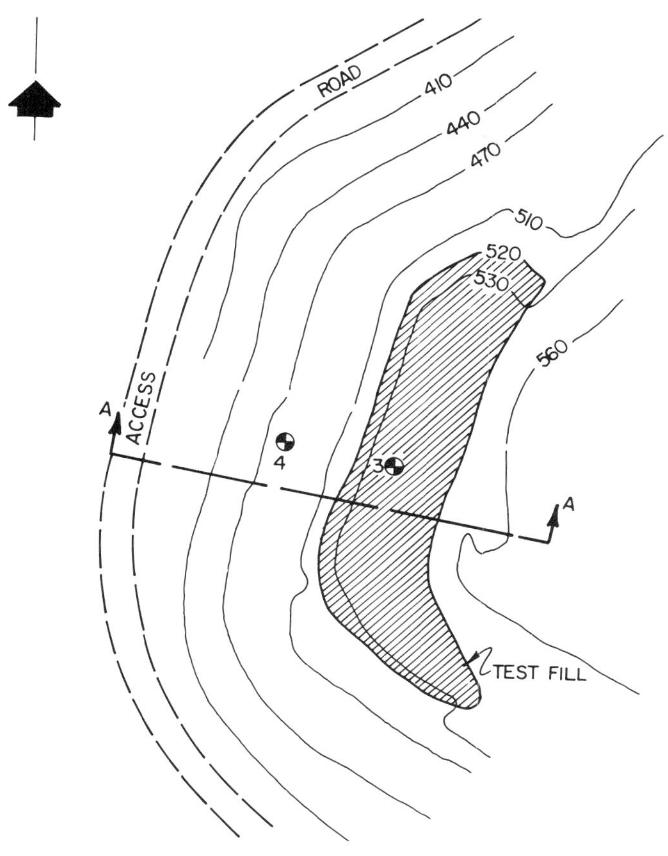

Figure 5.18 Plan of a test loading (Reproduced from Oweis *et al.*, 1985, by permission)

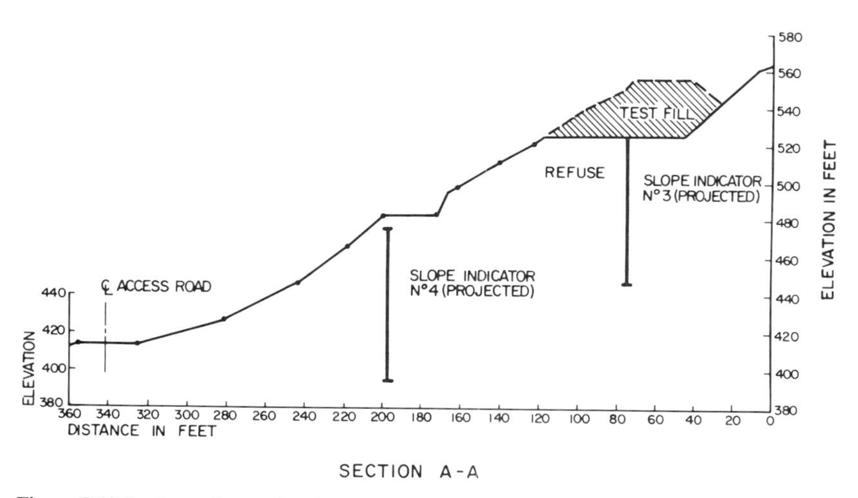

Figure 5.19 Section of a test loading (Reproduced from Oweis *et al.*, 1985, by permission)

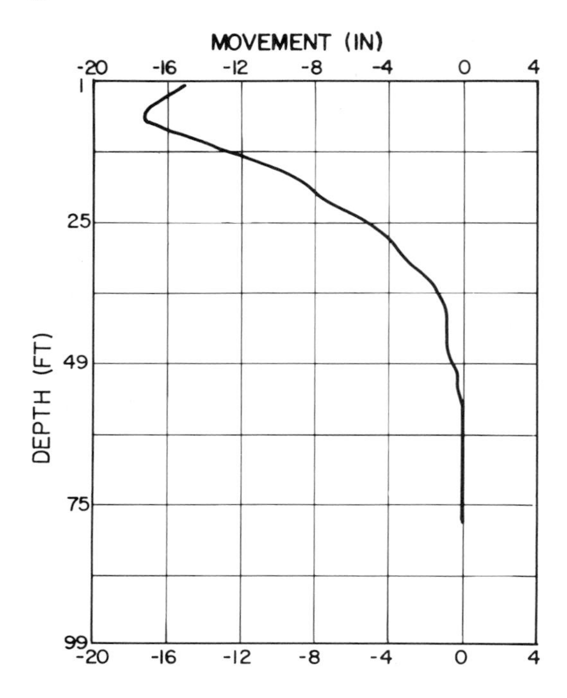

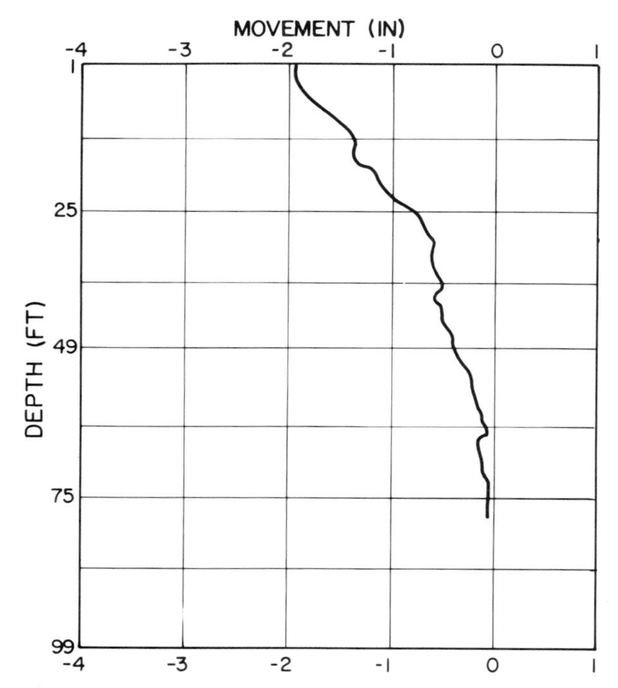

Figure 5.20 Lateral movement data from an inclinometer (Reproduced from Oweis *et al.*, 1985, by permission)

The refuse slope cited was apparently stable during and after placement of the test fill. In terms of conventional slope stability analysis, the factor of safety against failure would be at least 1.0. Assuming that the material can be represented by the conventional Coulomb parameters c and φ their values were calculated assuming a factor of safety of 1.0 and using the ordinary method of slices for stability analysis (Duncan and Buchignani, 1975). The results as depicted in Figure 5.21 represent pairs of c and φ which are less than the actual values. Since the slope had not failed these values are considered conservative.

Pairs of c and φ for the Global landfill are also shown in Figure 5.21. In the calculations a bulk unit weight of 60 lb/ft^3 (9.4 kN/m^3) was assumed (Woodward-Clyde, 1984).

Because of the lack of reliable means to assess the unit weight and strength parameters, the selection of these parameters is usually based on judgement. For example, numerous 'reasonable' combinations of the unit weight, cohesion and angle of friction would produce similar safety factors.

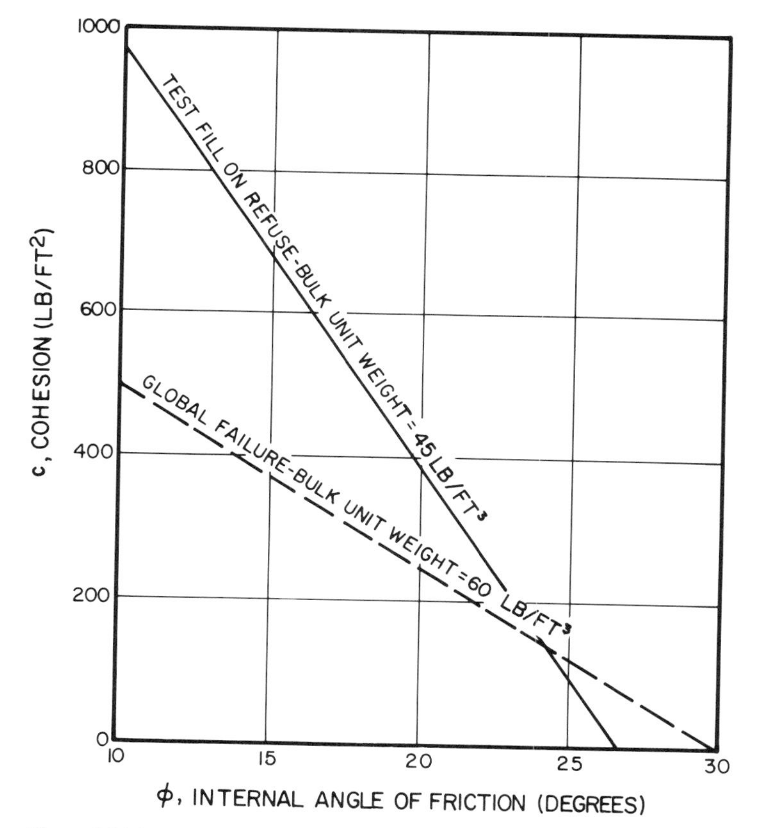

Figure 5.21 Slope stability (Reproduced from Oweis *et al.*, 1985, by permission)

5.9 Strength of mineral and industrial waste

Mine wastes, processes tailings, fly ash, FGD, paper and pulp waste, etc., are commonly disposed of as thin slurries in diked containment areas. Because of their high water content they occupy large volumes and their shear strength is low. To increase the life of a storage area it is necessary that volume of the disposed materials be reduced.

It is difficult to obtain undisturbed specimens from slurry disposal sites as these areas may be inaccessible owing to inability of the materials to support sampling equipment, the soil may contain a large proportion of cohesionless materials which are difficult to sample using conventional methods, or the deposit may be cemented and the sampling process may destroy the cementation. Therefore, laboratory prepared samples are often the only kind available for testing. However, when using the laboratory specimens, the effect of disturbance, ageing and applicability of testing technique on strength properties should be assessed.

5.9.1 Mixed coal refuse

Unconfined compression tests on compacted coal refuse indicate the strength to be dependent on the water content at the time of test. With increasing water content the strength decreases. Results of triaxial tests from coal mining waste from Moundsville, West Virginia, are shown in Figure 5.22. These specimens were consolidated to 45 lb/in^2 (310 kPa). Note that the response is similar to that of coarse-grained soils. Some of the loose specimens showed a sudden drop in stress level followed by a rapid increase. This was attributed to the collapse of structure followed by a partial recovery due to the lateral confining pressure (Saxena *et al.*, 1984).

The angle of friction for undrained conditions ranges between 10° and 28° with cohesion ranging from 0.1 to 0.4 ton/ft^2 (10–40 kPa). For effective stress the angle of friction ranges between 25° to 43° with cohesion values similar to those obtained in undrained tests. The large deviation in strength is caused by the variation in composition of coal waste materials.

In design the angle of friction may be taken as 30–35° for properly compacted coal refuse while neglecting cohesion. For uncontrolled placement materials the friction angle may range between 24° and 28° (Usmen, 1986). In uncontrolled loose fills a greater surface area is exposed to chemical weathering. The waste would be more susceptible to loss of strength at an accelerated rate.

5.9.2 Fly ash

Fly ash possesses pozzolanic properties, that is, in the presence of moisture its strength increases with time (Table 5.2). The increase in strength due to curing is much greater for class C fly ash, which contains lime and other compounds, than for class F fly ash. Class C fly ash possesses high cohesive strength while class F has no cohesion when in a saturated or dry state.

Cunningham *et al.* (1977) reported a decrease in strength with increasing depth for a slurry-deposited fly ash. Standard penetration tests on compacted ash consisting of primarily fly ash from an old lagoon are reported to yield erratic data. No correlation

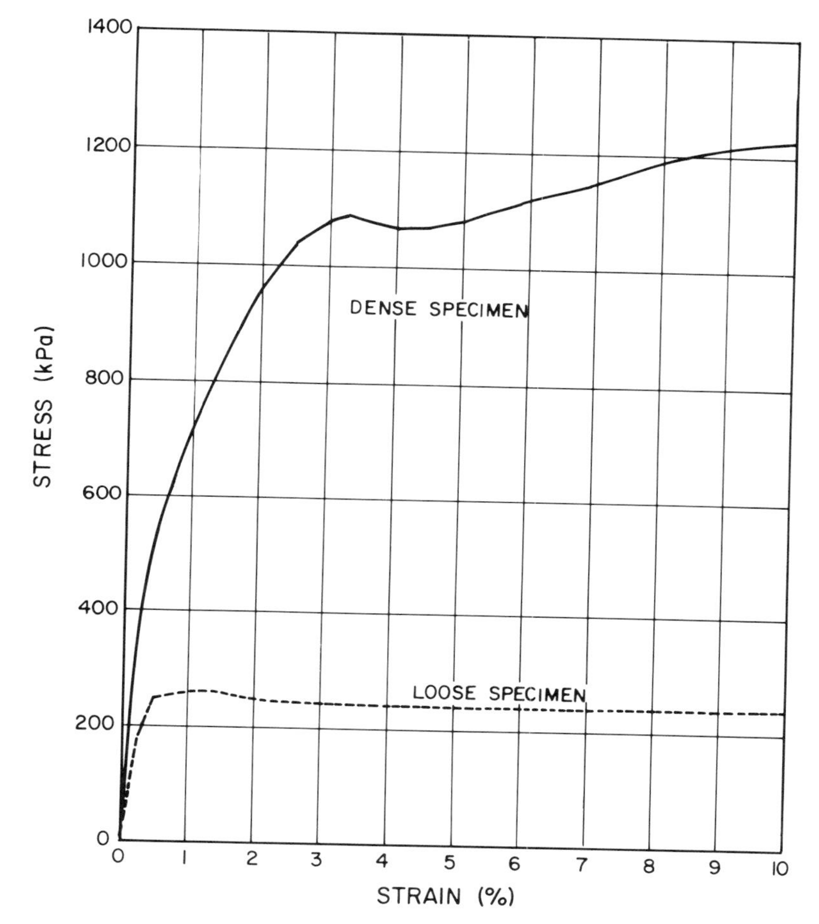

Figure 5.22 Stress–strain behaviour of coal waste (Data from Saxena *et al.*, 1984)

existed between N-values and shear strength. Consolidated undrained (CU) and consolidated drained (CD) tests showed $c=0$, $\varphi=34°$, and $c'=0$, $\varphi'=37.5°$, respectively (Seals *et al.*, 1977). CU tests on specimens of fresh or virgin fly ash prepared at a moisture content considerably higher than the liquid limit showed φ values between 36° and 37°. Cohesion ranged between 7 and 65 kPa with larger values associated with longer ageing periods. Compacted samples yielded $\varphi'=41°$, $c'=0.85-1.45$, and $\varphi'=37°$, $c'=0$ for CU and CD tests respectively (Gatti and Tripiciano, 1981). Note that the higher strength values are associated with fresh fly ash. It is clear that the strength behaviour of fly ash is most significantly affected by its composition, moisture content, method of sample preparation, age before specimen preparation, etc. Whenever fly ash is to be utilized or disposed of, representative samples must be carefully obtained and the laboratory method of specimen preparation must be

Table 5.2 Strength properties of mineral waste

Description	*Cohesion,* c (kPa)	*Friction angle,* φ (degrees)	*Undrained compressive strength,* q_u (kPa)	*Water content,* w (%)
Coal refuse	0–25	25–40		
Undrained condition	10–40	10–28		
Effective stress	0–40	25–43		
Fly ash, Arizona 7-day				
Unit wt, 12.6 kN/m³			223	
Unit wt, 13.4 kN/m³			331	
Unit wt, 13.8 kN/m³			587	
Fly ash (silica 46%, aluminium 34%, calcium 7%)				
Slurry samples	0	37–37		
Compacted, undrained effective stress	0	41		
Compacted, drained	0	37		
West Virginia fly ash	0–18	16–34		
Shelby tube samples (consolidated, undrained)	0	34		
Shelby tube samples (consolidated drained)	0	37.5		
West Virginia bottom ash		38–43		
FDG sludge				
Consolidated drained test	0	41.5		
Compacted	0–40	10–40		
Red mud				
Unleached			63	52
Leached			0	49
Mud : sand (5 : 1)			38	41

Sources: Gatti and Tripiciano, 1981; Hagerty *et al.*, 1977; Mabes *et al.*, 1977; Somogyi and Gray, 1977; Srinivasan *et al.*, 1977.

consistent with the field conditions. Every effort must be made to obtain *in situ* strength parameters, if the cost can be justified. For such measurements cone penetration resistance and pressuremeter tests appear to be more appropriate (Seals *et al.*, 1977).

The addition of lime or cement, even in small amounts, may result in substantial and rapid increase in the strength of fly ash. For ash A (Table 5.3), lime produced higher strength than the same proportion of cement. For ash B, cement was a better choice as far as strength was concerned. Similarly, fly ash can impart strength and stability to other materials, such as FGD sludge, pulp and paper waste (Table 5.2).

When fly ash is disposed of as slurry its strength may be lower than when it is compacted near its optimum moisture content. On old disposal sites the strength gain may be considerable, making it difficult to obtain samples even by coring. Generally the strength decreases with increasing moisture content and increases with increasing curing time. Strain at failure decreases with increasing time.

Table 5.3 Strength variation of fly ash and mixtures of fly ash with time

Material	q_u (kPa) (7 days)	q_u (kPa) (28 days)
Michigan	0.170×10^3	0.210×10^3
New Jersey	0.310×10^3	0.425×10^3
UK	0.550×10^3	0.660×10^3
West Virginia		
Ash A with 3% lime	19.6×10^3	31.6×10^3
Ash A with 3% cement	7.4×10^3	12.8×10^3
Ash B with 3% lime	2.8×10^3	4.0×10^3
Ash B with 3% cement	3.2×10^3	8.0×10^3
Lignite fly ash (fly ash 10%, sand 90%)	2.6×10^3	4.6×10^3
Lime : fly ash : FGD sludge		
6.3 : 43.7 : 50.0	3.3×10^3	5.5×10^3
2.5 : 55.8 : 41.7	3.0×10^3	4.0×10^3
0 : 41.2 : 58.8	0.3×10^3	0.4×10^3
Fly ash : lime : cement : dredge waste		
4 : 0 : 0 : 96		8
0 : 4 : 0 : 96		10.5
0 : 0 : 4 : 96		38
10 : 0 : 0 : 90		10
0 : 10 : 0 : 90		19
0 : 0 : 10 : 90		78

Sources: Bowders *et al.*, 1987; Chae and Gurdziel, 1976; Soliman *et al.*, 1986; Ullrich and Hagerty, 1987; Rizkallah, 1987.

5.9.3 Flue gas desulphurization

The angle of friction for compacted plain FGD sludge having dry density between 70 and 85 lb/ft^3 (11–13.4 kN/m^3) varies between 31° and 39° with negligible cohesion. The undrained compressive strength of double alkali sludge was reported to be between 200 and 1600 lb/ft^2 (10 and 80 kPa). The addition of fly ash to FGD sludge decreased the friction angle of FGD sludge under undrained conditions but showed no change under drained conditions (Krizek *et al.*, 1987). Addition of 40–50% of fly ash to sludge produced a considerable increase in angle of friction and cohesion (Hagerty *et al.*, 1977). The contradictory results may be due to the differences in type of fly ash or curing time.

The compressive strength of lime-treated sludge ranged from 100 kPa to 1 MPa. When FGD sludge containing 40–60% fly ash was treated with 1–3% lime its effective friction angle increased by several degrees provided the specimens were allowed a certain curing period (Krizek *et al.*, 1987). For uncured samples the strength was found to be dependent on the moulding water content (Ullrich and Hagerty, 1987). The strength of sludge–fly ash mixture was much greater for class C fly ash than for class F fly ash, which lacks self-cementing characteristics. Undisturbed samples and those compacted in the field had lower unconfined compressive strengths than the corresponding laboratory prepared samples. The higher strength of the laboratory prepared specimens was attributed to better mixing and quality control. The strength increased with curing time, reaching close to the maximum value between 20 and 40 days (Figure 5.23). The high water content of FGD makes it difficult to mix it with

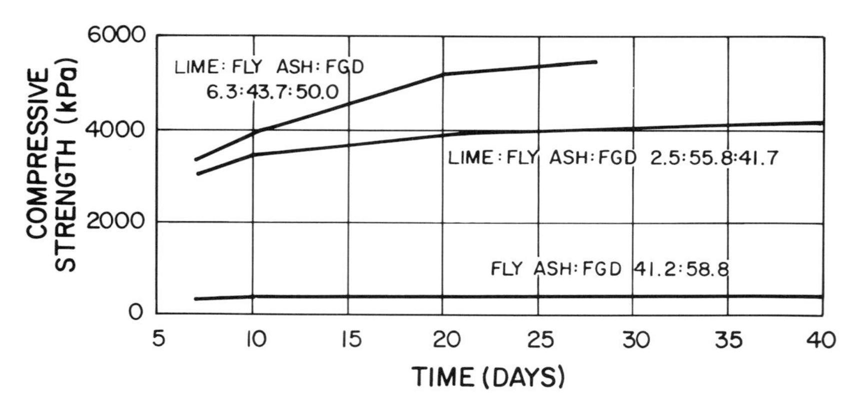

Figure 5.23 Effect of curing time on strength of FGD (Soliman *et al*, 1986)

additives in the field. A curing period of 28 days appears reasonable for materials treated with fly ash and/or lime.

In evaluating the possibility of disposing of FGD in the ocean brackish water was used during laboratory preparation of samples mixed with fly ash and lime. Such samples exhibited an unconfined compressive strength almost 90% greater than that of samples prepared without salt water. No relationship was reported to exist between confining pressure and shear strength. The drained compressive strength was lower than the unconfined strength (Soliman *et al.*, 1986).

5.9.4 Pulp and papermill waste

For fresh papermill waste the angle of friction, as measured in triaxial tests, decreased linearly from about 75° to 45° as the organic content decreased from 65 to 35%. Large strains are required to mobilize the shear strength fully (Charlie, 1977). At low organic contents the strength of the sludge approached that of its inorganic constituents. The decomposition of organics will have a similar effect. The low strength of existing sludge landfills is due to high placement water content, low permeability, long drainage distance, decomposition of a certain proportion of organics, and residual pore pressures.

The strength for the fresh waste was between 1116 lb/ft^2 (56 kPa) and 470 lb/ft^2 (23 kPa) for a moisture content range of 139–194%. When fly ash was added (10% by wet weight of sludge) to the fresh sludge, for comparable moisture content, the mixture showed a strength of only 638 lb/ft^2 (32 kPa). The reduced strength is attributed to the lowered organic content resulting from the addition of fly ash.

The shear strength of compacted samples obtained from the drying fields was 1140 lb/ft^2 (57 kPa) and 39 lb/ft^2 (2 kPa) at moisture contents of 45 and 134%, respectively. The material for these samples was blended mechanically before compaction. Note that for comparable moisture content the strength of 39 lb/ft^2 is considerably lower than the 1116 lb/ft^2 which is the strength for fresh sludge. The change in shear strength with change in moisture content was reported to be much greater at lower water content than at higher water content (Jedele, 1987).

5.9.5 Red mud

For red muds (bauxite residue) Somogyi and Gray (1977) reported a linear relationship between water content and undrained shear strength on a logarithmic scale. These soils also exhibited sensitivity between 3.6 and 7.4. The addition of sand to the red mud decreased its strength and increased its sensitivity. At water contents higher than 41% leaching resulted in complete loss of shear strength, but there was an increase in strength for lower initial water content.

5.9.6 Dredge waste

Extensive data on shear strength of dredged materials from the vicinity of Toledo, Ohio has been reported by Krizek and Salem (1977). The relationship between water content and undrained shear strength from laboratory and field tests is shown in Figure 5.24. The lower strength from the unconfined compression test was said to be from sampling disturbance, which cannot be avoided. The higher strength for laboratory vane tests, contrary to the lower expected values from sampling disturbance, is attributed partly to the confining effect of the sampling tube as opposed to the field vane where soil can move freely around the blades of the vane, and partly to the difference in the height-to-diameter ratios of the two vanes. For remoulded soil there appeared to be fairly good agreement between the field and the laboratory strength values.

The strength was found to increase with age and decrease with increasing distance from the inlet pipe. To determine the age of a deposit the starting time was taken to correspond to one-half the final volume of dredging at that site. The time-dependent strength was expressed (Krizek and Salem, 1977) as:

$$s_{avg} = (2.9 + 2x/l)(t_y - 1) \tag{5.29}$$

where s_{avg} is the average shear strength in kilopascals, x the distance of the specimen

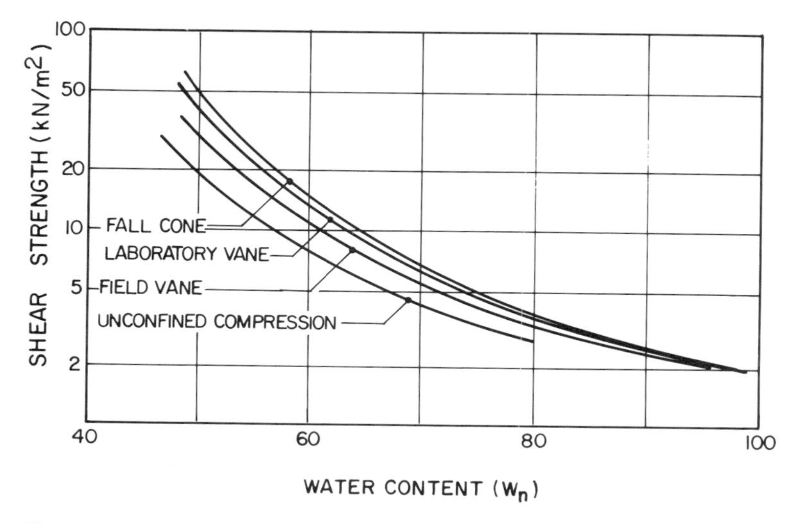

Figure 5.24 Time dependent development of strength in dredging (Krizek and Salem, 1977)

from the outflow weir, l the distance between the inlet pipe and the outflow weir, and t_y the time in years with a maximum value of 10.

The sensitivity values range between 2 and 10, putting this material in the class of 'sensitive' soils.

Laboratory vane shear tests were reported by Rizkallah (1987) on dredge waste stabilized with various proportions of cement, fly ash and lime. Addition of fly ash had virtually no effect on the strength. Some gain in strength was observed with lime, but a curing time of 3–28 days was of little consequence. The admix with cement showed the most increase in strength, with further improvement after 28 days of curing (Table 5.3).

Summary

Vertical stress changes due to surface loads are determined from elastic solutions which are available in the form of graphs. The total settlement of compressible soft soils and waste materials is estimated based on primary consolidation and secondary compression. The relevant parameters are determined from laboratory or field tests.

The strength parameters are determined assuming that the Mohr–Coulomb failure envelope is applicable. Several laboratory testing methods are available which can be used for most types of material. The selection of a test is determined by the type of soil, and loading and drainage conditions. In many instances the age of a deposit can have a significant effect on the strength properties. Because of the extreme heterogeneous nature of the materials constituting municipal waste the strength parameters are more difficult to evaluate. Field tests are highly desirable for most types of wastes.

Notation

a	projection of embankment slope
A	Skempton pore pressure parameter
b	base width of surface load
B	Skempton pore pressure parameter
b_t	top width of embankment
c	cohesion
c'	cohesion with respect to effective stress
C_c	compression index
CR	compression ratio
c_v	coefficient of consolidation for vertical flow
c_{vi}	coefficient of consolidation for layer i
c_{vn}	coefficient of consolidation for layer n
C_α	coefficient of secondary compression
D	constrained modulus
$d\sigma_3$	change in minor principal stress
$d\sigma_c$	applied change in confining pressure
$d\sigma_v$	increase in vertical stress
e	void ratio
e_0	initial void ratio
E_u	undrained modulus
H	thickness of compressible layer

H_d	shortest distance to the drainage boundary
H_n	actual thickness of layer n
H_p	thickness of compressible layer at the end of primary consolidation
h_w	static head of water
I	influence factor
I_p	plasticity index
k	coefficient of permeability
k_h	horizontal permeability
K_0	coefficient of lateral stress at rest
k_v	vertical permeability
l	distance between inlet pipe and outflow weir
m_v	coefficient of volume compressibility
p	intensity of surface load
q_u	undrained compressive strength
RR	recompression ratio
s_{avg}	average shear strength in kilopascals
t	time
t_p	time for completion of primary consolidation
T_v	time factor
t_y	time in years with a maximum value of 10
u	pore pressure
u_e	excess pore pressure at time t
u_i	initial excess pore pressure
u_0	neutral pressure
U_v	average degree of consolidation
U_z	consolidation ratio
w	water content
w_l	liquid limit
w_n	natural water content
x	distance of the specimen from the outflow weir
Z	depth parameter
z	depth measured from top
z_n	thickness of layer n
γ	unit weight of soil
γ_n	unit weight of soil in layer n
γ_w	unit weight of water
δ_c	primary consolidation settlement
δ_i	initial, immediate or distortional settlement
δ_s	secondary compression
δ_t	total settlement
ν	Poisson's ratio
σ	total stress
σ'	effective stress
σ_1	major principal stress
σ_3	minor principal stress
σ_c	confining pressure
σ'_c	consolidation pressure
σ_v	total vertical stress
σ'_v	effective vertical stress
σ'_{vc}	effective vertical consolidation stress

σ'_{vo} effective overburden stress
τ_f shear strength
φ angle of friction
φ' angle of friction with respect to effective stress

References

Berre, T. (1981) *Triaxial Testing at the Norwegian Geotechnical Institute*, NGI Pub. No. 134, pp.1–17

Bishop, A. W. and Henkel, D. J. (1962) *The Measurement of Soil Properties in Triaxial Test*, 2nd Ed., Edward Arnold, London

Black, D. K. and Lee, K. L. (1973) Saturating laboratory samples by back pressure, *Journal of the Soil Mechanics and Foundations Division, ASCE*, **99**, No. SM1, 75–94

Bowders, J. J., Usmen, M. A. and Gidley, R. A. (1987) Stabilized fly ash for use as low-permeability barriers, *Geotechnical Practice for Waste Disposal '87*, University of Michigan, Ann Arbor, Michigan, 15–17 June, Geot. Special Publ. No. 13, pp. 320–333

Chae, Y. S and Gurdziel, T. J. (1976) New Jersey fly ash as structural fill, *New Horizons in Construction Materials*, Vol. 1, Ed. Fang, H. Y., Envo Publishing Co. Inc., pp. 1–13

Charlie, W. A. (1977) Pulp and papermill solid waste disposal: a review, *Proc. Conf. on Geotechnical Practice for Disposal of Solid Waste Materials*, ASCE, Ann Arbor, Michigan, June 1977, pp. 71–86

Cunningham, J. A., Lukas, R. G. and Anderson, T. C. (1977) Impoundment of fly ash and slag – a case study, *Proc. Conf. on Geotechnical Practice for Disposal of Solid Waste Materials*, ASCE, University of Michigan, Ann Arbor, June 1977, pp. 227–245

Cunningham, J. A., Lukas, R. G. and Matthews, M. C. (1982) *Site Investigation*, Halsted Press, New York

Duncan, J. M. and Buchignani, A. L. (1975) *An Engineering Manual for Slope Stability Studies*, Department of Civil Engineering, University of California, Berkeley, March 1975

Fang, H. Y., Slutter, R. G. and Koerner, R. M. (1977) Load bearing capacity of compacted disposal materials, *Proc. Specialty Session: Geotechnical Engineering and Environmental Control, 9th ICSMFE*, Tokyo, pp. 265–277

Gatti, G. and Tripiciano, L. (1981) Mechanical behavior of coal fly ashes, *Proc. of the 10th ICSMFE*, Vol. 2, Stockholm, pp. 317–322

Hagerty, D. J., Ullrich, C. R. and Thacker, B. K. (1977) Engineering properties of FGD sludges, *Proc. Conf. on Geotechnical Practice for Disposal of Solid Waste Materials*, ASCE, Ann Arbor, Michigan, pp. 23–40

Holtz, R. D. and Kovacs, W. D. (1981) *An Introduction to Geotechnical Engineering*, Prentice-Hall, Englewood Cliffs, NJ

Jedele, L. P. (1987) Evaluation of compacted inert paper solids as a cover material, *Geotechnical Practice for Waste Disposal '87*, ASCE, Geot. Special Pub. No. 13, Ed. Woods, R. D., pp. 562–595

Krizek, R. J. and Salem A. M. (1977) Time-dependent development of strength in dredging, *Journal of the Geotechnical Engineering Division, ASCE*, **103**, No. GT3, 169–184

Krizek, R. J., Chu, S. C. and Atmatzidis, D. K. (1987) Geotechnical properties and landfill disposal of FGD sludge, *Geotechnical Practice for Waste Disposal '87*, Proc. Spec. Conf. Geot. Special Publ. No. 13, Ed. Woods, R. D., pp. 625–639

Lee, I. K. and Valliappan, S. (1974) Analyses of soil settlement. In *Soil Mechanics – New Horizons*, Ed. Lee, I. K., American Elsevier Publishing Co., New York, pp. 158–204

Mabes, D. L., Hardcastle, J. H. and Williams, R. E. (1977) Physical properties of Pb–Zn mine-process waste, *Proc. Conf. on Geotechnical Practice for Disposal of Solid Waste Materials*, ASCE, Ann Arbor, Michigan, June 1977, pp. 103–117

NAVFAC DM 7.1 (1982) *Soil Mechanics Design Manual*, Naval Facilities Engineering Command

Oweis, I. S., Mills, W. T., Leung, A. and Scarino, J. (1985) Stability of sanitary landfills, *Geotechnical Aspects of Waste Management*, Seminar, Met. Section, ASCE, December 1985

Saxena, S. K., Lourie, D. E. and Rao, J. S. (1984) Compaction criteria for eastern coal waste embankments, *Journal of Geotechnical Engineering, ASCE*, **110**, No. 2, 262–284

Rizkallah, V. (1987) Geotechnical properties of polluted dredged material, *Geotechnical Practice for Waste Disposal '87*, Proc Special Conf. Geot. Special Publ. No. 13, Ed. Woods, R. D., pp. 759–771

Seals, R. K., Moulton, L. K. and Kinder D. L. (1977) *In situ* testing of a compacted fly ash fill, *Proc. Conf. on Geotechnical Practice for Disposal of Solid Waste Materials*, ASCE, Ann Arbor, Michigan, June 1977, pp. 493–516

Skempton, A. W. (1954) The pore pressure coefficients *A* and *B*, *Geotechnique*, **4,** 143–147

Soliman, N., Houlik, Jr, C. W. and Schneider, M. H. (1986) Geotechnical properties of lime fixed fly ash and FGD sludge, *Int. Symp. on Environmental Geotechnology*, Vol. 1, pp. 549–563

Somogyi, F. and Gray, D. H. (1977) Engineering properties affecting disposal of red muds, *Proc. Conf. on Geotechnical Practice for Disposal of Solid Waste Materials*, ASCE, Ann Arbor, Michigan, June 1977, pp. 1–22

Srinivasan, V., Beckwith, G. H. and Burke, H. H. (1977) Geotechnical investigations of power plant wastes, *Proc. Conf. on Geotechnical Practice for Disposal of Solid Waste Materials*, ASCE, Ann Arbor, Michigan, June 1977, pp. 169–187

Stamatopoulos, A. C. and Kotzias, P. C. (1985) *Soil Improvement by Preloading*, Wiley-Interscience, New York

Taylor, D. W. (1948) *Fundamentals of Soil Mechanics*, John Wiley & Sons, New York

Terzaghi, K. (1943) *Theoretical Soil Mechanics*, John Wiley & Sons Inc., New York

Ullrich, C. R. and Hagerty, D. J. (1987) Stabilization of FGC wastes, *Geotechnical Practice for Waste Disposal '87*, Proc. Spec. Conf. Geot. Special Publ. No. 13, Ed. Woods, R. D., pp. 797–811

Usmen, M. A. (1986) Properties, disposal and stabilization of combined coal and refuse, *Int. Symp. on Environmental Geotechnology*, Vol. 1, pp. 515–526

Vallee, R. P. and Andersland, O. B. (1974) Field consolidation of high ash papermill sludge, *Journal of the Geotechnical Engineering Division, ASCE,* **100,** No. GT3, 309–328

Woodward-Clyde Consultants (1984) *Geotechnical Investigation Global Sanitary Landfill*, Report obtained from the New Jersey Department of Environmental Protection

Chapter 6

Site investigation

6.1 Introduction

Surface and subsurface conditions of soils and waste materials must be evaluated for proper design, operation and maintenance, and future utilization of a disposal facility. Prominent geotechnical design functions and the usually required geotechnical input are identified in Table 6.1. Some of the data needs are related to the physical and engineering characteristics of the subsoils while others relate to the characteristics of the wastes. The techniques used to collect the required data include remote sensing (air photo interpretation, infrared photography), geophysical methods (electrical methods, seismic methods), test pits, test boring and penetrometer resistance, sampling of soils and rocks, *in situ* tests for measuring properties of soils, rocks and waste materials, groundwater quality and profile, and geotechnical monitoring. Several of these methods are described in this chapter.

6.2 Health and safety

In any site investigation programme in an environment where refuse or potentially toxic waste is expected, proper environmental safety measures should be followed. The appropriate level of protection is usually determined from the available evidence and subsequently modified during work at the site.

Four levels of protection (A, B, C and D) have been identified by various agencies. Level D is used when sites are identified as having no toxic or hazardous waste. It consists of a work uniform (coveralls, hard hats, boots) with splash goggles in some cases. Level C protection includes disposable outer clothing, safety shoes or boots, hard hat, rubber gloves with cotton liner, and full-facepiece air-purifying respiratory protective equipment. Level B requires self-contained breathing apparatus (SCBA), disposable chemical-resistant outer clothing, overboots and gloves. Level A protection is where the worst possible conditions are considered. It requires wearing SCBA and a total encapsulating chemical protective suit with a positive seal, safety boots or shoes, rubber gloves, and hard hats.

For drilling in municipal landfill, level C protection is usually sufficient, although SCBA may be needed especially for drillers near the borehole. Explosive gases should be periodically monitored beneath the drilling rig. Methane, for example, where its concentrations reach 15% of the lower explosive limit (LEL) may require work stoppage or venting. The proper selection of protective tools and clothing should be determined by a qualified individual.

Table 6.1 Data needs for geotechnical design functions

	Design function												
Required investigation	*1*	*2*	*3*	*4*	*5*	*6*	*7*	*8*	*9*	*10*	*11*	*12*	*13*
Waste characteristics													
Disposal practice									×	×			×
Physical, chemical component				×					×	×	×		×
Unit weight				×	×				×	×			×
Strength									×	×			×
Age													×
Hydraulic properties				×	×	×			×	×	×		×
Deformation characteristics			×								×		×
Leachate build-up				×	×	×			×	×	×		
Leachate quality				×	×	×					×		
Site geology/hydrology													
Seismic history		×	×	×	×	×	×	×	×	×			×
Rock geology		×	×	×	×	×	×	×		×	×		×
Rock hydraulic properties				×		×	×					×	×
Aquifer characteristics				×		×	×					×	×
Groundwater flow												×	
Recharge and discharge area												×	
Seasonal high water levels				×		×							
Groundwater quality						×						×	
Water supply wells and reservoir characteristics						×						×	
Groundwater chemistry				×		×						×	
Index properties													
Grain size, Atterberg limits, moisture content	×	×	×			×	×	×	×	×			×
Soil chemistry		×	×	×		×		×			×		×
Foundation soils													
Borrow area, soils stratigraphy				×	×	×		×	×	×	×	×	×
Classification	×	×	×	×	×	×	×	×	×	×	×	×	×
Engineering properties													
Strength			×	×		×		×	×	×	×		×
Deformation			×	×		×		×	×	×	×		×
Hydraulic conductivity		×	×	×		×	×	×	×	×	×		×
Erodability	×	×	×										

1, daily and intermediate cover; 2, run-on/run-off system; 3, capping; 4, liner; 5, subsurface collection drain; 6, contaminant barrier; 7, gas venting; 8, leachate and gas management structures; 9, landfill slope stability; 10, landfill foundation stability; 11, geotechnical instrumentation; 12, groundwater monitoring; 13, construction on landfill.

6.3 Investigation phases

A geotechnical investigation for large sites such as disposal facilities is usually carried out in three phases: desk study, preliminary investigation, and detailed investigation.

6.3.1 Desk study

This phase covers review of available topographic and geological data, aerial photographs, site reconnaissance and data from previous investigations.

In the USA, topographic and geological data, including water supply papers, are available from the United States Geologic Survey and the States' Geological Survey. Useful data on near surface soils are available from the soil surveys published by the United States Department of Agriculture (USDA) and Soil Conservation Service (SCS). Soil surveys are primarily for agricultural purposes, although they do contain considerable information for geotechnical engineering use.

In the UK topographical and soil maps are available from Ordnance Survey and geological records and maps from the Geological Museum. Other useful data are available from highway departments, universities, public libraries, and many large cities around the world which have developed extensive geological information (Legget, 1973).

The value of topographical and geological maps is limited as they do not contain adequate details for geotechnical purpose. Greater detail can be obtained from aerial photographs, which are the most common type of remote sensing technique. An experienced air photo interpreter can determine valuable information on the site, such as depth to rock, drainage patterns, site morphology and surface soils (Way, 1973). Aerial photographs are usually available in 9 in frames in scales of 1:12 000 to 1:80 000. Other types of imagery (NAVFAC, 1982) include Skylab, NASA, SLAR and thermal infrared imagery. These are generally used for more detailed investigations such as fault studies and water resources.

The outcome of the desk study usually reveals the broad geological features at the site and helps to determine the scope of the preliminary investigation, such as the number of borings, and their location and depth.

6.3.2 Preliminary investigation

The areal extent and depth of investigation depend on the nature of the geological details identified in the desk study. For a new land disposal facility the minimum scope of the investigation is dictated by regulations (Table 6.2).

Test pits may be used for visual examination of soil strata or waste materials at shallow depths. The minimum depth of borings for sites where foundation instability is not a concern is usually less than 50 ft beneath the base of the disposal facility. Otherwise some or all the borings are advanced deeper to obtain data for preliminary stability assessment. The preliminary investigation defines the general site conditions,

Table 6.2 Minimum number of borings (NJDEP)

	No. of borings	
Acreage	*Total*	*Deep borings*
<10	4	1
10–49	8	2
50–99	14	4
100–200	20	5
>200	24 plus 1 boring each additional 10 acres	6 plus 1 boring each additional 10 acres

hydraulic conductivity and the groundwater flow. Sufficient data are collected to develop a preliminary design for the facility to seek regulatory preliminary approval.

6.3.3 Final investigation

Depending on the uniformity of site conditions the preliminary and final investigations may be combined into one phase. Various criteria have been proposed (NAVFAC, 1982; ASCE, 1976) for the extent and depth of investigations for various engineering structures. Table 6.3 presents guidance for scoping out an exploration programme utilizing test borings.

6.3.4 Drilling methods

Several methods are available for advancing bore holes to obtain samples of underlying materials (Clayton *et al.*, 1982). For land disposal facilities the two commonly used techniques are the hollow-stem continuous flight auger and rotary drilling. In extremely difficult conditions a percussion or cable tool drilling method may be used.

Table 6.3 Guidance for borings

Areas of investigations	*Boring layout*	*Boring depth*
New or expanded land disposal facility	As per Table 6.2 add borings to establish subsurface sections at most critical areas. Provide 4–6 borings along each section to establish data for stability analysis. Supplement by test pits for examination of near surface soils and waste	Advance borings into a relatively incompressible soil to depths where the increase in stress is 10% or less of the existing effective overburden stress, where settlement is not a key design factor. Advance to below active or potential failure surface or to a depth where foundation failure is unlikely
Facility structure	Minimum of four borings at corners plus one in the interior for 5000 ft^2 (465 m^2). Supplement by test pits for examination of near surface soils and waste	Advance borings into a relatively incompressible soil or where vertical stress induced by the structure is less than 10% of the overburden effective stress. Advance for a minimum of 30 ft (9.1 m)
Retention ponds	Advance preliminary borings at 100–300 ft centres (30.5–91.4 m). Add intermediate borings along the centre line at critical facilities (cut-off, outlet and inlet structures). Supplement by test pits for examination of near surface soils and waste	Add borings to a minimum of 0.5–1 times the width of the embankment or to the relatively hard stratum
Containment barriers	Advance preliminary borings at 500 ft (152.4 m) intervals. Advance intermediate borings for a final spacing of 300 ft. Supplement by test pits for examination of near surface soils and waste	Extend a minimum of 10 ft (3 m) into stratum with the design hydraulic conductivity

Hollow-stem continuous flight auger

The hollow-stem auger (Figure 6.1) serves as a casing and has a continuous outer spiral. It contains an inner rod with a plug at the lower end which keeps the soil from entering the auger during advancing of the hole. The drill is operated by power from a truck or drilling rig.

For sampling the plug is replaced by a sampler. In soils such as sand and silt below the water table, when the plug is removed the high water pressure outside the stem may cause the soils to be washed into the casing. In such soils, if representative samples are called for, the plug should not be used during augering; instead the casing should be cleaned from within, or the hole should be stabilized by mud or by keeping a positive head of water above the groundwater level.

Rotary drilling

Rotary drilling (Figure 6.2) employs power to rotate the drilling rods and the cutting bit through which water or drilling fluid is circulated to remove cuttings from the hole. The drilling fluid is prepared from bentonite and water and has a density of 68–72 lb/ft^3. If the stability of the hole cannot be maintained by the drilling fluid, a casing is driven and its inside is cleaned before taking a sample. This method is applicable to all soils except those containing large boulders and cobbles.

Figure 6.1 Drilling using a hollow-stem auger

Figure 6.2 Rotary drilling

Percussion or cable tool drilling
The hole is advanced by alternately raising and dropping a heavy drilling bit (Figure 6.3). A small amount of slurry is kept in the hole, which keeps the cuttings in suspension. The slurry is removed by a bailer or a pump as it reaches its soil carrying capacity. This method can be used in most instances where obstructions are anticipated. The hole diameter is 4 in or larger. When possible the bore hole is advanced ahead of the casing.

Drilling in refuse
In advancing a test boring through refuse difficulties may be encountered requiring one or more relocations of the bore holes before the refuse layer is penetrated.

Augering is the preferred method of advancing a hole through refuse and is usually attempted first. Textiles and other objects may prevent advancement of the auger. In some cases rotary drilling may be successful if circulation of the drilling fluid can be maintained and no large and hard objects are encountered. The most difficult drilling is usually below the leachate or water level since collapse of the hole becomes a problem. For large holes over 6 or 8 in in diameter, the conventional drilling rigs usually do not have enough power to advance the hole through either augering or

Figure 6.3 Cable tool drilling

rotary action. In such cases a larger machine usually used for caisson construction may be considered. An alternative is to use a pile-driving rig to advance the casing or to use a bucket auger which has proved to be successful above leachate level. Another effective method, although slow, is that of cable tool drilling, which can penetrate almost all types of obstruction. Its rate of penetrating through refuse may vary from 4 to 8 ft per 8 h day.

6.3.5 Sampling

Sampling procedures can be broadly grouped into 'disturbed' or 'undisturbed' sampling. Disturbed samples are usually adequate for classification and physical property testing such as Atterberg limits (liquid limit, plastic limit, moisture content) and chemical testing (chemical corrosivity, contamination) and laboratory resistivity tests for assessing potential for galvanic corrosion. Undisturbed samples are required to conduct laboratory strength, permeability and compressibility testing. A survey of the current practice of soil sampling is contained in the 1979 *Proceedings of the International Society of Soil Mechanics and Foundation Engineering*.

Disturbed samples

Disturbed samples are usually recovered from a test boring at typically 5 ft intervals, using a standard split barrel (ASTM D 1586) or a split spoon sampler (Figure 6.4). The standard barrel has a 2 in outside diameter and 1.375 in inside diameter and it consists of two interlocking halves which are held together by end caps. After the sampler is withdrawn from the ground, the two ends are removed, the barrel is split open and the sample is retrieved. The sample is then visually classified utilizing an appropriate classification system and transported to the laboratory. Larger diameter samples are obtained using non-standard samplers fitted with an inner tube. The blows driving these samplers such as the Converse sampler or Dames and Moore sampler are usually converted to equivalent standard penetration test (SPT) blows if required. Other techniques for obtaining disturbed samples are described in Table 6.4.

Undisturbed samples

Better quality or undisturbed samples can be obtained in soils using techniques summarized in Table 6.5. Proper sample handling after removal from the test boring is important in order to minimize sample disturbance (ASTM D1587-74).

In sampling soils at designated hazardous waste sites, some of the soil samples retrieved may be used for environmental testing for contaminants. Certain protocols mandated by the regulatory agencies should be followed to avoid cross-contamination between samples and achieve the intent of sampling. The protocols relate to 12 requirements (NJDEP, 1986) and include selection of the sampler and container, and sample handling and preservation. The sampling device (e.g. split barrel) is required to be laboratory cleaned using a non-hazardous cleaning compound (e.g. methanol) and wrapped in aluminium foil and identified. Ideally, each sampler should be used for taking one sample only. Because of the large number of samples it is impractical to use this procedure. To re-use the sampler in the field, cleaning of samplers and containers requires the following steps:

1. Soap and tap water wash.
2. Tap water rinse.

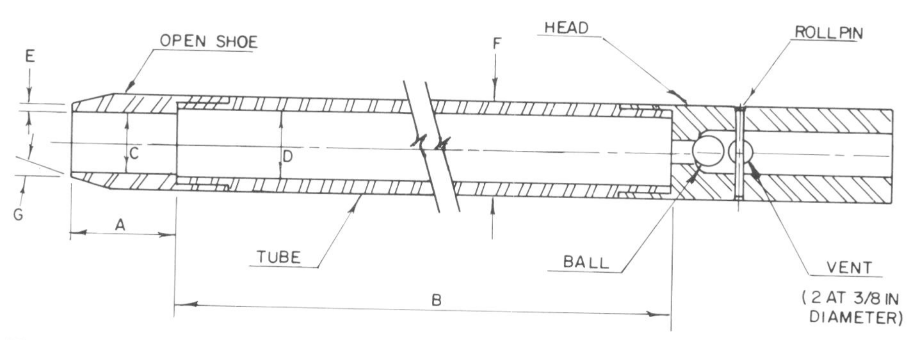

Figure 6.4 The standard split spoon sampler. The 1½ in (38 mm) inside diameter split barrel may be used with a 16 gauge wall thickness split liner. The penetrating end of the drive shoe may be slightly rounded. Metal or plastic retainers may be used to retain soil samples. A, 1.0–2.0 in (25–50 mm); B, 18.0–30.0 in (0.457–0.762 m); C, 1.375 ± 0.005 in (34.93 ± 1.3 mm); D, 1.50 ± 0.05–0.00 in (38.1 ± 1.3–0.0 mm); E, 0.10 ± 0.02 in (2.54 ± 0.25 mm); F, 2.00 ± 0.05–0.00 in (50.8 ± 1.3–0.0 mm); G, 16.0–23.0°

Table 6.4 Common samplers for disturbed soil samples and rock cores

Sampler; method of penetration	*Sampler dimensions*	*Most suitable for soil or rock types*	*Causes of disturbance or low recovery*	*Remarks*
Split barrel hammer driven (ASTM D 1586-84)	2 in OD, 1.375 in ID is standard. Penetrometer sizes up to 4 in OD, 3.5 in ID available	All fine-grained soils in which sampler can be driven. Gravels, refuse invalidate drive data	Vibration	SPT is made using 140# hammer falling 30 in. Better quality samples are obtained with liners. Some sample disturbance does occur
Augers: continuous helical flight; rotation (ASTM D 1452-8)	3–16 in diameter can penetrate to depths in excess of 50 ft	For most soils above water table. Will not penetrate hard soils or those containing cobbles or boulders	Hard soils, cobbles, boulders	Rapid method of determining soil profile. Bag samples can be obtained. Log and sample depths must account for lag between penetration of bit and arrival of sample at surface
Single tube; rotation		Primarily for strong, sound and uniform rock	Fractured rock, rock too soft	Drill fluid must circulate around core, rock must not be subject to erosion. Not often used
Diamond core barrels (ASTM D 2113-83)	Standard sizes $1\frac{1}{2}$–3 in OD, $\frac{7}{8}$–$2\frac{1}{8}$ in core. Barrel lengths 5–10 feet for exploration	Hard rock. All barrels can be fitted with insert bits for coring soft rock or hard soil		
Double tube	Non-uniform, fractured, friable and soft rock		Improper rotation or feed rate in fractured or soft rock	Has inner barrel or swivel which does not rotate with outer tube. For soft, erodible rock. Best with bottom discharge bit
Triple tube		Same as double tube	Same as double tube	Differs from double tube by having an additional inner split tube liner. Intensely fractured rock core best preserved in this barrel
Bucket; rotation	Up to 48 in diameter is common. Larger diameter available. With extensions, depths greater than 80 ft are possible	For most soils above water table. Can dig harder soil and penetrate refuse, soils with cobbles and small boulders when equipped with a rock bucket	Soil too hard to dig	Several types of buckets are available including those with ripper teeth and chopping buckets. Progress is slow when extensions are used
Hollow stem; rotation	Generally 6–8 in OD with 3–4 in ID hollow stem	Same as bucket	Soil too hard to dig	A special type of flight auger with hollow centre allows taking of undisturbed samples or standard penetration test

Table 6.5 Common samplers for undisturbed samples

Sampler; method of penetration	*Dimensions*	*Best results soil types*	*Causes of disturbance*	*Remarks*
Shelby tube; pressing with fast, smooth stroke. Can be carefully hammered (ASTM D1587-85)	3 in OD, 2.875 in ID most common. Available from 2 to 5 in OD sampler standard length is 30 in. Standard sampler length is 30 in	For cohesive fine-grained or soft soils. Gravelly soils will crimp the tube	Erratic pressure applied during sampling hammering, gravel particles crimping tube edge, improper soil types for sampler	Simplest sampler for undisturbed samples. Boring should be clean before lowering sampler. Little waste area in sampler. Not suitable for hard, dense or gravelly soils
Stationary piston; pressing with continuous, steady stroke (ASTM D1587-85)	3 in OD most common, available 3–5 in OD. Standard sampler length is 30 in	For soft to medium clays and fine silts. Not for sandy soils	Erratic pressure during sampling, allowing piston rod to move during press. Improper soil types for sampler	Piston at end of sampler prevents entry of fluid and contaminating material. Requires heavy drill rig with hydraulic drill head. Generally less disturbed samples than Shelby. Not suitable for hard, dense or gravelly soil. No positive control of specific recovery ratio
Hydraulic piston (Osterberg); hydraulic or compressed air pressure	3 in OD most common, available from 2 to 4 in OD. Standard sampler is length 36 in	For silts, clays and some sandy soils	Inadequate clamping of drill rods, erratic pressure	Needs only standard drill rods. Requires adequate hydraulic or air capacity to activate sampler. Generally less disturbed samples than Shelby. Not suitable for hard, dense or gravelly soil. Not possible to limit length of push or amount of sample penetration
Denison; rotation and hydraulic pressure	Samplers from 3.5 to $7\frac{3}{4}$ in OD. (2.375–6.3 in size samples). Standard sampler length is 24 in	Can be used for stiff to hard clay, silt and sands with some cementation, soft rock	Improperly operating sampler. Poor drilling procedures	Inner tube face projects beyond outer tube which rotates. Amount of projection can be adjusted. Generally takes good samples. Not suitable for loose sands and soft clay
Pitcher sampler; same as Denison	Sampler 4.125 in OD uses 3 in Shelby tubes. 24 in sample length	Same as Denison	Same as Denison	Differs from Denison in that inner tube projection is spring controlled. Often ineffective in cohesionless soils
Hand cut block or cylindrical sample	Sample cut by hand	Highest quality undisturbed sampling in cohesive soils, cohesionless soil, residual soil, weathered rock, soft rock	Change of state of stress by excavation	Requires accessible excavation. Needs dewatering if sampling below groundwater

3. 10% acidic (HNO_3) solution rinse (perform only if sample is to be analysed for metals).
4. Distilled/deionized water rinse (perform only if sample is to be analysed for metals).
5. Acetone (pesticide grade) rinse only if sample is to be analysed for organics.
6. Total air dry or nitrogen blow out.
7. Distilled/deionized water rinse.
8. Steps 5, 6 and 7 above could be eliminated if samples are not analysed for organics. The sampling devices (e.g. split barrel) should be cleaned after each sample. In between test borings, the drilling equipment is required to be steam cleaned. This includes the casing, drill rods, drill bits and auger. If the bucket is used to obtain the material to be sampled then the bucket must be steam cleaned, scrubbed with non-phosphate detergent and water and then rinsed by steam cleaning again. Steam cleaning of the arm and bucket of the backhoe is required between test pits.

6.4 *In situ* measurement of soil and rock properties

Although there are a large number of methods available for *in situ* measurements of properties of geotechnical materials only a few, which are pertinent, are described here.

6.4.1 Standard penetration test

In the standard penetration test (SPT) the penetration resistance of the subsurface materials is determined by driving the split barrel sampler with a 140 lb hammer dropping 30 in. The number of blows, *N*, required to drive the sampler a distance of 12 in (300 mm) after an initial penetration of 6 in (150 mm) is referred to as SPT *N* value (ASTM D 1586). This test is typically conducted at 5 ft interval in a test boring. Some of the factors that affect the measured *N* value are as follows.

Low *N* values are obtained by driving the sampler through disturbed soils, by improper cleaning of the hole, failure to keep the water level inside the bore hole above the groundwater level at all times to prevent sand boiling at the bottom, and overwashing ahead of the casing.

High *N* values are generated by restricting the fall of the drive weight, a plugged sampler, the presence of coarse gravel, or overdriving the sampler.

Another major factor is the technique by which the energy is delivered to the drill rod. The usual US practice is to use a safety or doughnut hammer with two wraps of rope around a rotating cathead (pulley). The energy delivered to the drill rod can vary from 20 to 90% of the theoretical maximum (4200 in lb) (Kovacs and Salomone, 1982).

The size of the hole and drilling procedure also affects the *N* value. Larger holes produce lower *N* values for granular soils. Mudded holes generally produce lower *N* values for granular soils compared with cased holes.

Despite the non-standard nature of the SPT, this field test has been used extensively and the *N* values have been correlated with engineering parameters. The *N* value is influenced by the effective vertical stress, soil density, gradation, and stress history. The Gibbs and Holtz (1957) relationship of *N* to relative density D_r could be approximated by the following relationships:

$$D_r = 30\ (N'/5)^{0.42} \text{ for } N' > 5 \qquad (6.1)$$

$$N' = \frac{50N}{\sigma'_{vo} + 10} \qquad (6.2)$$

where σ'_{vo} is the effective overburden stress in lb/in^2 at the depth where N is measured.

The angle of internal friction φ for granular soils may be expressed as (Meyerhoff, 1956):

$$\varphi = 25 + 0.15D_r \qquad (6.3)$$

$$\varphi = 30 + 0.15D_r \qquad (6.4)$$

Equation 6.3 is for sands containing greater than 5% silt and fine sand by weight. Equation 6.4 is for sand with less than 5% silt and fine sand.

A crude estimate of the dynamic shear modulus G (F/L^2), may be made based on (NAVFAC, 1982):

$$G\ (\text{ton/ft}^2) = 120N^{0.8} \qquad (6.5)$$

When utilized on sanitary landfills, the SPT yields results with wide scatter. Figure 6.5 shows SPT results from landfills with similar waste composed of household garbage and light industrial waste. The N values for soaked refuse are usually much lower than those for refuse above leachate or groundwater level. In order to achieve spoon penetration, larger hammers are often used to drive the standard sampler.

The standard penetration resistance of slurried fly ash very often tends to indicate the material to be much lower in strength and density than the other relevant comparative tests would do. Cunnigham *et al.* (1977) reported a relative density of about 25% when based on standard penetration tests. Based on unit weight of the undisturbed specimen it was estimated at about 50%. The penetration resistance at greater depths was less than at shallower depth. Although this response could not be clearly explained, one of the reasons given for the lower penetration resistance was the possible upward flow of water due to the slight artesian condition detected at the underlying rock. Other examples of standard penetration tests yielding values that could not be correlated with the observed behaviour of the structure are given in Chapters 5 and 10.

6.4.2 Cone penetration test

The static cone or the Dutch cone penetration test (CPT) consists of forcing a cone into the ground at a rate of 10–20 mm/s and measuring the pressure needed for each increment of penetration (ASTM D 3441). The cone has a point angle of 60° and an area of 10 cm^2. The measurements of the soil resistance to the penetration of the cone are taken at 200 mm intervals or less. Where both friction and point resistance are needed a Begamann friction cone is used. The point resistance, q_c, is obtained while the cone is extended 80 mm, then the total resistance of the cone and friction sleeve is obtained during the last 120 mm of the total 200 mm that the outer rods are advanced. The frictional resistance, f_r, of the sleeve is the total penetration resistance less q_c. Errors may be introduced by friction between telescoping rods, friction between soil and the mantle immediately above the cone, excessive compression of the inner rod at high resistance (10 ton) or by its large length at greater depths (30 m). Some of the difficulties are overcome in an electric cone, which is more rapid, allows automatic data acquisition, reduction and plotting, and has greater accuracy and repeatability.

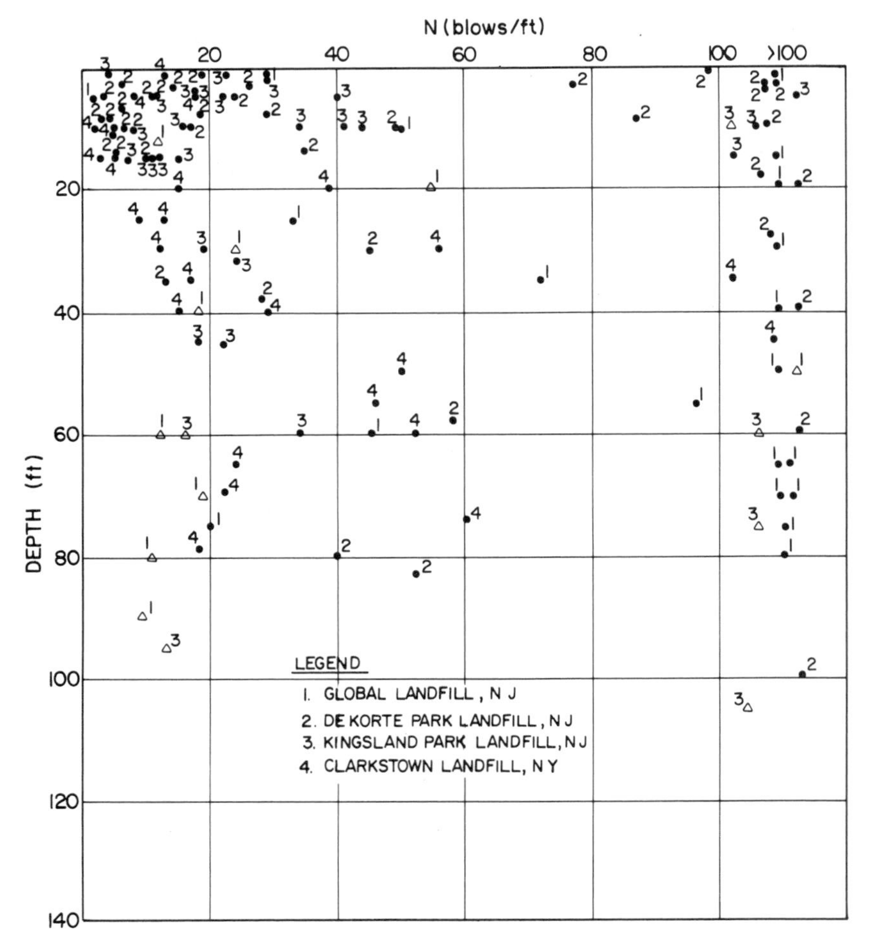

Figure 6.5 *N* values in refuse

As no soil samples can be recovered for identification, cone resistance, q_c, and friction ratio, f_r/q_c, are used to differentiate between various soil types. Clean sands generally exhibit low friction ratios and it increases with increase in clay content. A soil classification chart based on the standard friction electric cone is given in Figure 6.6 (Olsen and Farr, 1986). To determine the true value of q_{cn} start with the minimum n value of 0.6 and compute q_{cn}. Using the computed value of q_{cn} and the normalized value of f_{rn} determine a new value of n from the graph. Each successive n value will be larger than the previous one. Usually less than five repetitions are adequate for convergence.

The cone penetration resistance, q_c, has been correlated with various engineering parameters. A useful relationship is:

$$s_u = (q_c - \sigma_{vo})/N_k \tag{6.6}$$

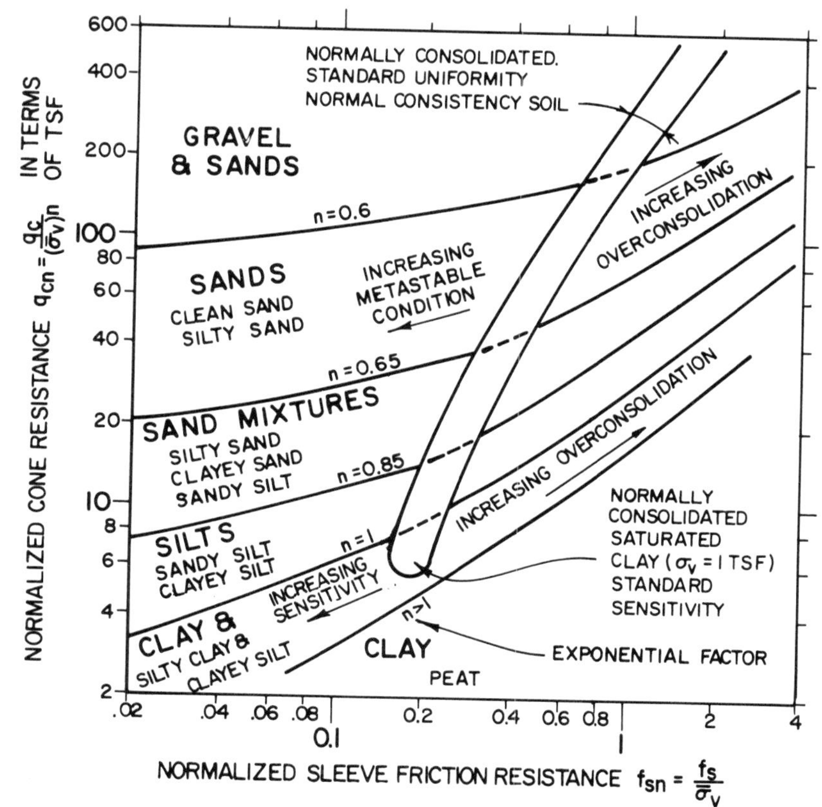

Figure 6.6 Soil classification based on standard friction electric cone tests (Reproduced from Olsen and Farr, 1986, by permission)

where s_u is the undrained shear strength, σ_{vo} is the total vertical stress at the point of measurement, and N_k is the cone factor.

N_k ranges between 10 and 20. Use an N_k value of 15 for a preliminary estimate of s_u and a value of 10 for sensitive soils (Robertson and Campanella, 1983b).

Young's modulus of sands, E, is estimated as follows:

$$E = \alpha q_c \tag{6.7}$$

The empirical parameter α depends on stress history and the fraction of failure stress level at which E is determined. At 25% of the failure stress level, for normally consolidated sand the value of α varies from 1.5 to 3.0 and for overconsolidated sand α is 4–10 times the value for normally consolidated sands. A relationship between q_c and E is given in Figure 6.7.

The ratio of cone resistance q_c (kg/cm^2) to the SPT N value has been correlated with the mean grain size D_{50} (mm). The ratio q_c/N increases with increased grain size as illustrated in Table 6.6. The value of D_{50} may be estimated from Figure 6.6 if actual soil is not available for classification (Robertson and Campanella, 1983a).

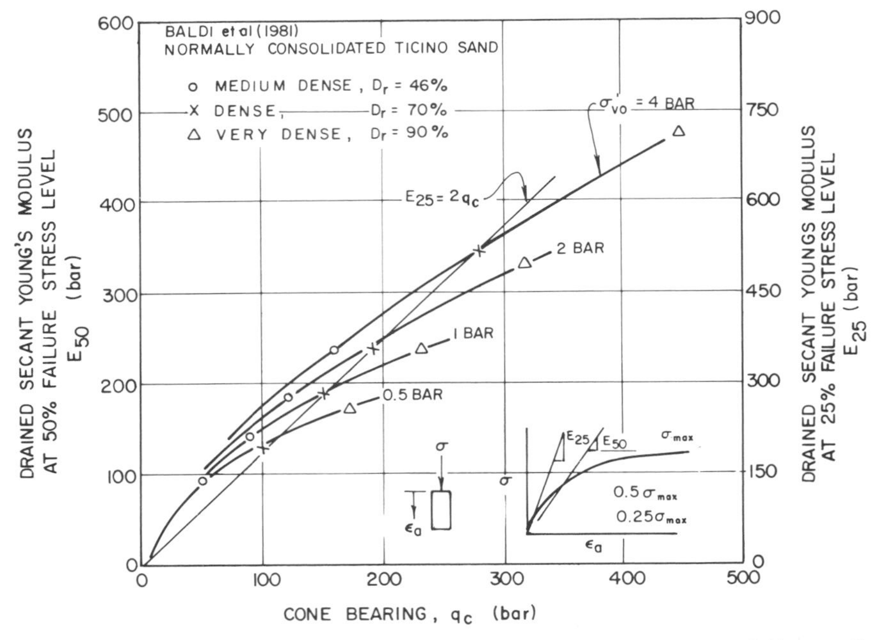

Figure 6.7 Relationship between q_c and E (Reproduced from Robertson and Campanella, 1983, by permission)

Table 6.6 Variation of q_c/N with mean grain size

Mean grain size, D_{50} (mm)	*Soil*	q_c/N
0.001	Clay	1
0.01	Sandy silts and silts	2
0.1	Silty sand	4
0.5	Medium sand	5–6
1.0	Gravelly sand	6–8

6.4.3 Piezocone test

The conventional cone has been modified to include a filter element for measuring pore pressure. The filter may be located at the cone tip, cone face or immediately behind the cone. The cone penetration resistance, pore pressure and the frictional resistance of the sleeve are recorded continuously using a data logger. The position and the size of the filter element affect the point resistance. The recorded q_c value is less than that obtained using a cone without the filter element and must be corrected as follows:

$$q_T = q_c + u(1 - a) \qquad (6.8)$$

where q_T is the corrected cone penetration resistance for a jointless cone, u the measured pore water pressure, and a the area ratio for the given piezocone.

The undrained strength is given by:

$$s_u = (q_T - \sigma_{vo})/N_{kT} \quad (6.9)$$

For normally consolidated clays, N_{kT} increases linearly with plasticity index and may be taken as 13 ± 2 for $I_p = 0$ and 18.5 ± 2 for $I_p = 50$ (Aas *et al.*, 1986).

6.4.4 Pressuremeter

The pressuremeter test is an on-site lateral loading test performed by means of a cylindrical probe. Under increments of pressure, radial expansions are measured and deformation moduli calculated. The Menard pressuremeter (Winter, 1982) is used in a pre-drilled hole. The self-boring pressuremeter test (Baguelin *et al.*, 1986) is performed with the sides of the hole supported as the instrument is pushed into the ground by augering or jetting, thus minimizing disturbance. The self-boring device is useful in soft soils where sample disturbance may result in significant strength loss. Its use for characterization of refuse such as encountered in sanitary landfill has been limited by expected large scatter results. Figure 6.8 shows results of tests made in a sanitary landfill (STS, 1985) at various depths. Data for medium sands and soft peat are shown for comparison. As shown in Figure 6.8 the modulus derived for refuse varies almost by an order of magnitude because of the extremely variable nature of the material. The tests, however, could be used to assess the improved modulus of deformation after stabilization.

6.4.5 Field vane tests

Field vane shear tests are useful in soft cohesive deposits where much of the soil strength could be lost by sample disturbance. The test is not suitable for stiff clays, or soft clay containing layers or varves of cohesionless soils, pieces of gravel, wood, shells, etc.

A standard vane consists of four blades with a height-to-diameter ratio of two. Depending on the size of casing, the height of the vane may be about 75–185 mm and blade thickness 1.6–3.2 mm (ASTM D 2573-72). According to the British Standard, the vane shear test is not suitable for shear strengths greater than 75 kPa.

To perform a test the blade is pushed into the ground at the bottom of the bore hole. The tip of the vane is advanced to the undisturbed soil at least five times the diameter of the hole. The vane is rotated at 6°/min or less and the torque is measured. The test is performed within 5 min of insertion of the vane. The time required for failure to occur may range between 2 and 5 min except for very soft clays where it may be as much as 10–15 min (ASTM D 2573-72). In Norway the required time is 1–3 min (Aas *et al.*, 1986). Once the maximum torque is reached, the value of remoulded strength may be obtained after giving between 10 and 25 turns to the vane. The soil's shear strength is calculated by assuming a cylindrical failure surface corresponding to the periphery of the plate. The measured shear strength is corrected based on the soil plasticity index (Bjerrum, 1972) as per Table 6.7. Charlie (1977) reported that for paper mill sludge, which had an average I_p of 150, the correction factor had a value of 0.6. The corrected shear strength yielded a factor of safety close to one at failure at paper mill sludge waste sites where failure was investigated.

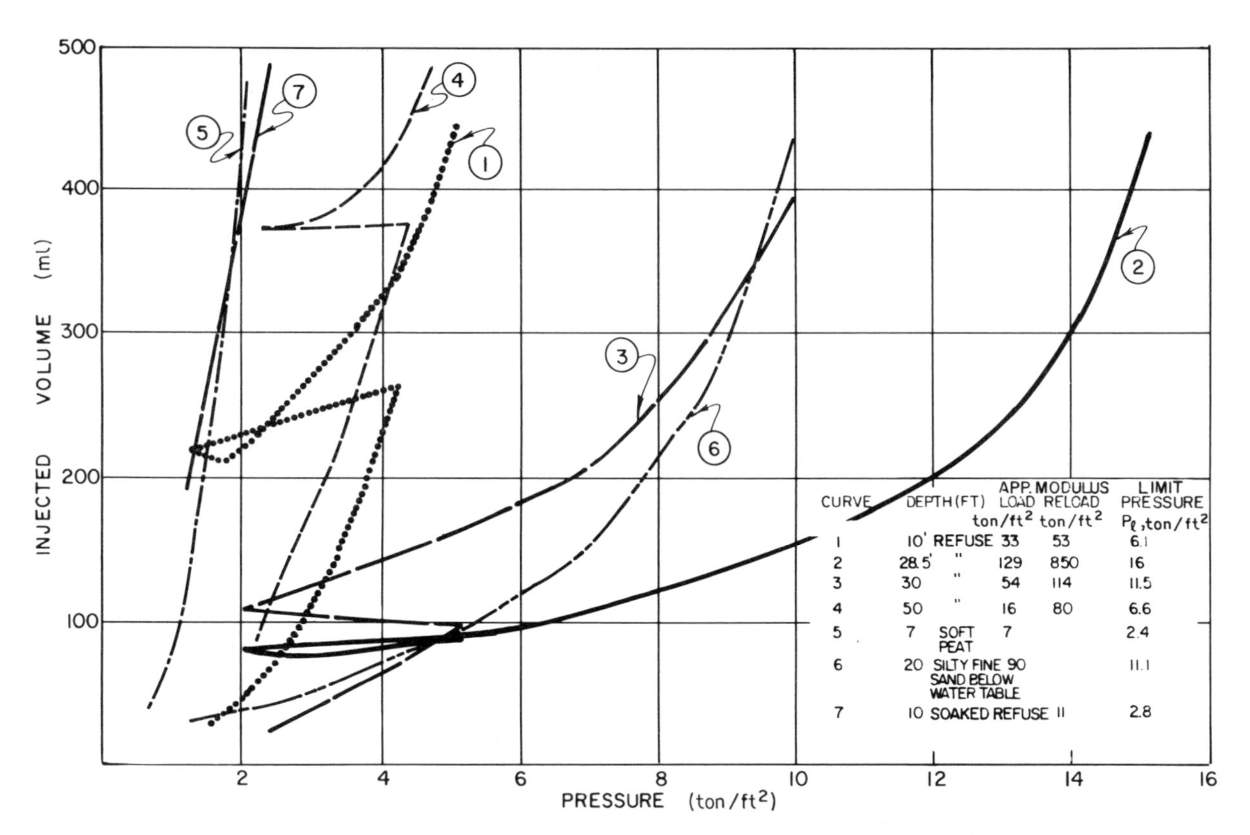

Figure 6.8 Results of pressure tests on refuse and soils at various depths in a sanitary landfill (Reproduced from STS Consultants, 1985, by permission)

Table 6.7 Correction factors for field vane undrained shear strength

Plasticity index	*Corrected shear strength/vane shear strength*
20	1.0
40	0.85
60	0.74
80	0.67
100	0.63

6.4.6 Plate load test

Plate load tests are performed to determine the bearing capacity of materials that are difficult to sample or where field strength tests cannot be readily performed such as in municipal landfills. A rigid circular plate is seated on the location of the test and loaded in increments while the settlement is recorded. The load is generally applied with a calibrated jack reacting against dead load or tension piles. The load increments are usually one-tenth of the ultimate bearing capacity of the area being tested. After the rate of settlement under a given load increment has become equal to or less than a specified value and remained so for a given length of time, the next load increment is applied (ASTM D 1194-72). The test is continued until failure occurs or the computed ultimate bearing capacity is reached.

The required minimum plate diameter is 1 ft (300 mm) but it should not be less than six times the largest particle size. The stress at a depth of 1–1.5 times the plate diameter becomes essentially insignificant. The largest practical plate diameter should be selected so that the compressibility and strength properties of the underlying materials are properly evaluated.

6.5 Groundwater level measurements

Groundwater measurements are made using three common categories of construction: the open stand pipe piezometer, the porous element piezometer, and the air actuated piezometer. Other types include electrical piezometers, oil pneumatic and water pressure type piezometers. The open type piezometer is illustrated in Figure 6.9. It is used primarily through reasonably free draining materials. Where several strata exist, the stratum of interest can be isolated (Figure 6.12).

The porous element piezometer (Figure 6.10) is used in fine-grained soils where a more rapid response is needed. The porous tip is driven or inserted into the fine-grained soil when measurement of the pore water pressure is desired. A metallic tip is less susceptible to damage than a ceramic tip. In the unsaturated zones if size of the pores in the porous tip is small, that is, it has high air entry value, then only water will enter the tip and not the air. This allows the measurement of pore water pressure without interference by air.

In the air-actuated piezometer (Figure 6.11) the ceramic tip and the two air tubes are connected with a flexible diaphragm between them. Compressed air or nitrogen is introduced at increasing pressure through one of the air tubes. The applied air pressure is resisted by the pore water pressure acting on the other side of the

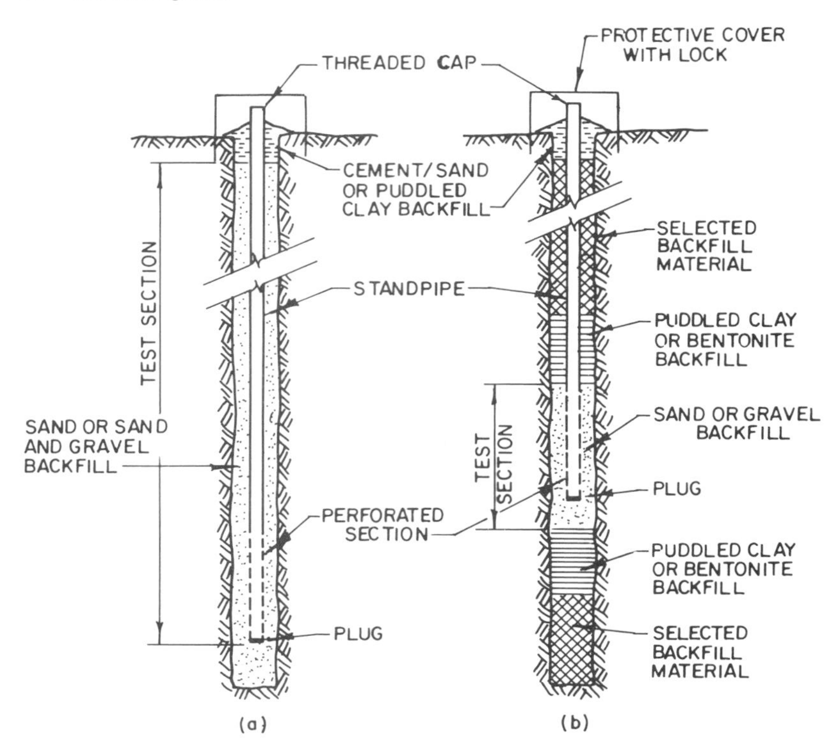

Figure 6.9 Open standpipe piezometer. Test sections may be perforated with slots or drilled holes (Reproduced from NAVFAC, 1982)

diaphragm. When the air pressure is equal to the water pressure, the membrane is pushed away allowing the air to flow up through the other tube to an indicator where the air bubbles become visible. The pneumatic device is sensitive to pressure changes as low as 1 cm.

6.6 Field permeability tests

The purpose of the field permeability tests is to measure the hydraulic conductivity of materials *in situ*. The test is usually conducted in a test boring or in a monitoring well. The test is influenced by the position of the water level, type of material, depth of test zone, hydraulic conductivity of the test zone, heterogeneity and anisotropy of the test zone. The tests are grouped into three broad categories: the constant head test, variable head test, and pump test.

The test zone could be isolated in several ways as illustrated in Figure 6.12. Where flow rates are very high or very low, the falling head test is more suitable. In reasonably pervious soils below the water table a rising head test is recommended. In unsaturated zones, the falling head is applicable.

Since the flow surface is very limited for the isolation methods of Figure 6.12, the test results are sensitive to the conditions of the bore hole (the hole should be clean), and drilling methods. The use of circulating drilling fluid or driving the casing during

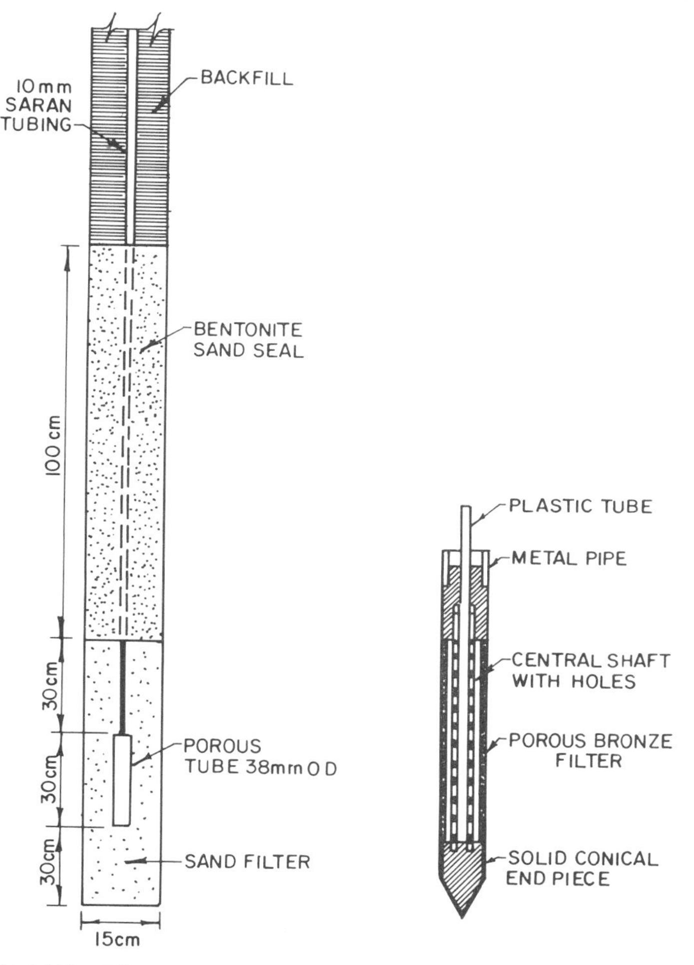

Figure 6.10 Porous element piezometer (Reproduced from NAVFAC, 1982)

advancing of the bore hole negatively affects the results of the tests. In stable formations the use of augering or air rotary drilling techniques is usually preferable if permeability tests are to be performed. Figures 6.13–6.15 can be used for interpreting the variable head tests (NAVFAC, 1982).

Because of the several factors influencing the results in bore hole tests, the hydraulic conductivity is frequently determined from the laboratory tests or estimated based on the index properties. Table 9.1 may be used for this purpose or an empirical relation between effective particle diameter D_{10} (mm) and coefficient of permeability (cm/s) (Hazen, 1892):

$$k = C(D_{10})^2 \tag{6.10}$$

where $C = 100$ (it varies between 40 and 150).

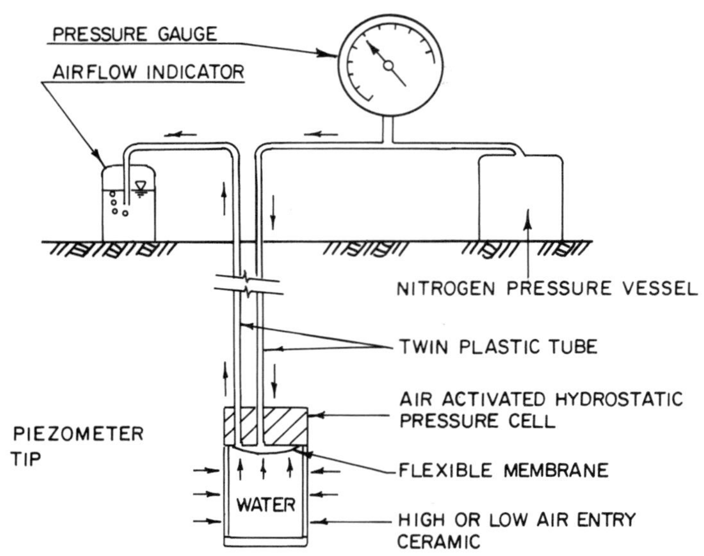

Figure 6.11 Air-actuated piezometer

A more reliable assessment of the hydraulic conductivity is obtained from pumping tests. However, they are much more expensive to perform and require much greater time to complete. An excellent treatment of the subject is presented by Driscoll (1986).

In municipal refuse the corrosive nature of the materials as well as the probable high temperatures inside the landfill affect the selection of materials for the well. Figure 6.15 shows the cross-section of a test well installed through a sanitary landfill (Oweis *et al.*, 1990).

High gas flow rates from the observation wells, which were screened partly in the vadose zone, along with high fluid conductivity and persistent foaming, caused considerable difficulties with the accurate measurement of fluid levels. This was resolved by inserting $\frac{3}{4}$ in diameter high-temperature CPVC pipe into the bottom of the well screens before commencement of the pumping test. This allowed measurements to be taken within the CPVC pipes using M-scopes with conductivity gain controls. In addition pressure transducers were attached to the tip of the CPVC pipes for automatic recording of head changes.

Special well caps were constructed to allow access for the monitoring equipment and prevent free gas discharge at the surface. This made working in the area tolerable and prevented any possible explosion hazard. In addition, sealing the well head reduced the amount of bubbling and foaming in the well, which allowed reliable water level measurements. A threaded nipple in the cap permitted gas pressures to be periodically measured in each of the observation wells using a water manometer. Maximum measured gas pressures in three observation wells were 1.92, 1.53 and 0.48 ft.

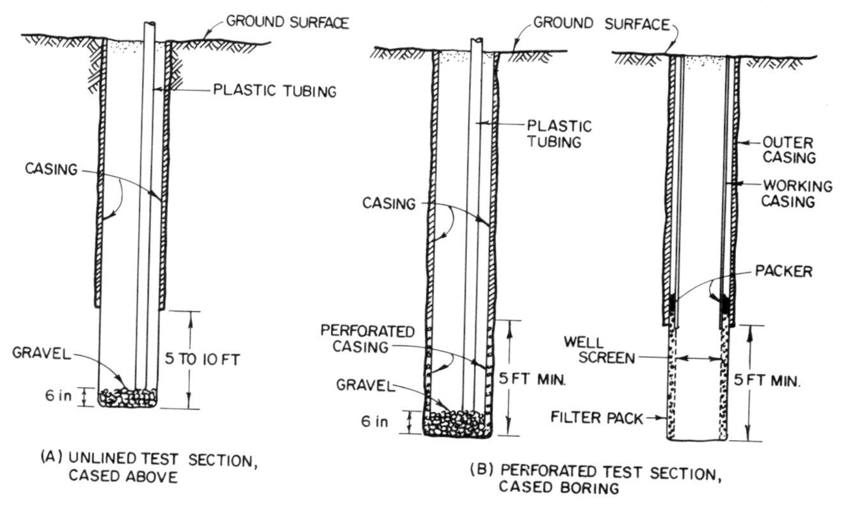

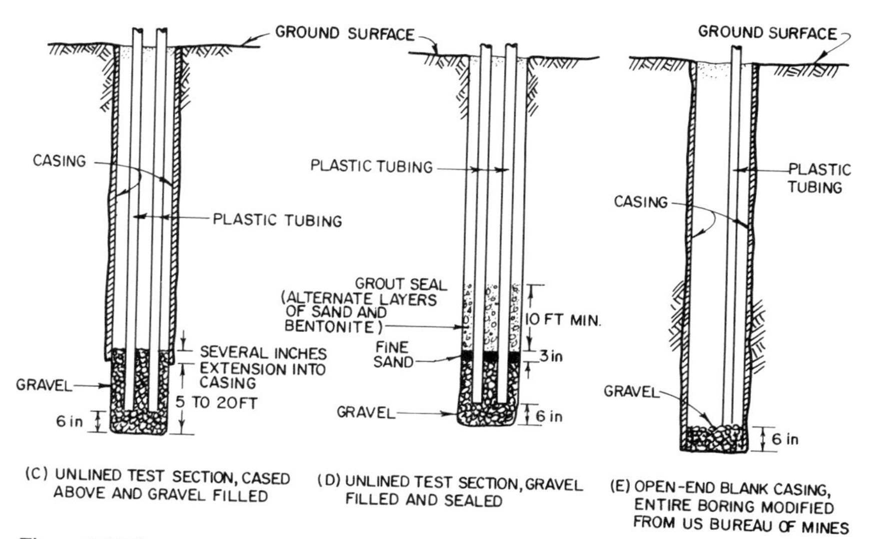

Figure 6.12 Test zone isolation methods

The temperature was measured by lowering a thermometer into the observation wells and it ranged from 140 to 150°F. Water temperature during the pumping tests was measured at 132°F. Use of a flow meter to measure discharge rates during pumping was aborted because of plugging by fibres during the early portion of the test. Flow calculations were subsequently based on measured discharge temporarily diverted into drums.

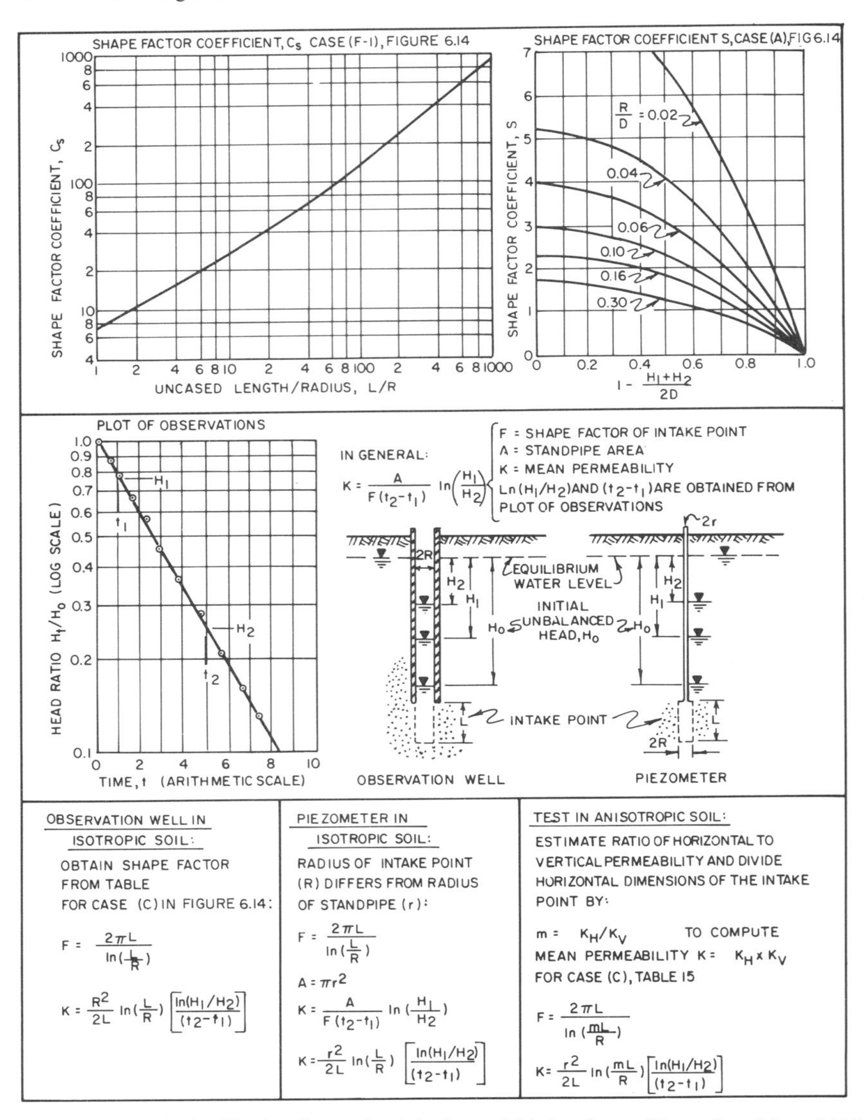

Figure 6.13 Analysis of hydraulic conductivity by variable head tests (Reproduced from NAVFAC, 1982)

6.6.1 Monitoring wells

Monitoring wells are installed to sample and assess the quality of the groundwater for environmental purposes. Figures 6.16 and 6.17 show the type of construction specified by the New Jersey Department of Environmental Protection. The larger hole in rock allows pumping from the well, if necessary, to induce flow into the well.

Evacuation of three to six well volumes is recommended before sampling at rates of 5–10 gal/min (19–38 l/min). Evacuation could be accomplished by hand bailing or pumping.

For installing monitoring wells at a contaminated site, drilling and sampling methods should be carefully selected. A well defined protocol is specified for field measurements and sampling (NJDEP, 1986).

6.7 Field monitoring

Field observations are made for the following reasons:

1. To provide information on the behaviour of the structure during construction and loading.
2. To determine if there are any indications of failure of a structure under service loads.

Some of the measurements that are made consist of vertical displacements, lateral displacements, rotation, pore water pressure, loads and contract pressure.

Methods for monitoring the pore water pressure are described in the previous sections.

Some disposal facilities built on marginal sites may require monitoring of soil movements and excess pore water pressure to monitor stability during landfilling phases.

To monitor movements, reliable benchmarks are established which are sufficiently removed from the site that they are not affected by construction activities or subsequent stress changes.

6.7.1 Displacements

Surface monuments can be used for monitoring horizontal and vertical displacements (Figure 6.18). Settlements are obtained from level readings and the lateral displacements by surveying.

Where the area to be monitored is loaded by a fill, settlement platforms are installed. They consist of a steel pipe attached to a rigid base plate, which is seated on a prepared horizontal surface. The top of the pipe is monitored for vertical displacement. With increasing height of fill the pipe is extended from time to time (Figure 6.19).

6.7.2 Pore pressure measurements

Properly installed piezometers (Figures 6.9 and 6.10) are used for pore water pressure measurements. This information is used in assessing the stability of the structure (see Chapter 10).

6.7.3 Inclinometers

Soil inclinometers are excellent devices for measurement of lateral deformations, which is needed in evaluating the stability of slopes and retaining structures. The key elements to an inclinometer system are a probe, a guide casing, a cable and a readout unit (Figure 6.20).

	CONDITION	DIAGRAM	SHAPE FACTOR, F	PERMEABILITY, K BY VARIABLE HEAD TEST	APPLICABILITY
OBSERVATION WELL OR PIEZOMETER IN SATURATED ISOTROPIC STRATUM OF INFINITE DEPTH	(A) UNCASED HOLE	2R, H_2, H_1, D	$F = 16\pi DSR$	(FOR OBSERVATION WELL OF CONSTANT CROSS SECTION) $K = \frac{R}{16DS} \times \frac{(H_2 - H_1)}{(t_2 - t_1)}$ FOR $\frac{D}{R} < 50$	SIMPLEST METHOD FOR PERMEABILITY DETERMINATION. NOT APPLICABLE IN STRATIFIED SOILS.
	(B) CASED HOLE, SOIL FLUSH WITH BOTTOM	CASING, 2R, H_2, H_1, D	$F = \frac{11R}{2}$	$K = \frac{2\pi R}{11(t_2 - t_1)} \ln\left(\frac{H_1}{H_2}\right)$ FOR $6'' \leq D \leq 60''$	USED FOR PERMEABILITY DETERMINATION AT SHALLOW DEPTHS BELOW THE WATER TABLE. MAY YIELD UNRELIABLE RESULTS IN FALLING HEAD TEST WITH SILTING OF BOTTOM OF HOLE.
	(C) CASED HOLE, UNCASED OR PERFORATED EXTENSION OF LENGTH "L"	CASING, 2R, H_2, H_1, L	$F = \frac{2\pi L}{\ln\left(\frac{L}{R}\right)}$	$K = \frac{R^2}{2L(t_2 - t_1)} \ln\left(\frac{L}{R}\right) \ln\left(\frac{H_1}{H_2}\right)$ FOR $\frac{L}{R} > 8$	USED FOR PERMEABILITY DETERMINATIONS AT GREATER DEPTHS BELOW WATER TABLE.
	(D) CASED HOLE, COLUMN OF SOIL INSIDE CASING TO HEIGHT "L"	CASING, 2R, H_2, H_1, L	$F = \frac{11\pi R^2}{2\pi R + 11L}$	$K = \frac{2\pi R + 11L}{11(t_2 - t_1)} \ln\left(\frac{H_1}{H_2}\right)$	PRINCIPAL USE IS FOR PERMEABILITY IN VERTICAL DIRECTION IN ANISOTROPIC SOILS.

	CONDITION	DIAGRAM	SHAPE FACTOR, F	PERMEABILITY, K BY VARIABLE HEAD TEST	APPLICABILITY
OBSERVATION WELL OR PIEZOMETER IN AQUIFER WITH IMPERVIOUS UPPER LAYER				(FOR OBSERVATION WELL OF CONSTANT CROSS SECTION)	
	(E) CASED HOLE, OPENING FLUSH WITH UPPER BOUNDARY OF AQUIFER OF INFINITE DEPTH	CASING, 2R, H_2, H_1	$F = 4R$	$K = \frac{\pi R}{4(t_2 - t_1)} \ln\left(\frac{H_1}{H_2}\right)$	USED FOR PERMEABILITY DETERMINATION WHEN SURFACE IMPERVIOUS LAYER IS RELATIVELY THIN. MAY YIELD UNRELIABLE RESULTS IN FALLING HEAD TEST WITH SILTING OF BOTTOM OF HOLE.
	(F) CASED HOLE, UNCASED OR PERFORATED EXTENSION INTO AQUIFER OF FINITE THICKNESS: (1) $\frac{L_1}{T} \leqq 0.2$ (2) $0.2 < \frac{L_2}{T} < 0.85$ (3) $\frac{L_3}{T} = 1.00$ NOTE: R_0 EQUALS EFFECTIVE RADIUS TO SOURCE AT CONSTANT HEAD	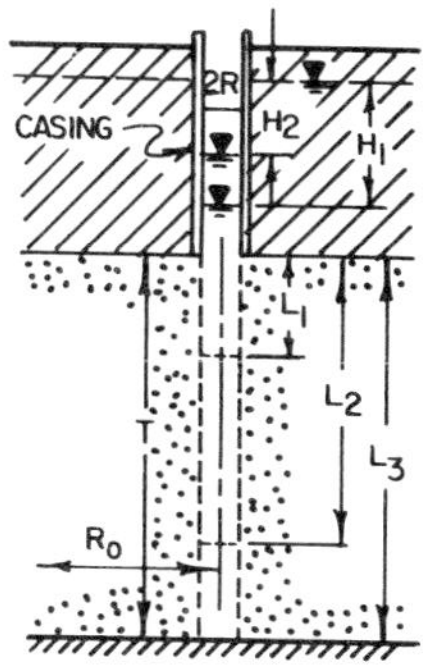 CASING, 2R, H_2, H_1, L_1, L_2, L_3, T, R_0	(1) $F = C_s R$	$K = \frac{\pi R}{C_s(t_2 - t_1)} \ln\left(\frac{H_1}{H_2}\right)$	USED FOR PERMEABILITY DETERMINATIONS AT DEPTHS GREATER THAN ABOUT 5FT.
			(2) $F = \frac{2\pi L_2}{\ln(L_2/R)}$	$K = \frac{R^2 \ln\left(\frac{L_2}{R}\right)}{2L_2(t_2 - t_1)} \ln\left(\frac{H_1}{H_2}\right)$ FOR $\frac{L}{R} => 8$	USED FOR PERMEABILITY DETERMINATIONS AT GREATER DEPTHS AND FOR FINE-GRAINED SOILS USING POROUS INTAKE POINT OF PIEZOMETER.
			(3) $F = \frac{2\pi L_3}{\ln\left(\frac{R_0}{R}\right)}$	$K = \frac{R^2 \ln\left(\frac{R_0}{R}\right)}{2L_3(t_2 - t_1)} \ln\left(\frac{H_1}{H_2}\right)$	ASSUME VALUE OF $\frac{R_0}{R}$ = 200 FOR ESTIMATES UNLESS OBSERVATION WELLS ARE MADE TO DETERMINE ACTUAL VALUE OF R_0.

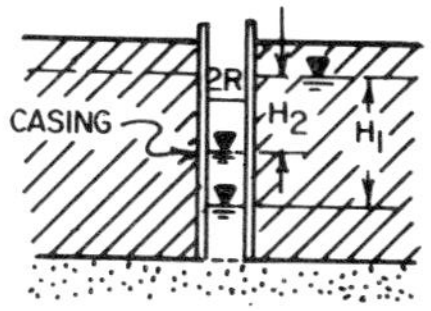

Figure 6.14 Shape factors for calculation of hydraulic conductivity from variable head tests (Reproduced from NAVFAC, 1982)

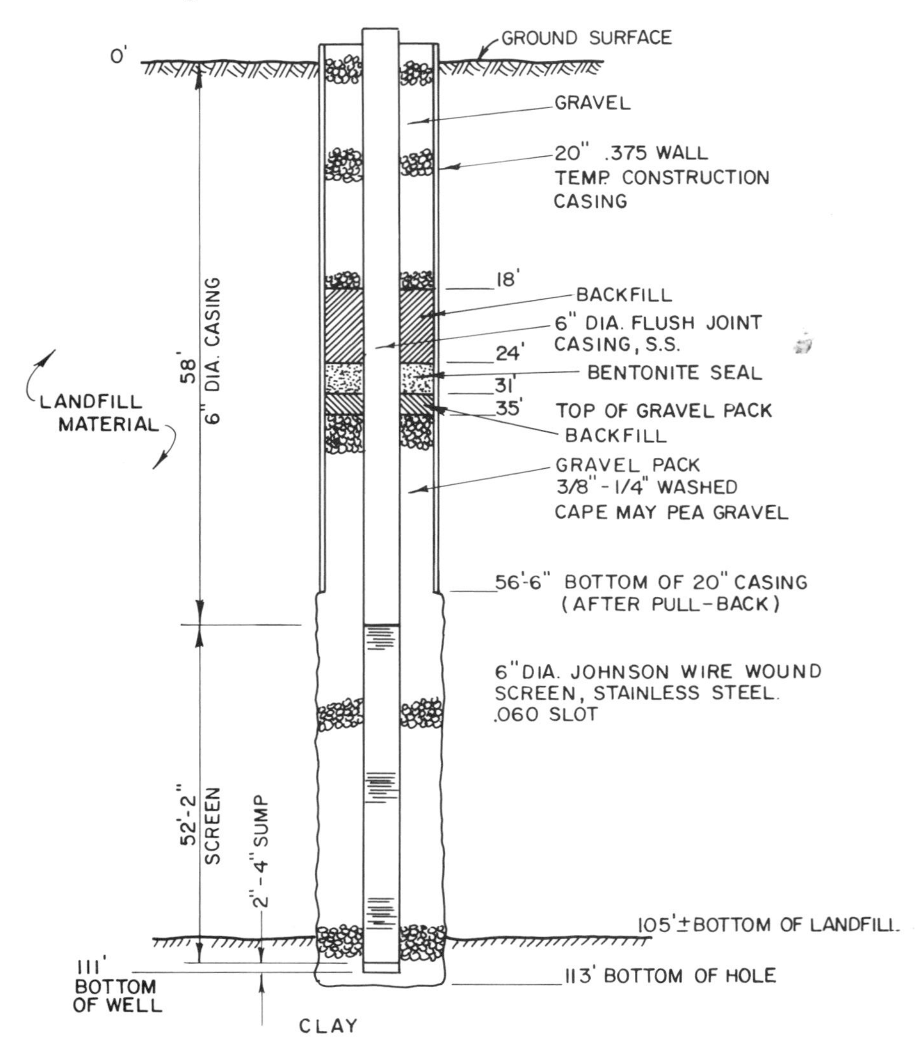

Figure 6.15 A test well in refuse

The probe is mounted on guide wheels which travel up and down in the guide casing. The electrical cable is used to transmit the signals from the sensing element to a readout unit at the surface. In addition, the cable has graduated marks on it to control the depth at which the sensing element is operating. The inclinometer measures the average inclination of the casing between the upper and lower sets of wheels. Multiplying the distance between the two sets of wheels by the sine of the angle of inclination of the probe, the displacement of the upper set of wheels, with respect to the lower set, is obtained. Starting at the bottom of the casing and summing up these displacements up to the top, the total displacement of the top of the casing with respect to the bottom is computed. Repeating such measurements periodically

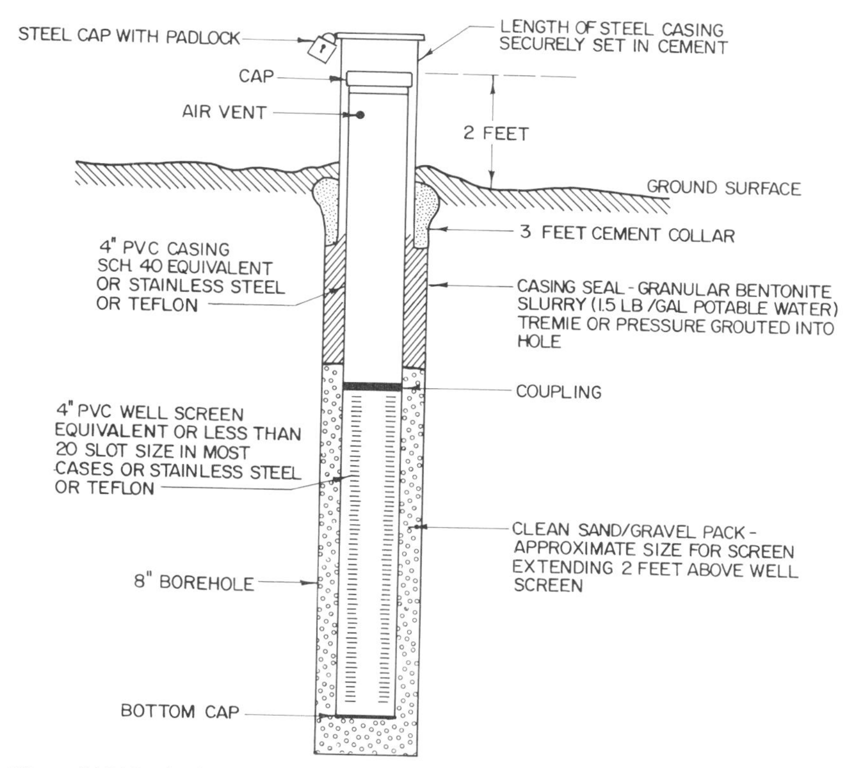

Figure 6.16 Monitoring well in soil

provides data on the depth, magnitude, direction and rate of casing deflection. The casing should either extend into a firm stratum not expected to deform significantly or extend to well below any critical failure surface.

Summary

In an investigation programme where refuse or potentially toxic waste is expected, proper environmental safety measures should be followed. The appropriate level of protection (A–D) should be pursued, with A being applicable to the worst conditions. The proper selection of protective tools and clothing should be determined by a qualified individual.

Investigation of large sites consists of a desk study, preliminary investigations, and detailed investigations. Applicable regulations and guidelines are provided. For hazardous wastes special conditions apply regarding cleaning of the equipment, record keeping and storage of the samples, etc. *In situ* measurements are made using standard penetration tests, cone penetration tests, pressuremeters, vane shear tests, load tests, and field permeability tests.

Field monitoring provides information on the behaviour of the structure during construction and in service. The equipment is available for measuring vertical and lateral displacement, pore pressure, etc.

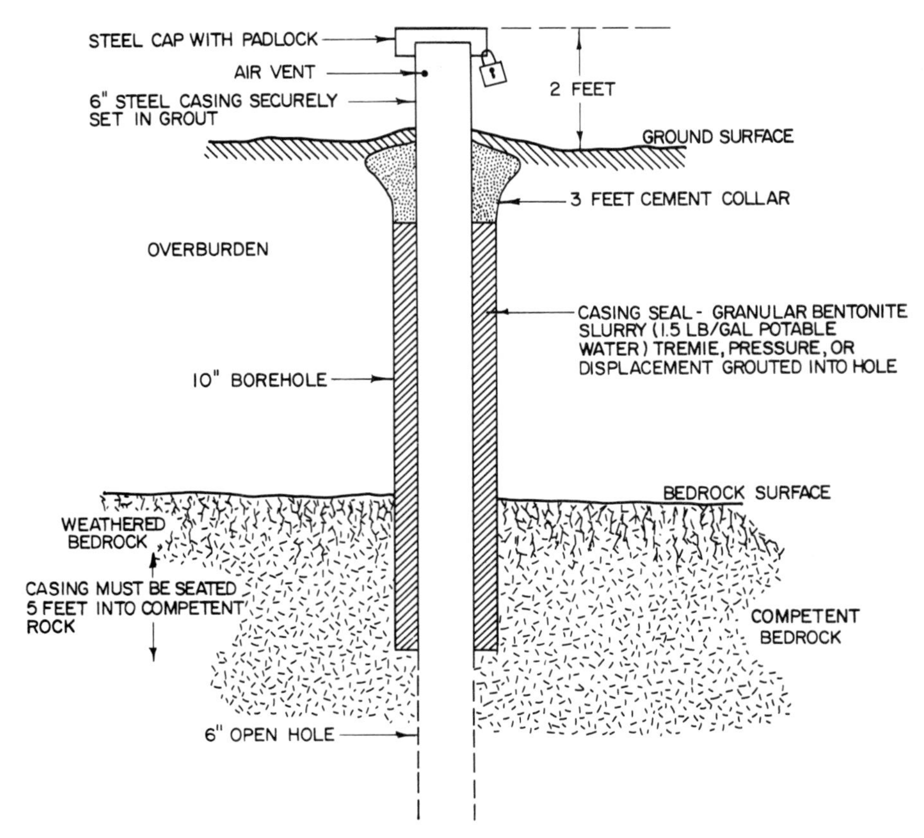

Figure 6.17 Monitoring well in rock

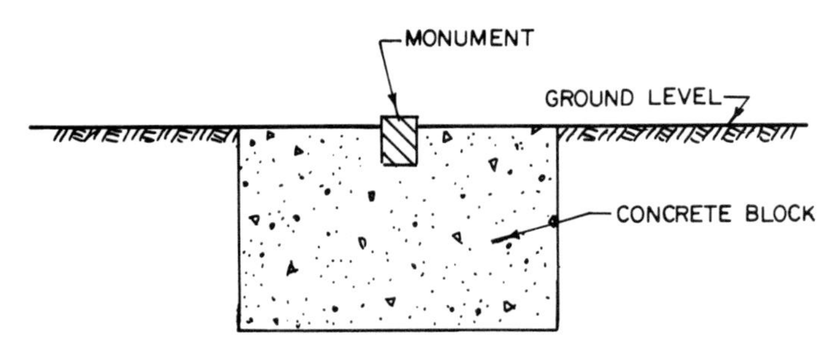

Figure 6.18 Surface monuments

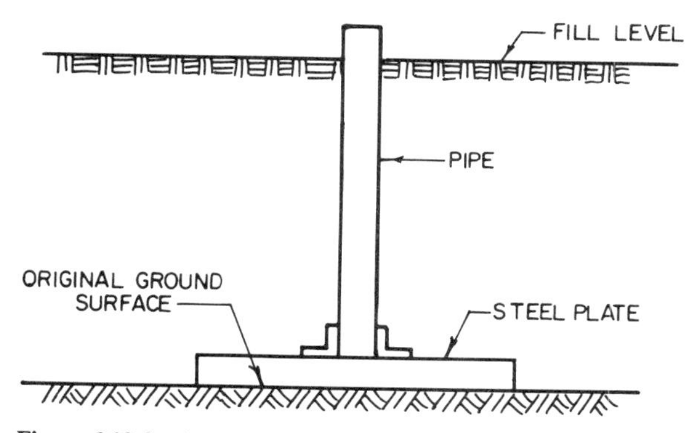

Figure 6.19 Settlement plate

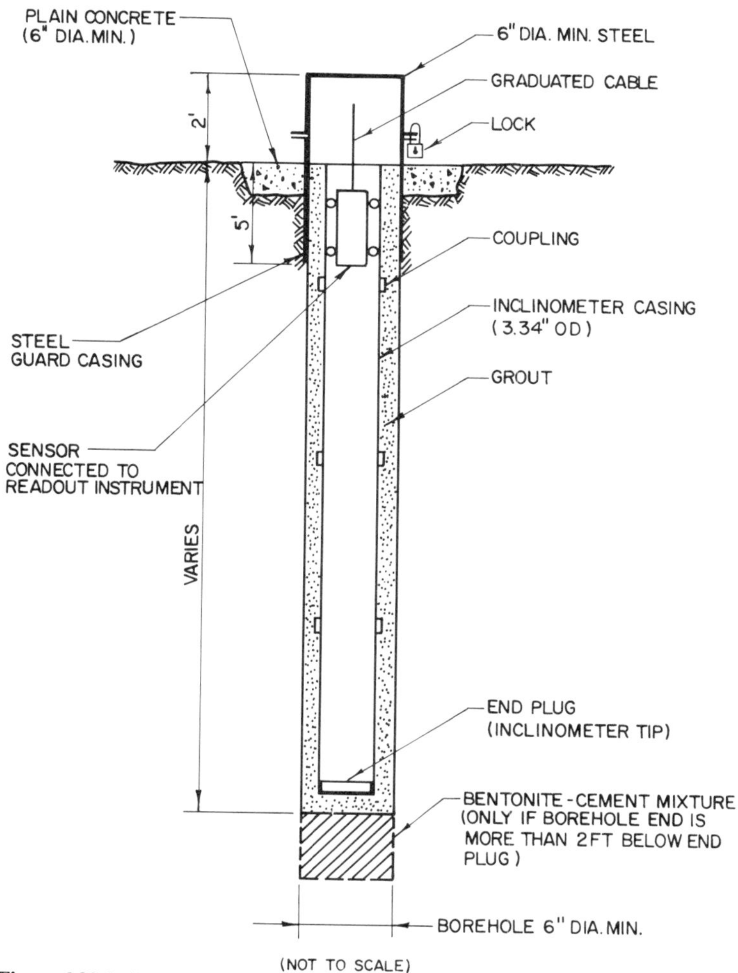

Figure 6.20 Soil inclinometer

Notation

a	area ratio for the given piezocone
C	constant for permeability computation in sand
D_{10}	effective particle diameter
D_r	relative density
E	Young's modulus
G	dynamic shear modulus
I_p	plasticity index
k	coefficient of permeability
N	standard penetration resistance
N'	corrected N for overburden
N_k	cone factor
N_{kT}	cone correction factor for soil plasticity
q_c	cone penetration resistance
q_T	corrected cone penetration resistance for a jointless cone
s_u	undrained shear strength
u	measured pore water pressure
α	empirical parameter
σ'_{vo}	effective overburden stress
σ_{vo}	total vertical stress
φ	angle of friction

References

Aas, G., Lacasse, S., Lunne, T. and Hoeg, K. (1986) Use of in situ tests for foundation design on clay. *In situ 86*, Ed. Clemence, S. P. ASCE, Geot. Special Publication No. 6, pp. 1–30

ASCE (1976) *Subsurface Investigation for Design of Foundations of Buildings*, ASCE Manual and Report on Engineering Practice, No. 56

Baguelin, F. J., Bustsmantem, M. G. and Frank, R. (1986) The pressuremeter for foundations: French experience. *In Situ 86*, Ed. Clemence, S. P., ASCE, Geot. Special Publication No. 6, pp. 321–346

Bjerrum, L. (1972) Embankments on soft clay, *Proc. ASCE Specialty Conf. on Earth and Earth Supported Structures*, Vol. II, Purdue University, USA, pp. 1–54

Charlie, W. A. (1977) Pulp and papermill solid waste disposal: a review, *Proc. Conf. on Geotechnical Practice for Disposal of Solid Waste Materials*, ASCE, Ann Arbor, Michigan, June 1977, pp. 71–86

Clayton, C. R. I., Simons, N. E. and Matthews, M. C. (1982) *Site Investigation*, Halsted Press, New York

Cunnigham, J. A., Lukas, R. G. and Anderson, T. C. (1977) Impoundment of fly ash and slag: a case study, *Proc. Conf. on Geotechnical Practice for Disposal of Solid Waste Materials*, ASCE, Ann Arbor, Michigan, June 1977, pp. 227–245

Driscoll, F. G. (1986) *Groundwater and Wells*, Chapter 16, Johnson Division, St. Paul, Minnesota

Gibbs, H. J. and Holtz, W. G. (1957) Research in determining the density of sands by spoon penetration testing, *Proc. 4th Int. Conf. Soil Mechanics and Foundation Engineering*, Vol. 1, London, pp. 35–39

Hazen, A. (1892) *Physical Properties of Sands and Gravels with Reference to Their Use in Filtration*, Rept. Mass. State Board of Health, p. 539

Hvorslev, M. J. (1962) *Subsurface Exploration and Sampling of Soils for Civil Engineering Purposes*, ASCE, Engineering Foundation

Kovacs, W. D. and Salomone, L. A. (1982) SPT hammer energy measurement, *ASCE, Journal of the Geotechnical Engineering Division*, **108,** No. GT4, pp. 599–620

Legget, R. F. (1973) *Cities and Geology*, McGraw-Hill, New York

Meyerhoff, G. G. (1956) Penetration tests and bearing capacity of cohesionless soils, *ASCE, Journal of the Soil Mechanics and Foundation Engineering Division*, **82,** No. SM1

NAVFAC DM 7.1 (1982) *Soil Mechanics Design Manual*, Naval Facilities Engineering Command, Alexandria, Va

NJDEP *Field Sampling Procedures Manual*, New Jersey Department of Environmental Protection, Division of Hazardous Site Mitigation

Olsen, R. S. and Farr, J. V. (1986) Site characterization using the cone penetrometer test, *Use of In Situ Tests In Geotechnical Engineering*, *Proc. In Situ 86*, Geot. Special Pub. No. 6, Ed. Clemence, S. P., pp. 854–868

Oweis, I. S. and Khera, R. (1986) Criteria for geotechnical construction on sanitary landfills, *Int. Symp. on Environmental Geotechnology*, Vol. 1, Ed. Fang, H. Y., pp. 205–222

Oweis, I., Smith, D., Ellwood B. and Greene, D. (1990) Hydraulic properties of refuse, *Journal of the Geotechnical Engineering Division, ASCE*

Robertson, P. K. and Campanella, R. G. (1983a) Interpretation of cone penetration tests. Part I: Sand, *Canadian Geotechnical Journal*, **20,** No. 4, 718–733

Robertson, P. K. and Campanella, R. G. (1983b) Interpretation of cone penetration tests. Part II: Clay, *Canadian Geotechnical Journal*, **20,** No. 4, 734–745

State of the Art on Current Soil Sampling (1979) The Sub-Committee on Soil Sampling, International Society of Soil Mechanics and Foundation Engineering

STS Consultants (1985) *Geotechnical Data Report for the Delcorte Park Landfill*, Report to the Hackensack Meadowland Commission, Lyndhurst, New Jersey

Way, S. G. (1973) *Terrain Analysis: A Guide to Site Selection Using Aerial Photographic Interpretation*, Dowden, Hutchinson, and Russ Inc., Straudsburg, Pennsylvania

Winter, E. (1982) Suggested practice for pressuremeter testing in soils, *ASTM*, *Geotechnical Testing Journal*, **5,** No. 3/4

Chapter 7

Leachate and gas formation

7.1 Introduction

The waste in a sanitary landfill is a mixture of organic materials (e.g. garbage, paper, cardboard, textiles, plastic, wood, rubber) and inorganic waste, which may include metals such as cans and wires. As the refuse is placed in landfill, it undergoes oxidation and decomposition in the presence of oxygen, moisture and appropriate temperature. Water, which is essential for decomposition, is derived from the waste itself and is about 10–20% by volume or 100–200 mm^2 of water for each 1 m^2 of refuse (Fenn *et al.*, 1975). The rate of decomposition is influenced by type of refuse, ambient temperature, oxygen supply and water content. When infiltration exceeds the total evapotranspiration plus the moisture retention capacity of the refuse in the gravitational field, the water percolates through the refuse removing dissolved and/or suspended products of biological and chemical decomposition.

The biological degradation of the waste may occur in the presence of oxygen (aerobic bacteria), in an environment devoid of oxygen (anaerobic bacteria), or with very little oxygen (facultative anaerobic bacteria). Aerobic bacteria require oxygen to attack the organic material if the appropriate nutrients and moisture are present. Anaerobic bacteria remain mostly dormant in the presence of oxygen. Facultative anaerobic bacteria are adaptable to the presence of oxygen.

In all cases organic waste is broken down by enzymes produced by bacteria in a manner comparable to food digestion. Considerable heat is generated by these reactions with methane, carbon dioxide, and other gases as the by-products. A satisfactory design of landfill must consider measures for collecting and treating leachate, a leakage detection system, and gas ventings.

7.1.1 Aerobic decomposition

In the presence of oxygen, appropriate nutrients and moisture, aerobic bacteria usually generate water, carbon dioxide, organic acids and inorganic minerals. The major constituents of organic wastes in a sanitary landfill are foodstuffs, cellulose, and plastics, rubber and leather. Foodstuffs principally contain proteins, carbohydrates and fats. Materials containing cellulose include paper, rags, fruit skins, etc. Proteins consist of complex nitrogenous compounds and sulphur compounds that may be reduced to ammonia gas, nitric acid, sulphuric acid, carbon dioxide and water. Carbohydrates (sugar, starch) successively change to glucose, lactic acid and acetic acid, which finally oxidizes yielding carbon dioxide and water. Fats are split into fatty acids and glycerine. Methane gas and carbon dioxide are the by-products of the oxidation of fatty acids.

Rubber and most plastics are usually resistant to biochemical degradation. Plastic polymers decompose into fragment molecules or monomers (Zerlaut and Stake, 1975) depending on their structure and conditions of heat, air, radiation, and mechanical methods used in manufacturing. Vinyl polymers such as polyethylene are resistant to monomer formation by virtue of their structure. Because of the high temperature associated with the aerobic phase some plastics may be altered.

The heat produced from aerobic decomposition elevates the initial temperature. Peak temperatures of 160°F (71°C) can be achieved in a few days to a few weeks after application of the cover (Noble, 1976). The high temperature (roughly 70°F (21°C) above ambient) may cause combustion of dry waste and generate fire.

7.1.2 Anaerobic decomposition

The consumption of nutrients and depletion of oxygen and moisture tend to inhibit the aerobic process and initiate the anaerobic decomposition process (Figure 7.1). In

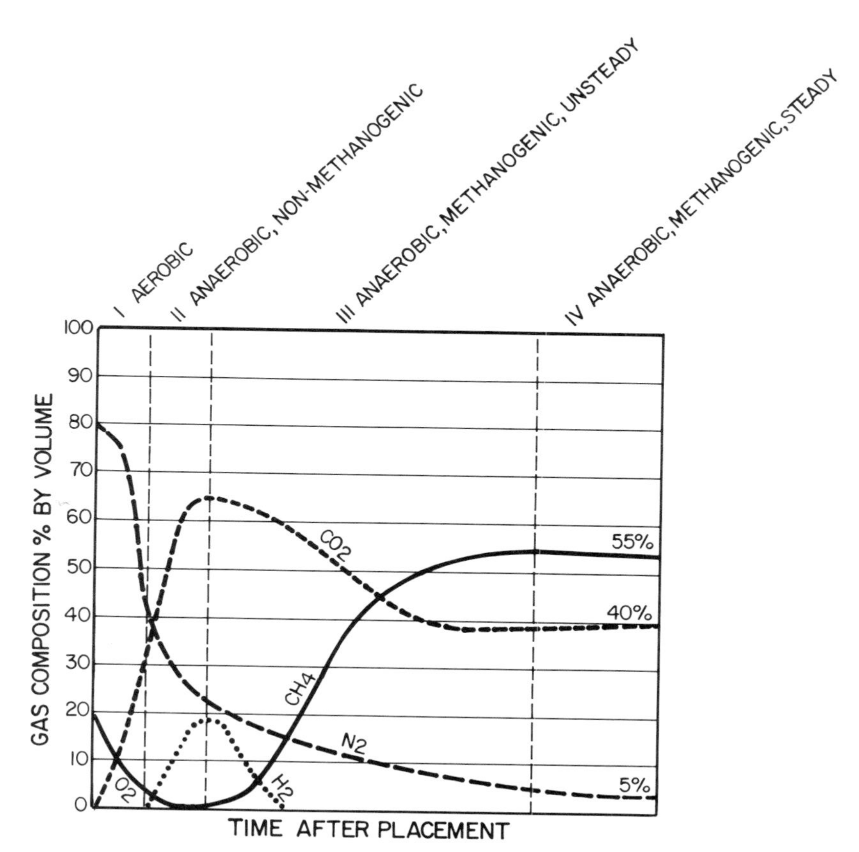

Figure 7.1 Gas composition and evolution in a typical landfill (Reproduced from Gas Generation Institute, 1981, by permission)

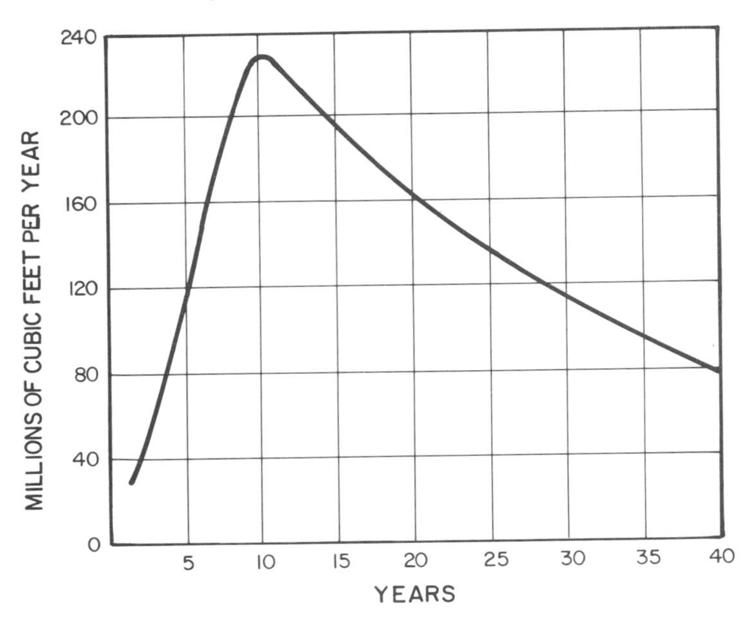

Figure 7.2 Typical gas generation profile for 1 million tons of solid waste (Reproduced from Gas Generation Institute, 1981, by permission)

the presence of moisture and appropriate nutrients, the anaerobic process and gas generation could extend over a long period after completion of the landfill as illustrated in Figure 7.2. Organic materials are broken down by facultative and anaerobic bacteria. In the initial phase of anaerobic decomposition, the principal gas produced is carbon dioxide. With time, the amount of carbon dioxide decreases and methane increases, each reaching a plateau. The concentration of methane can reach 50–60%. A landfill may continue to produce methane at this rate for 10 or more years.

Carbohydrates and cellulose are converted into sugar then alcohol and various acids. Carbon dioxide is released and the pH drops, thus inhibiting the acid-forming process. The soluble acids are converted into methane and carbon dioxide. This raises the pH and the acid fermentation bacteria resume activity. The process would continue in the presence of nutrients, appropriate pH, temperature, and moisture. The heat generated by the anaerobic process is much less than that of the aerobic phase.

With the exception of metals, inorganic materials are not subject to corrosion. Metals are oxidized. Strong inorganic acids produced as by-products of decomposition could corrode metals. Chemical reactions could cause the galvanic action that accelerates the corrosion of metals. Plastics, glass and synthetic rubber remain essentially inert, and natural rubber breaks down extremely slowly.

7.2 Leachate

Leachate refers to the highly contaminated water that emanates from a disposal site.

The percolation of rainwater through a landfill or surface runoff from an ineffective landfill cover generates leachate. Leachate from a decomposing landfill usually contains various amounts of organic and inorganic chemicals. Tables 7.1 and 7.2 show the range of inorganic and organic constituents found in leachate from many landfills (Subtitle D Study, 1986). These chemicals and their concentrations impact the groundwater quality beneath and beyond the landfill. Table 7.3 illustrates the type of waste expected from various industries. Leachate generated from the disposal of hazardous waste may contain elevated amounts of heavy metals (e.g. mercury, lead), toxic substances (e.g. arsenic), and organic compounds. As illustrated in Tables 7.1 and 7.2 the concentrations of various organic and inorganic constituents vary. The leachate constituents significantly impact the design of the leachate containment and collection systems.

In the formation of leachate initially the solid waste particles and soluble materials are carried by percolating water. Later, the soluble components that enter the leachate stream are the result of complex series of biological and chemical reactions that generate both liquids and gases.

The constituents and character of the landfill affect the quality of the leachate (O'Leary and Tansel, 1986). An acidic pH condition increases the solubility of chemical constituents, decreases the sorptive capacity of refuse and increases the ion exchange between leachate and organic matter. Other parameters affecting the activity of leachate are the redox potential, adsorption, temperature and biological mechanism. The decomposition process could be inhibited by large changes of pH, unfavourable ionic concentrations and deficiency of nutrients (Lu *et al.*, 1985; Chen and Bowerman, 1975).

Table 7.1 Leachate constituents from municipal waste landfills

Constituent	*Concentration* (mg/l)	*Constituent*	*Concentration* (mg/l)
Chemical oxygen demand	50–90 000	Hardness (as $CaCO_3$)	0.1–36 000
Biochemical oxygen demand	5–75 000	Total phosphorus	0.1–150
Total organic carbon	50–45 000	Organic phosphorus	0.4–100
Total solids	1–75 000	Nitrate nitrogen	0.1–45
Total dissolved solids	725–55 000	Phosphate (inorganic)	0.4–150
Total suspended solids	10–45 000	Ammonia nitrogen	0.1–2000
Volatile suspended solids	20–750	Organic nitrogen	0.1–1000
Total volatile solids	90–50 000	Total Kjeldahl nitrogen	7–1970
Fixed solids	800–50 000	Acidity	2700–6000
Alkalinity (as $CaCO_3$)	0.1–20 350	Turbidity (Jackson units)	30–450
Total coliform bacteria (c.f.u./100 ml)	0–10	Chlorine	30–5000
Iron	200–5500	pH (dimensionless)	3.5–8.5
Zinc	0.6–220	Sodium	20–7600
Sulphate	25–500	Copper	0.1–9
Sodium	0.2–79	Lead	0.001–1.44
Total volatile acid	70–27 700	Magnesium	3–15 600
Manganese	0.6–41	Potassium	35–2300
Faecal coliform bacteria (c.f.u./1000 ml)	0–10	Cadmium	0.0.375
Specific conductance (mho/cm)	960–16 300	Mercury	0–0.16
Ammonium nitrogen	0–1106	Selenium	0–2.7
		Chromium	0.02–18

Source: Subtitle D Study, 1986.

Table 7.2 Organic constituents of leachate from various municipal waste landfills

Constituent	*Minimum* (ppb)	*Maximum* (ppb)	*Median* (ppb)
Acetone	140	11 000	7500
Benzene	2	410	117
Bromomethane	10	170	55
1-Butanol	50	360	220
Carbon tetrachloride	2	398	10
Chlorobenzene	2	237	10
Chloroethane	5	170	7.5
bis(2-Chloroethoxy)methane	2	14	10
Chloroform	2	1 300	10
Chloromethane	10	170	55
Delta BHC	0	5	0
Dibromomethane	5	25	10
1,4-Dichlorobenzene	2	20	7.7
Dichlorodifluoromethane	10	369	95
1,1-Dichloroethane	2	6 300	65.5
1,2-Dichloroethane	0	11 000	7.5
cis-1,2-Dichloroethane	4	190	97
trans-1,2-Dichloroethane	4	1 300	10
Dichloromethane	2	3 300	230
1,2-Dichloropropane	2	100	10
Diethyl phthalate	2	45	31.5
Dimethyl phthalate	4	55	15
Di-*n*-butyl phthalate	4	12	10
Endrin	0	1	0.1
Ethyl acetate	5	50	42
Ethyl benzene	5	580	38
bis(2-Ethylhexyl) phthalate	6	110	22
Isophorene	10	85	10
Methyl ethyl ketone	110	28 000	8300
Methyl isobutane ketone	10	660	270
Naphthalene	4	19	8
Nitrobenzene	2	40	15
4-Nitrophenol	17	40	25
Pentachlorophenol	3	25	3
Phenol	10	28 800	257
2-Propanol	94	10 000	6900
1,1,2,2-Tetrachloroethene	7	210	20
Tetrachloroethene	2	100	40
Tetrahydrofuran	5	260	18
Toluene	2	1 600	166
Toxaphene	0	5	1
1,1,1-Trichloroethane	0	2 400	10
1,1,2-Trichloroethane	2	500	10
Trichloroethane	1	43	3.5
Trichlorofluoromethane	4	100	12.5
Vinyl chloride	0	100	10
m-Xylene	21	79	26
p-Xylene + *o*-xylene	12	50	18

Source: Subtitle D Study, 1986.

Table 7.3 Typical constituents of industrial leachate

Industry	*As*	*Cd*	*CH*[a]	*Cr*	*Cu*	*CN*	*Pb*	*Hg*	*MO*[b]	*Se*	*Zn*
Battery		×		×	×						×
Chemical manufacturing			×	×	×			×	×		
Electrical and electronic			×		×	×	×	×		×	
Electroplating and metal finishing		×		×	×	×					×
Explosives	×				×		×	×	×		
Leather				×					×		
Mining and metallurgy	×	×		×	×	×	×	×		×	×
Paint and dye		×		×	×	×	×	×	×	×	
Pesticide	×		×			×	×	×	×		×
Petroleum and coal	×		×				×				
Pharmaceutical	×			×				×	×		
Printing and duplicating	×				×		×		×	×	
Pulp and paper								×	×		
Textile				×	×				×		

[a] Chlorinated hydrocarbons and polychlorinated biphenyls. [b] Miscellaneous organics such as acrolein, chloropicrin, dimethyl sulphate, dinitrobenzene, dinitrophenol, nitroaniline and pentachlorophenol.
Source: Matrecon, 1980.

The leachate components from leather, garbage and cloth products may be carbon dioxide, aldehydes, ketones, organic acids, sulphates, phosphates, ammonium, nitrates and nitrites. Paper and paper products have similar constituents except for ketones, phosphates and carbonates. Metals may leach as sulphates of calcium and magnesium; bicarbonates of iron, calcium and magnesium; and oxides of tin, zinc and copper. Ashes in landfills release soluble minerals. Leachate may also contain phenol.

7.2.1 Other wastes

The leachate generated by fly ash is very alkaline with high concentrations of sodium, sulphate and total dissolved solids. The field pH values range between 5 and 12.5. Trace elements such as arsenic, selenium, molybdenum, chromium and lead are also present and may result in contamination of groundwater. Very high concentrations of these pollutants have been reported by Groenewold *et al.* (1985). In an unsaturated disposal environment the concentration of sodium and sulphate decreased significantly after 2–3 years but under saturated conditions the pollutant concentration in groundwater was high for several years after disposal.

7.2.2 Estimating leachate volume

The amount of leachate generated is dependent on the available water, landfill constituents, its surface and foundation soils.

The available water is affected by moisture in the refuse itself, precipitation, surface runoff, irrigation, groundwater moving through the landfill, rise in an otherwise low groundwater table, water originally present on site before the placement of refuse, and water generated from the decomposition process.

Table 7.4 Water holding capacity of refuse

Point in time	*% by volume*	*Equivalent* mm *of water per* m *of refuse*	*Equivalent gallons of water per* yd³ *of refuse*
At placement time	10–20	150	30
At field capacity	20–35	300	60
At saturation (porosity = 0.4)		550	110

Source: Fenn *et al.*, 1975.

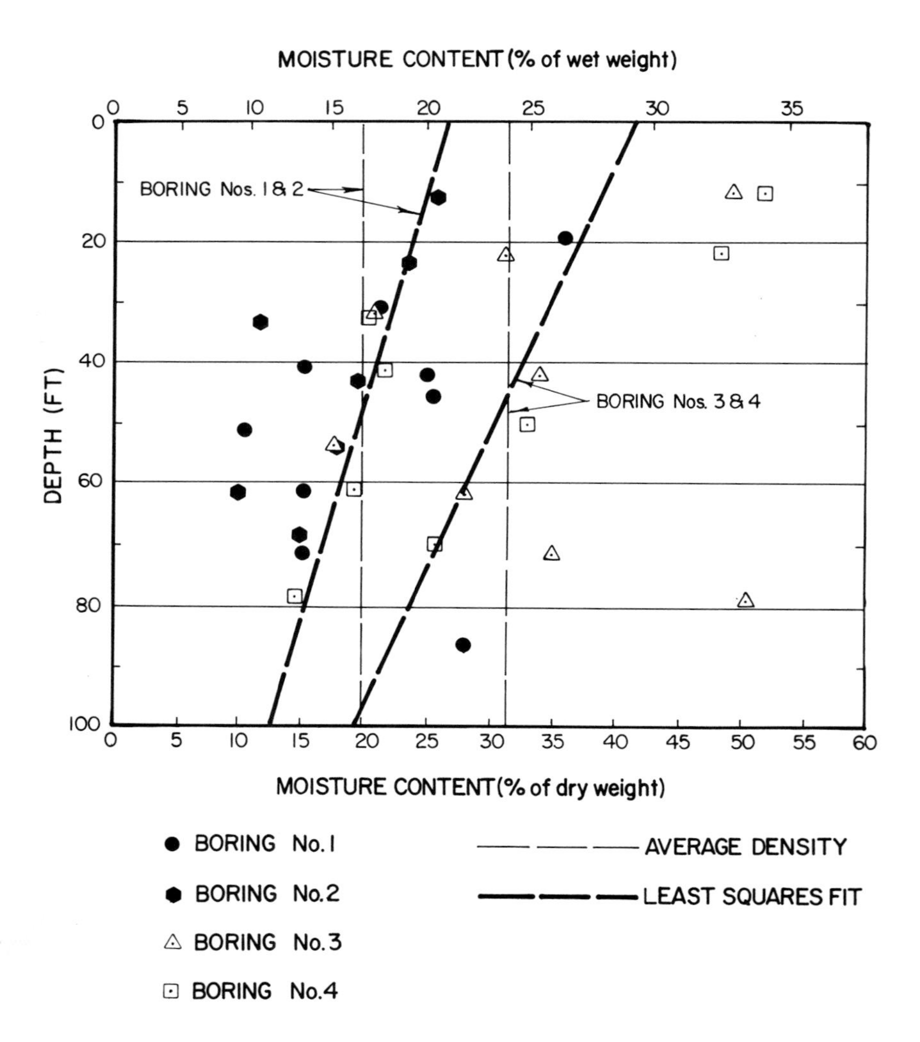

Figure 7.3 Variation of moisture content with depth for a landfill in Southern California

The water reaching the landfill is affected by the surface runoff, evapotranspiration, and the field capacity of the soil cover. The field capacity is the maximum amount of water a soil or refuse can retain in a gravitational field without percolation. The field capacities for fine sand and clay are 1.5 in (125 mm) and 5.5 in (460 mm) of water per foot (1 m) of soil, respectively (Fenn *et al.*, 1975). A portion of the field capacity water is not available to the plants. The upper limit of this moisture content is known as the wilting point. The available water is the field capacity minus the wilting point. The average amount of water in refuse is 10–20% by volume, as illustrated in Table 7.4.

Figure 7.3 shows the variation of moisture content with depth for a landfill in southern California. The refuse bulk samples were retrieved using a 12 in coring bucket at frequent intervals. Representative portions of 100–200 g were used for moisture content determination and placed in a 40°C oven for the 72 h to determine moisture content. The low temperature prevented burning of organic materials while the lengthy drying time allowed most of the moisture to be removed.

7.2.3 Water balance method

The water balance method allows the estimation of rate of percolation and is most commonly used to predict the amount of leachate generated from a landfill. The method assumes a one-dimensional flow, conservation of mass, and known retention and transmission characteristics of the soil cover and refuse. The basic equations (Figure 7.4) are:

$$P + \mathrm{SR} + \mathrm{IR} = I + R_0 \tag{7.1}$$

where P is the input water from precipitation, SR the input water from surrounding surface runoff, IR the input water from irrigation, I the infiltration, and R_0 the surface runoff.

The portion of infiltrating liquid, I, that will percolate in the soil cover, $\mathrm{PER_S}$, is given by:

$$\mathrm{PER_S} = I - \mathrm{AET} - \mathrm{d}S_{\mathrm{TC}} \tag{7.2}$$

where AET is the actual evapotranspiration (i.e. the sum of water loss from

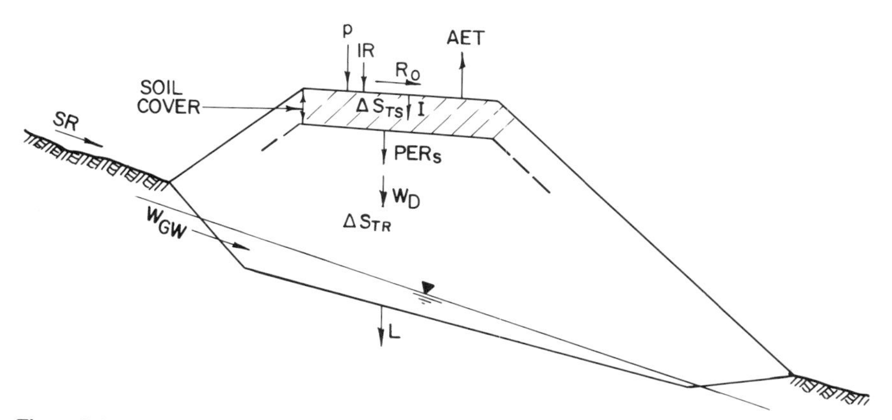

Figure 7.4 Leachate generation parameters

evaporation and plant water use), and dS_{TS} is the change in moisture storage in soil cover.

Similarly, the portion of I percolating in the refuse, PER_R, which represents leachate volume, is given by:

$$PER_R = I - AET - dS_s + W_D - dS_{TR} \quad (7.3a)$$

$$= PER_S + W_D + dS_{TR} \quad (7.3b)$$

where W_D is the water from decomposition of solid waste, and dS_{TR} is the change in moisture storage in refuse.

If there is groundwater intrusion, the amount of leachate generated, L, is modified:

$$L = PER_R + W_{GW} \quad (7.4)$$

where W_{GW} is the input water from groundwater underflow.

To use the water balance equation it is necessary to evaluate precipitation, infiltration and evapotranspiration.

The precipitation, P, is the amount of water (in inches or millimetres) that accumulates on a sealed level surface. It is determined in the field and is available from the US Weather Bureau. Gauges installed near the landfill are generally used to take an average monthly reading over a long period of time.

Surface runoff, R_0, can be determined by actual measurements, graphical methods, or empirical methods such as the rational formula. A fraction of the precipitation becomes the runoff. The runoff is given by:

$$R_0 = CP \quad (7.5)$$

where C is the runoff coefficient and is a function of type of surface, vegetation, slope, etc. Its values are given in Tables 7.5 and 7.6.

Monthly potential evapotranspiration may be computed from the Thornthwaite equation. This equation uses the temperature efficiency index, TE, which is the sum of 12 monthly values of the heat index, I_t, which is given by:

$$I_t = (t/5)^{1.514} \quad (7.6)$$

where t is the mean monthly temperature in °C.

Table 7.5 Parry's runoff coefficient as affected by cover material and slope

Type of area	*Runoff coefficient, C*		
	Flat slope (<2%)	*Rolling slope* (2–10%)	*Hilly slope* (>10%)
Grassed area	0.25	0.30	0.30
Earth areas	0.60	0.65	0.70
Meadows and pasture lands	0.25	0.30	0.35
Cultivated land			
Impermeable (clay)	0.50	0.55	0.60
Permeable (loam)	0.25	0.30	0.35

Source: Lu *et al.*, 1985.

Table 7.6 Runoff coefficients after salvato for leachate estimation

Surface condition	*Slope (%)*	*Surface runoff coefficients*		
		Sandy loam	*Clay or silt loam*	*Clay*
Pasture or meadow (surface with cover crop)	0–2 (flat)	0.10	0.30	0.40
	5–10 (rolling)	0.16	0.36	0.55
	10–30 (hilly)	0.22	0.42	0.60
No vegetation (raw soil surface)	0–5 (flat)	0.30	0.50	0.60
	5–10 (rolling)	0.40	0.60	0.70
	10–30 (hilly)	0.52	0.72	0.82

Source: Lu *et al.*, 1985.

The potential evapotranspiration is given by:

$$\mathrm{PET(mm)} = 16(10t/\mathrm{TE})^{a}\ \text{(unadjusted)} \tag{7.7}$$

where

$$a = 0.000\,000\,675(\mathrm{TE})^3 - 0.000\,077\,1(\mathrm{TE})^2 + 0.017\,92\mathrm{TE} + 0.492\,39$$

The above value of the unadjusted evapotranspiration (for 12 h standard duration of sunlight) is corrected for unequal day lengths and months. Table 7.7 (Chow, 1964) presents the correction factors for various latitudes. The factors for 50° are used for poleward from 50°. The other parameters are the soil moisture storage at field capacity and the soil moisture retained after evapotranspiration has occurred. Figure 7.5 shows the water-holding characteristics of various soils with the USDA Classification and the approximate Unified Classification system. Table 7.8 is an abbreviated version of an expanded table (Thornthwaite and Mather, 1957) for estimating the soil moisture retention after the potential evapotranspiration has occurred.

If the moisture content in refuse is kept below the field capacity, leachate will not be generated from rainfall provided there is no other source. An effective cover could produce such conditions. Generation of leachate at a rate similar to the rate of percolation can be achieved only following soil cover saturation, which requires substantial amounts of water.

Table 7.7 Adjustment factors for potential evapotranspiration computed by the Thornthwaite equation

Latitude	*Jan.*	*Feb.*	*Mar.*	*Apr.*	*May*	*June*	*July*	*Aug.*	*Sep.*	*Oct.*	*Nov.*	*Dec.*
0	1.04	0.94	1.04	1.01	1.04	1.01	1.04	1.04	1.01	1.04	1.01	1.04
10	1.00	0.91	1.03	1.03	1.08	1.06	1.08	1.07	1.02	1.02	0.98	0.99
20	0.95	0.90	1.03	1.05	1.13	1.11	1.14	1.11	1.02	1.00	0.93	0.94
30	0.90	0.87	1.03	1.08	1.18	1.17	1.20	1.14	1.03	0.98	0.89	0.88
35	0.87	0.85	1.03	1.09	1.21	1.21	1.23	1.16	1.03	0.97	0.86	0.85
40	0.84	0.83	1.03	1.11	1.24	1.25	1.27	1.18	1.04	0.96	0.83	0.81
45	0.80	0.81	1.02	1.13	1.28	1.29	1.31	1.21	1.04	0.94	0.79	0.75
50	0.74	0.78	1.02	1.15	1.33	1.36	1.37	1.25	1.06	0.92	0.76	0.70

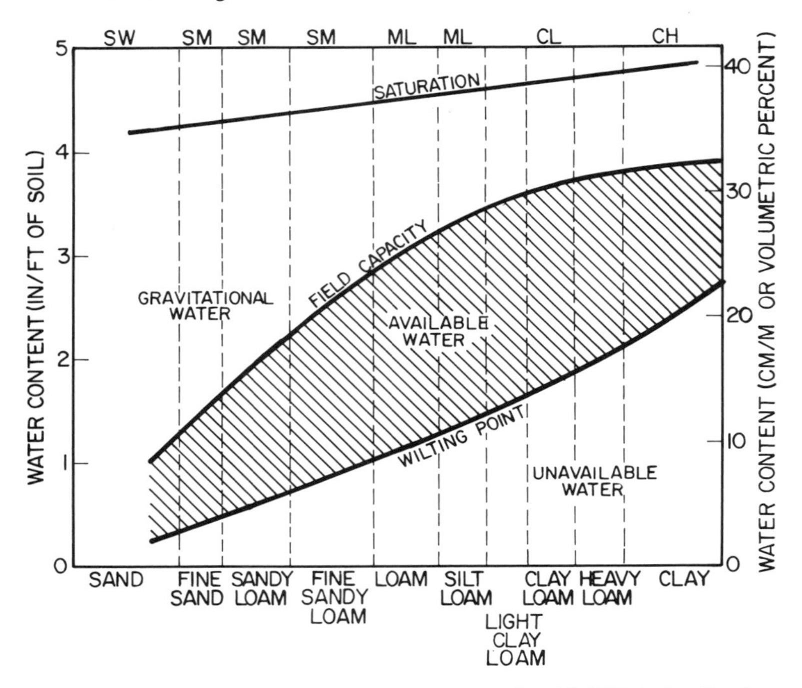

Figure 7.5 Water-holding characteristics of various soils with USDA classification

Use of the water balance method for a site in Lyndhurst, New Jersey, is shown in Table 7.9. The mean monthly temperatures are given in line 1; the heat index I_t, computed from Equation 7.6, is shown for each month in row 2. The sum of row 2 is the temperature efficiency index, TE. Using TE and Equation 7.7 the potential evapotranspiration (PET) is computed in line 3. The adjustment factors are obtained for the latitude of 40° from Table 7.7 and are shown in line 4. The adjusted PET values are computed in line 5. The monthly precipitation data (US Weather Bureau) are listed in line 6.

The runoff coefficient for a sandy clay cover with 2% slope is 0.25 (Table 7.5). Because in January the temperature is below freezing, to account for the impermeable nature of the essentially frozen ground the runoff coefficient is increased to 0.5. The runoff in line 8 is computed by multiplying the runoff coefficients in line 7 by the monthly rainfall in line 6. The annual runoff is 288.55 mm. The infiltration, I, is computed in line 9 (line 6 minus line 8).

When annual value of ($I-$Adj PET) is positive, as is the case here (i.e. 44.34), the area is known as a humid area. The negative values in May to September represent potential moisture deficiency where infiltration fails to supply water needed for vegetation. These deficiencies are summed up on a running basis in row 11. A sum of zero is assigned to the last month having a positive value of ($I-$Adj PET). This is because at the end of the wet season the soil moisture is at its field capacity.

Table 7.8 Soil moisture retention after potential evapotranspiration has occurred

	S_T (mm)[b]								
ΣNEG (I − PET)[a]	25	50	75	100	125	150	200	250	300
0	25	50	75	100	125	150	200	250	300
10	16	41	65	90	115	140	190	240	290
20	10	33	57	81	106	131	181	231	280
30	7	27	50	74	98	122	172	222	271
40	4	21	43	66	90	114	163	213	262
50	3	17	38	60	83	107	155	204	254
60	2	14	33	54	76	100	148	196	245
70	1	11	28	49	70	93	140	188	237
80	1	9	25	44	65	87	133	181	229
90	1	7	22	40	60	82	127	174	222
100		6	19	36	55	76	120	167	214
150		2	10	22	37	54	94	136	181
200		1	5	13	24	39	73	111	153
250			2	8	16	28	56	91	130
300			1	5	11	20	44	74	109
350			1	3	7	14	34	61	92
400				2	5	10	26	50	78
450				1	3	7	20	41	66
500				1	2	5	16	33	56
600					1	3	10	22	40
700						1	6	15	28
800						1	4	10	20
1000							1	4	10

[a] NEG(I − PET) is lack of infiltration water needed for vegetation. See Table 7.9 (row 12) for an example.
[b] S_T, soil moisture storage at field capacity.

Example 7.1

Considering a grassed sandy soil cover with a thickness of 2 ft (0.6 m), the available water by volume from Figure 7.5 is 7.8% (i.e. 10.8 − 3.0) or 78 mm/m. The root depth is limited by cover thickness (0.6 m). The potentially available soil moisture is 0.6 × 78 or about 50 mm. This value of soil moisture storage, S_T, is assigned to the initial wet months. For months with negative values of (I − Adj PET) the storage is derived from Table 7.8. After the dry period when (I − Adj PET) becomes positive the storage S_T in any given month is the sum of the storage in the previous month plus the value of (I − Adj PET) for that month, but this value cannot exceed the storage capacity (i.e. 50 mm in this case). Where positive (I − Adj PET) occurs between two negative values, S_T is calculated by direct addition of (I − Adj PET) to the previous S_T. Change in storage, dS_T, is the storage from this month less the storage from the previous month.

The actual loss due to evapotranspiration (AET) in line 14 during wet months is equal to Adj PET as the soil is at its storage capacity and there is more than adequate moisture available. For dry months, i.e. the months with negative sums of (I − Adj PET), the infiltration drops to below the Adj PET and the actual evapotranspiration becomes less than the potential. AET cannot exceed the infiltration plus the change in soil moisture storage ($I + |dS_T|$) calculated in line 14. For months with maximum moisture storage (50 mm), any excess moisture becomes percolation into the refuse

Table 7.9 An example of the use of the water balance method

Line	*parameter*	*Jan.*	*Feb.*	*Mar.*	*Apr.*	*May*	*June*	*July*	*Aug.*	*Sep.*	*Oct.*	*Nov.*	*Dec.*	*Annual total*
1	Temperature (°C)	−0.39	0.22	4.90	10.80	16.70	21.70	24.70	23.80	19.80	13.80	7.80	1.72	
2	I_t	0.00	0.01	0.97	3.21	6.21	9.23	11.23	10.61	8.03	4.65	1.96	0.20	56.31
3	PET	0.00	0.18	13.21	39.21	71.44	102.45	122.44	116.34	90.31	54.95	25.05	3.13	638.71
4	Adj Fac	0.84	0.83	1.03	1.11	1.24	1.25	1.27	1.18	1.04	0.96	0.83	0.81	
5	Adj PET	0.00	0.15	13.61	43.52	88.59	128.06	155.50	137.28	93.92	52.75	20.80	2.53	736.71
6	P (mm)	84.60	73.90	99.60	88.90	92.50	84.10	97.30	105.70	96.00	76.00	87.40	83.60	1069.60
7	C (R/O)	0.50	0.25	0.25	0.25	0.25	0.25	0.25	0.25	0.25	0.25	0.25	0.25	
8	R_o (mm)	42.30	18.48	24.90	72.23	23.13	21.03	24.33	26.43	24.00	19.00	21.85	20.90	288.55
9	I	42.30	55.43	74.70	66.68	69.38	63.08	72.98	79.28	72.00	57.00	65.55	62.70	781.05
10	I-Adj PET	42.30	55.27	61.09	23.15	−19.21	−64.99	−82.52	−58.01	−21.92	4.25	44.75	60.17	44.34
11	NEG (I − Adj PET)				0.00	−19.21	−84.20	−166.72	−224.73	−246.65				
12	S_T (Table 7.8)	50.00	50.00	50.00	50.00	35.00	9.00	2.00	1.00	1.00	5.25	50.00	50.00	
13	dS_t	0.00	0.00	0.00	0.00	−15.00	−26.00	−7.00	−1.00	0.00	4.25	44.75	0.00	0.00
14	AET (mm)	0.00	0.15	13.61	43.52	84.38	89.08	79.98	80.28	72.00	52.75	20.80	2.53	539.06
15	PERC (mm)	42.30	55.27	61.09	23.15	0.00	0.00	0.00	0.00	0.00	0.00	0.01	60.17	241.99

underlying the cover and is shown in line 15. For negative dS_T the percolation is equivalent to zero. For positive dS_T the percolation is equal to $(I-AET-dS_T)$. As one would expect the annual percolation (PERC) in Table 7.9 is equal to the annual infiltration (I) less the actual evapotranspiration (AET) $(241.99=781.05-539.06)$. The change in storage is zero.

During land filling the cover is usually thin and is unlikely to be more than 150 mm thick. The surface is not sloped, resulting in less runoff and higher infiltration. Maximum storage (line 12) is unlikely to exceed 10–20 mm. Ponding of rainwater is common during landfilling and the runoff coefficient in line 7 becomes less, leading to more infiltration (line 9). Since refuse does not support vegetation the Adj PET values in line 5 (or AET values in line 14) should be reduced. Correction factors to AET of 0.5 for dry months and 0.75 for wet months have been used. With all these adjustments the percolation into the refuse would be calculated at about 550mm, which is higher than the value calculated in Table 7.9.

Leachate, L, as illustrated in Figure 7.4 will first appear when the refuse attains its field capacity moisture content. Considering a placement moisture content of 15% and a field capacity of 30% (see Table 7.4) the refuse could store an additional 15% before leachate formation. Some leaching would occur prior to the soil mass attaining its field capacity. This is due to channelling. The approximate time of leachate appearance can be estimated based on:

$$t_1 = D_R(w_{FR} - w_R)/PER_R \qquad (7.8)$$

where t_1 is the time of first appearance of leachate (months), D_R the depth of refuse (L), PER_R the percolation (L) monthly average, w_{FR} the moisture content at field capacity (L^3/L^3), and, w_R the placement moisture content (L^3/L^3).

As illustrated in Equation 7.8, the leachate appearance time, t_1, is sensitive to the parameter $(w_{FR} - w_R)$. The composition of the refuse influences this parameter as illustrated in Table 7.10. If the refuse composition and its initial moisture content are known, the field capacity may be estimated.

Modern landfills are designed to be several feet above the groundwater. Surface run-on is prevented and the leachate produced by decomposition is usually relatively small (Lu *et al.*, 1985). If the change in refuse storage, dS_{TR}, is assumed to be zero, the leachate generated would be equal to PER_s. If this leachate is not collected and the landfill is founded on an impervious layer, then a leachate mound would develop.

Table 7.10 Moisture-holding capacity of refuse as determined in the laboratory

Component	*Water absorption capacity (% dry weight basis)*[a]
Newsprint	290
Cardboard	170
Miscellaneous paper	100–400
Leaves and grass	60–200
Tree pruning, shrubs	0–100
Garbage (food waste)	0–100
Textiles	100–300
Wood, plastic, sand, ashes	0

[a] Field absorptive capacity is the value shown less the moisture as delivered.
Source: Lu *et al.*, 1985.

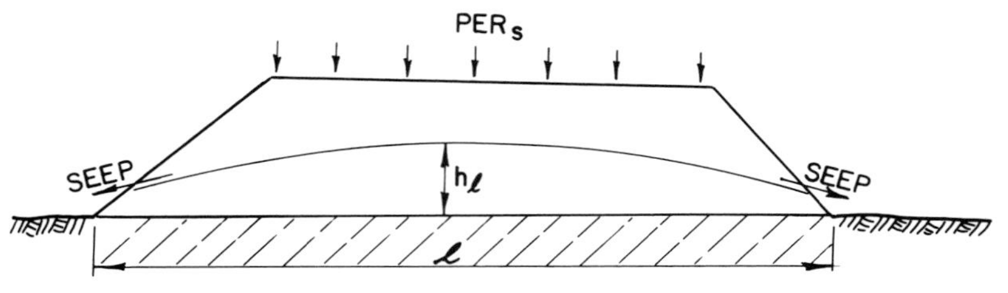

Figure 7.6 Leachate mound over a relatively impervious base

Considering an effective drain only around the toe of the landfill and an impervious base, the leachate build-up h_l (Figure 7.6) is estimated based on Equation 7.9 (USEPA, 1983):

$$h_l = \frac{l(\mathrm{PER_R}/k_\mathrm{R})^{0.5}}{2} \tag{7.9}$$

where l is the width of the landfill, k_R the hydraulic conductivity of refuse, and h the porosity of refuse.

It must be noted that the method presented here and those in use at the present time do not always yield reliable estimates of the amount of leachate generated or its first appearance. Computer aided simulation such as the HELP model could be used for estimates of leachate generated (Schroeder *et al.*, 1984). This model accounts for cover type, drainage conditions of the cover and other factors that affect the runoff and evapotranspiration (see Chapter 11).

7.3 Gas generation and migration

7.3.1 Composition

The decomposition process generates various gases that may affect construction on and near landfills. Table 7.11 (Subtitle D Study, 1986) shows the range of gas concentration found in municipal landfills. In addition, various volatile organic compounds (e.g. benzene, trichloroethylene) have been identified in gases generated in landfill simulators.

The presence of these gases even in very small concentrations could impact the health and safety of occupants of structures founded on or near a landfill. The gas production increases with increased moisture and temperature and decreases with increased density of the waste (Chian and DeWalle, 1979). The optimum moisture content for gas production is between 75 and 100% of dry weight. Carbon dioxide predominates in the initial phase of the decomposition process; the percentage of methane increases with time and appears to stabilize after the fourth year. The time for various phases is variable and depends on the availability of moisture and oxygen. Impeding moisture availability by covering the landfill with a reasonably impervious cover can substantially reduce the volume of gas produced.

Table 7.11 Range of landfill gas composition

Gas	*Percentage*
Methane	44–53
Carbon dioxide	34–47
Nitrogen	4–21
Hydrogen	<1
Oxygen	<2
Hydrogen sulphide	<1
Carbon monoxide	<1/10
Trace compounds[a]	<5/10

[a] Includes sulphur dioxide, benzene, toluene, methylene chloride, perchlorethylene and carbon sulphide in concentrations less than 50 parts per million.

7.3.2 Gas production

The reported volume of gas produced from waste varies over a wide range. A site-specific investigation is essential if such data are required. Constable *et al.* (1979) cited a gas production range of 2.2–250 l/kg dry weight ($0.0022-0.25\ m^3/kg$). Tchobanoglous *et al.* (1977) computed a theoretical production of $0.41\ m^3/kg$ for typical refuse with 70% achievable in 5 years under optimum conditions. Laboratory and field evidence indicates that gas production continues long after the active life of the landfill. Construction may alter the rate of gas production. York *et al.* (1977) presented a case where the gas production per unit area dropped from 4 to 1.5 $ft^3/1000\ ft^2$. Covering the area with pavement and discontinuing a sub-drain system were given as tentative explanations.

7.3.3 Construction implications

The presence of methane and volatile organic vapours are of concern because of their impact on health and safety. Gas venting is usually required for construction on landfills or in close proximity (up to several hundred feet) to landfills. In concentration of 5–15% by volume in air, and in the presence of sufficient oxygen, methane is explosive. The 5% concentration is usually referred to as the lower explosive limit (LEL) and defined as the lowest percentage of volume of a mixture of explosive gases which will propagate a flame in air at 25°C and atmospheric pressure. Concentrations as low as $\frac{1}{4}$ LEL have been suggested as an action level for gas venting and other remedial measures (USEPA, 1980).

The problem of gas generation and the need to protect structures has been recognized by engineers for many years (Sowers, 1968). MacFarland (1970) and Moore *et al.* (1979) cited case histories of explosions and loss of life. In the North Carolina National Guard Armory at Winston-Salem explosions killed three guardsmen and injured 25 (USEPA, 1975). The closest section of the armory concrete block construction was about 30 ft away from the landfill. The facility was built on 6 in reinforced concrete slab.

Methane migration resulted in the abandonment of an almost completed school in northern New Jersey. Landfill gas migration in Cherry Hill, Cinnaminson, Pennsauken and Glassboro, New Jersey has killed vegetation and entered buildings, causing chronic odour problems and fires in some cases (Duell and Flower, 1986).

Methane could enter a structure through cracks in floor slabs and cracks or leaks in underground utilities.

7.3.4 Investigations for gas

In order to determine the presence of landfill gases a specimen of gases is extracted from the landfill. In the usual method a hole is first made in the deposit. Commercial hole makers are available in lengths up to 7 ft. At least 1 h before sampling a probe is inserted into the hole, which is then closed with a stopper on the upper end. This allows the gases in the hole to displace the air that may have entered the hole. The gases from the hole are withdrawn with a suction pump into a portable detection apparatus. An apparatus calibrated for methane indicates the percentage LEL of methane.

Where gas detection is a part of an overall geotechnical investigation, a vacuum probe as shown in Figure 7.7 could be used to extract a gas sample. The probe consists of a stainless steel rod with a drive tip and drive head. It has machine-cut very fine slots connecting to an axial tube $\frac{3}{8}$ in in diameter. This tube in turn connects to $\frac{3}{8}$ in Teflon tubing which extends to the ground surface through a 5 ft length of rod. The probe is advanced by driving with a standard 140 lb hammer. To sample the gas, the lines are evacuated by pumping for a few minutes. The gas is injected directly into the detection device for measurement of the LEL for methane or concentrations of other gases or vapours. Where future laboratory analyses are called for, samples are collected in gas bags.

7.3.5 Gas migration

Gas flows from one point to another because of differential in gas pressure or by diffusion, which is the result of the differences in the gas concentrations. The impediments to gas flow by diffusion are collisions with other gas molecules (molecule diffusion) and collisions with the walls of the soil particles (Knudson diffusion). The Knudson diffusion may be significant in very fine porous materials when the pore dimensions are of the order of 10^{-5} mm, which is very small compared with pore dimensions of ordinary soils (American Petroleum Institute, 1986). Gas flow by diffusion and differential in total gas pressure are both affected by blockage by soil particles and tortuosity which results in an increase in the effective length of flow. The area available for flow is characterized by the effective porosity:

$$n' = n(1 - S) \tag{7.10}$$

when n' is the effective porosity (fraction), n the porosity of soil, and S the degree of saturation (fraction).

Moore *et al.* (1979) presented a detailed analytical model accounting for the flow of multi-component gas and developed design charts for assessing migration of methane in granular soils beneath and around a landfill. The model predicts that gas generated during a 5 year period can continue to migrate outwards for as much as 150 years.

In an effort to classify solid waste facilities EPA developed criteria in terms of the depth of refuse below grade, the type of material (pervious or impervious) underlying and surrounding the refuse, and the nature of the soil surface surrounding the landfill. The basic case for a landfill 25 ft below grade with pervious soil around it is shown in Figure 7.8. The 'clay' curve may be used for clayey gravel (GC), clayey sands (SC),

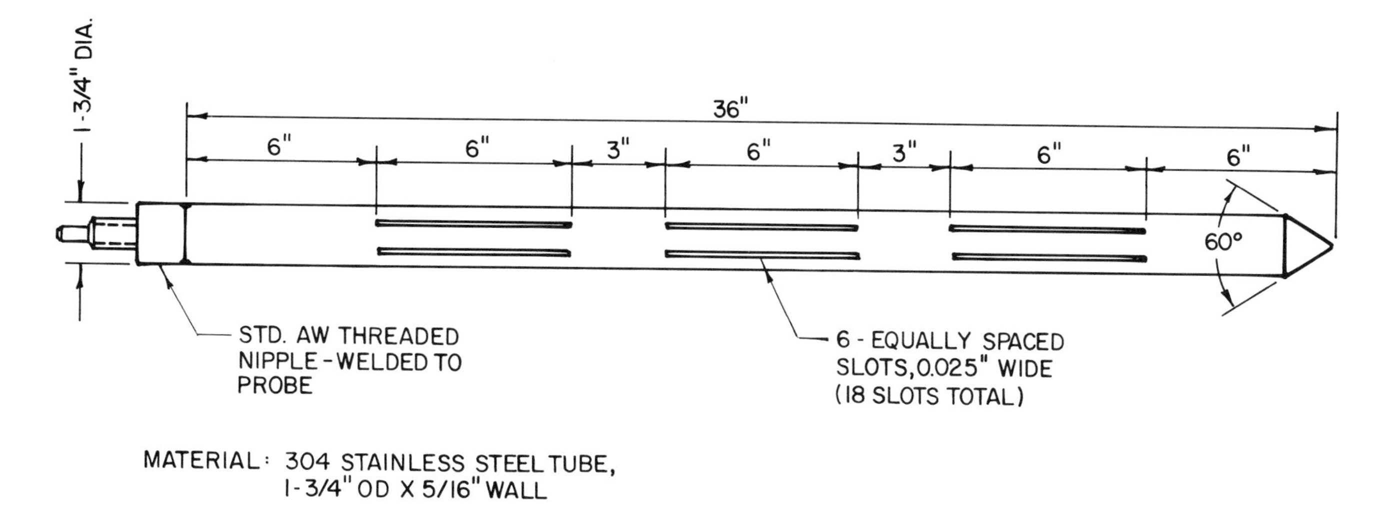

Figure 7.7 Vacuum probe

lean clay (CL), fat clay (CH) and organic clay (OH). Similarly the 'sand' curve is used for clean gravels (GW, GP) and sands (SW, SP). Interpolation between the clay and sand curves is recommended for border soils such as silty gravels and sands (GM–SM), silt, silty and sandy loam and organic silt (ML, MH, OL). For stratified soils, the most pervious unsaturated thickness is used. In conjunction with Figure 7.8, correction charts are provided in Figures 7.9 and 7.10 for different types of refuse below grade and different types of soil surrounding the landfill.

Considering molecular diffusion alone, the equation describing gas flow is similar to the heat flow equation and is expressed as:

$$J = -D\partial C_g/\partial x - D\partial C_g/\partial y - D\partial C_g/\partial z \qquad (7.11)$$

and

$$\partial C_g/\partial t = D\partial^2 C_g/\partial x^2 + D\partial^2 C_g/\partial y^2 + D\partial^2 C_g/\partial z^2 \qquad (7.12)$$

where J is the quantity of gas diffusing per second through a unit area perpendicular to the concentration gradient, C_g is the gas concentration by volume, and D the diffusion coefficient, which depends on the properties of the medium.

Experimentally determined values of D for air dry sandy silt range from 29.2 ft^2/s at 0.001 psi gas pressure to 22.2 ft^2/s at 0.05 psi (Laguros and Robertson, 1977). The rate of diffusion of gases in soil is expected to be less than that in air and may be approximated by (Lutton *et al.*, 1979):

$$D = 0.66n'D_0 \qquad (7.13)$$

where D_0 is the free space diffusion at 20°C (21 ft^2/day for methane and 15 ft^2/day for carbon dioxide).

Equation 7.13 implies that the diffusion constant is a function of the tortuosity introduced by a reduced cross-sectional area. Since air and water content are

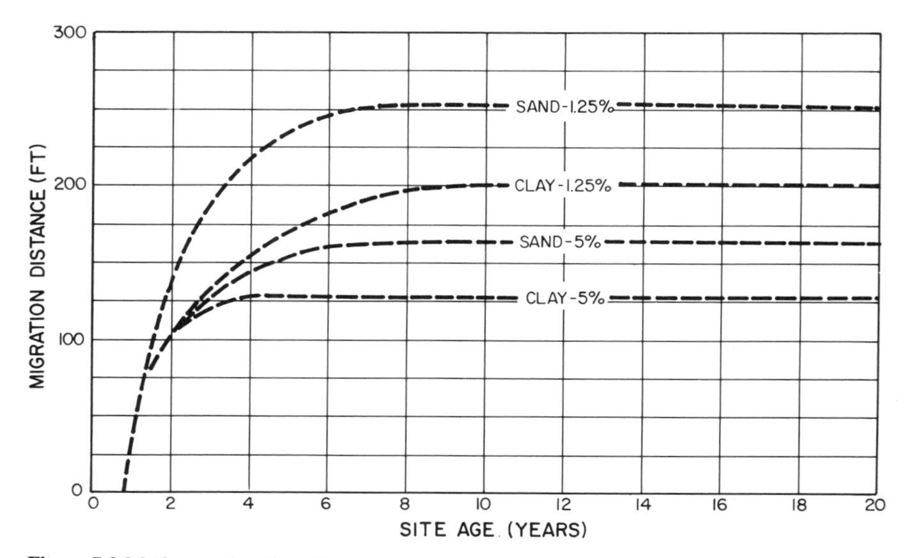

Figure 7.8 Methane migration distance

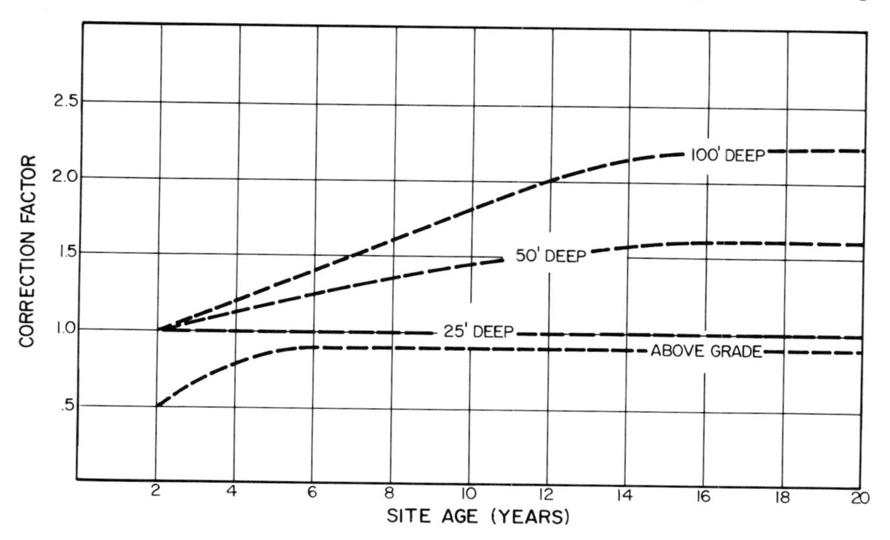

Figure 7.9 Correction factors for landfill depth below grade (Reproduced from EPA, 1980)

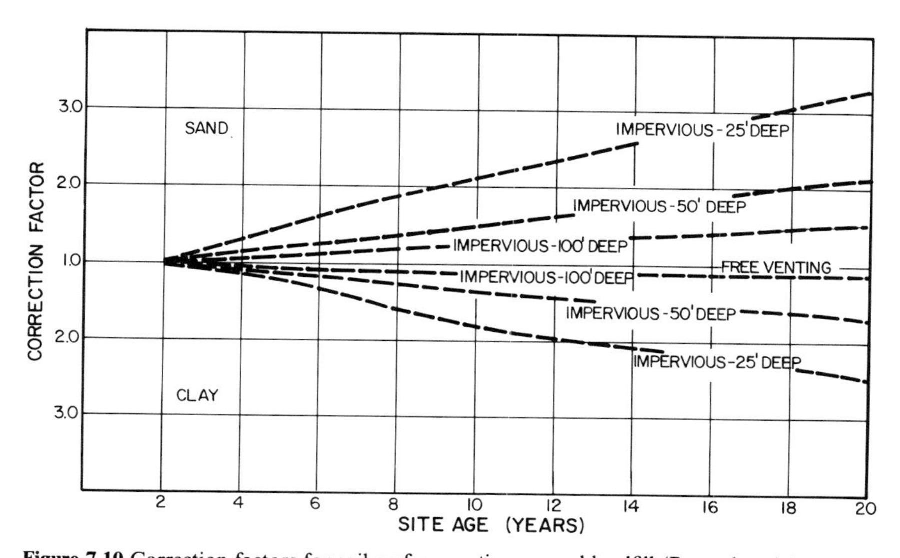

Figure 7.10 Correction factors for soil surface venting around landfill (Reproduced from EPA, 1980)

functions of porosity, the relationship is meaningful for only very dry soil where chemical adsorption is insignificant.

Equation 7.11 expresses the mass continuity condition considering a one-dimensional problem (z-direction). Equation 7.11 reduces to:

$$J = -D\, \partial C_g / \mathrm{d}z \qquad (7.14)$$

Considering for example a sandy cover with $n = 0.4$, $S = 0.5$, and a methane concentration of 40% in refuse and zero on the surface, D for methane is

$$21 \times [0.66 \times 0.4(1 - 0.5)]$$

or 2.77 ft²/day. For 2 ft cover, the volume of methane migration is $J = 2.77(0.4)/2$ or 0.55 ft³/ft² per day.

The flow under a total gas pressure gradient may be described by an equation similar to the familiar Darcy equation:

$$q = -k\Gamma i_p/\mu \tag{7.15}$$

$$i_p = (1/\Gamma) \times (dp/dz) \tag{7.16}$$

where q is the rate of flow per unit area (L^3/L^2 per T), k the intrinsic permeability (L^2), Γ the specific weight (F/L^3), i_p the pressure head gradient (head difference divided by distance), dp/dz the gradient in terms of total gas pressure (F/L^3), and μ the viscosity (FT/L^2).

Table 7.12 shows the unit weights and viscosities for gases found at sanitary landfills. The specific weight, Γ, is related to pressure and temperature by the equation:

$$\Gamma = mp/Rt \tag{7.17}$$

where p is the absolute pressure (F/L^2), t the temperature in Kelvin (K) (°F + 460 or °C + 273.15), R the gas constant, $L^3(F/L^2)(1/t)(1/mol) = \dfrac{8.3144\ Nm}{K\ mol}$ and m the molecular weight (F/mol).

Table 7.12 Viscosity and specific weights of gases found in sanitary landfills

Gas	*Temperature* (°C)	*Viscosity* (μP)	*Specific weight* (0°C, 1 atm) g/l	lb/ft³
Air	0	170.8	1.2928	0.0808
	18	182.7		
	54	190.4		
Carbon dioxide	0	139	1.9768	0.1235
	20	148		
	40	157		
Hydrogen sulphide	0		1.5392	0.0961
	17	124.1		
Nitrogen	0		1.2507	0.0782
	10.9	170.7		
	27.4	178.1		
Oxygen	0	189	1.4289	0.0892
	19.1	201.8		
Carbon monoxide	0	166	1.2501	0.0781
	21.7	175.3		
Ammonia	0	91.8	0.7708	0.0482
	20	98.2		
Hydrogen	0	83.5	0.0898	0.0052
	20.7	87.6		
	28.1	89.2		
Methane	0	102.6	0.7167	0.0448
	20	108.7		

$1\mu P = 10^{-6}$ poise, 1 Poise = 1 dyn s/cm² = 0.0672 lb/(s ft) = 1g/(s cm) = 0.00209 (lbf s)/ft².
Sources: Tchobanoglous *et al.*, 1977; Weast and Melvin, 1980.

The viscosity is dependent on temperature, and tables are available for viscosities at various temperatures (Weast and Melvin, 1980).

Equation 7.15 assumes laminar flow. The kinematic viscosity for gas ug/Γ (g being acceleration due to gravity) is about one order of magnitude higher than that of water. For the same flow velocity V and characteristic length L, the range of applicability of Equation 7.15 is larger for gases compared with water because of the smaller Reynolds number (VL/n).

From Equations 7.15 and 7.16 the rate of gas flow in terms of the total pressure is given by:

$$q = -(k/\mu) \times (dp/dz) \quad (7.18)$$

The intrinsic permeability, k, is independent of the type of fluid or gas flowing through porous medium (Chow, 1964). The parameter k/μ with gas as the permeant is about 50 times larger than for water as the permeant ($\mu = 10^{-2}$ P).

Various equations available for flow into wells may be used in combination with Equation 7.18. Consider for example a gas well with a radius r_1 placed in an area sealed at the surface to provide an effective containment, then the flow into the well, Q, is given by:

$$Q = \frac{2\pi hk\, dp}{u \times \ln(r_1/r_w)} \quad (7.19)$$

where h is the depth to the water table (lower containment), and r_w the radius of influence.

In the above analysis, it is assumed that the unit weight is independent of the pressure. This allows the use of the solutions for groundwater flow into wells. If the gas pressure head (i.e. cm of gas) is used in lieu of the hydraulic head, then the gas conductivity KI/μ is used in lieu of the hydraulic conductivity k. The intrinsic permeability could be determined by back-calculations if the hydraulic conductivity is known since the unit weight of water and its viscosity are known. If the intrinsic permeability is expressed in darcies then the hydraulic conductivity in cm/sec is 0.001 times the intrinsic permeability (1 darcy is approximately 10^{-8} cm^2). Another procedure is to express the intrinsic permeability as:

$$k = 0.125\, r^2 \quad (7.20)$$

The average pore radius r for sands and gravels may be estimated based on (Sherard *et al.*, 1984)

$$r = 0.1\, D_{15} \quad (7.21)$$

where D_{15} is the particle size which 15% of the soil is finer.

Other parameters such as the storage coefficient in groundwater flow is equivalent to mn'/Rt (t, temperature in Kelvin; n', effective porosity and m is the molecular weight (F/mol)). Time dependent gas pressure drawdowns resulting from pumping could be estimated using analyses developed for groundwater flow into wells (Theis 1935, Jacob 1946).

For example, for $k = 10^{-11}$ ft^2, $\mu = 3.5 \times 10^{-7}$ lb s/ft^2, $r_w = 150$ ft, $r_1 = 1.5$ ft, $h = 20$ ft, if the imposed dp over the distance r_w is 40 lb/ft^2, then the discharge Q is

$$\frac{2\pi \times 20 \times 1.0 \times 10^{-11} \times 40}{3.5 \times 10^{-7}\ \ln(150/1.5)} = 0.031 \text{ ft}^3/\text{s}$$

Some of the factors such as moulding water content that affect the hydraulic conductivity also influence the conductivity of gases. For a given moisture content densification resulted in lower conductivity. Densification of clays and sandy silts of low plasticity by kneading compaction produced lower conductivity than by dynamic compaction, with the differences becoming larger as the optimum moisture content approached. The changes in conductivity were lower for moisture content dry of optimum. The conductivity was reduced by several orders of magnitude at moisture content wet of optimum (Langfelder *et al.*, 1968).

Summary

From chemical and biological decomposition of waste, leachate and gases are generated. Leachate refers to the highly contaminated water containing organic and inorganic chemicals that emanates from a waste disposal site. The volume of leachate is estimated by a water balance method which assumes one-dimensional flow, conservation of mass, and known retention and transmission characteristics of the soil cover and refuse. None of the methods available always yield reliable estimates of the leachate generated or its first appearance.

Primarily methane and carbon dioxide are produced at the municipal waste sites. Carbon dioxide predominates in the initial phase of the decomposition process and methane increases in the later phase. The presence of these gases has impact on health and safety and requires proper venting. If not properly vented, methane can migrate laterally in the soil and damage vegetation and enter structures located near the disposal site. Some of the factors that affect the hydraulic conductivity, such as moulding water content, method of soil compaction, etc., also influence the conductivity of the gases.

Notation

AET	actual evapotranspiration
C	runoff coefficient
C_g	gas concentration by volume
D	diffusion coefficient
D_0	free space diffusion at 20°C
dp/dz	gradient in terms of total gas pressure
D_R	depth of refuse
dS_{TR}	change in moisture storage in refuse
dS_{TS}	change in moisture storage in soil
h	depth to water table (lower containment)
h_l	height of leachate build-up
I	infiltration
i_p	pressure head gradient
IR	input water from irrigation
I_t	heat index
J	quantity of gas diffusing per second through a unit area perpendicular to the concentration gradient
k	intrinsic permeability
k_R	hydraulic conductivity of refuse
L	amount of leachate generated
l	width of the landfill
m	molecular weight

mR	1545 (foot-pound-second system)
n	porosity of refuse
n'	effective porosity (fraction)
P	precipitation
p	absolute pressure
PER_R	percolation in the refuse
PER_S	percolation in the soil
Q	flow into the well
q	rate of flow per unit
R	gas constant
R_0	surface runoff
r	average pore radius
r_w	radius of influence
s	degree of saturation (fraction)
SR	input water from surrounding surface runoff
t	mean monthly temperature
T	temperature
t_1	time of first appearance of leachate (months)
u	viscosity
W_D	water from decomposition of solid waste
w_{FR}	moisture contents at field capacity
W_{GW}	input water from groundwater underflow
w_R	refuse placement moisture content
Γ	specific weight

References

American Petroleum Institute (1986) *Examination of Venting for Removal of Gasoline Vapors from Contaminated Soils*, API Publication No. 4429, p. 24

Chen, K. Y. and Bowerman, F. R. (1975) Mechanism of leachate formation in sanitary landfills. In *Recycling and Disposal of Solid Waste*, Ed. Yen, T. F., Ann Arbor Science Publications, Ann Arbor, Michigan

Chian, E. S. K. and DeWalle, F. B. (1979) Effect of moisture regime and temperature on municipal solid waste stabilization, *Municipal Solid Waste: Land Disposal*, Proc. of the 5th Annual Research Symp., Orlando, Florida, March 1979, pp. 32–40

Chow, T. V. (1964) *Handbook of Applied Hydrology*, McGraw-Hill, New York

Constable, T. W., Farquhar, G. J. and Clement, B. N. C. (1979) Gas migration and modeling, *Municipal Solid Waste: Land Disposal*, Proc. of the 5th Annual Research Symp., Orlando, Florida, March 1979, p. 397

Duell, R. W. and Flower, F. B. (1986) Biological reclamation of landfill areas, Seminar, *Sanitary Landfill Closure*, New Jersey Department of Environmental Protection, May 1980

Fenn, D. G., Hanley, K. J. and DeGeare, T. U. (1975) *Use of the Water Balance for Predicting Leachate Concentration from Solid Waste Disposal Sites*, EPA-530/SW-168, US Environmental Protection Agency, Cincinnati, Ohio.

Gas Generation Institute (1981) *Landfill Methane Recovery*, Part 1

Groenewold, G. H., Hassett, D. J., Koob, R. D. and Manz, O. (1985) Disposal of western fly ash in the northern Great Plains, *Fly Ash and Coal Conversion by By-products: Characterization, Utilization and Disposal*, I, Vol. 43, Materials Research Soc. pp. 213–226

Laguros G. J. and Robertson, J. M. (1977) Generation, movement and control of gas in sanitary landfills, *Proc. of the Specialty Session: Geotechnical Engineering and Environmental Control, 9th Int. Conf SMFE*, Tokyo, pp. 296–304

Langfelder, L., Chen, C. F. and Justice, J. A. (1968) Air permeability of compacted cohesive soils, *ASCE, Journal of the Soil Mechanics and Foundation Division*, **94,** No. SM4, 981–1002

Lu, C. S., Eichenberger, B. and Stearns R. J. (1985) *Leachate for Municipal Landfills*, Noyes Publication

Lutten, R. J., Regan, G. L. and Jones, L. W. (1979) *Design and Construction of Covers for Solid Waste Landfills*, EPA 600/12-79-165. United States Environmental Protection Agency

MacFarland, I. C. (1970) Gas explosion hazards in sanitary landfills, *Public Works*, May, 76

Martecon Inc. (1980) *Lining of Waste Impoundment and Disposal Facilities*, EPA 530/SW-870

Massmann, J. W. (1989) Applying groundwater flow models in vapor extraction system design, *Journal of Environmental Engineering*, ASCE, **115**, no. 1

Moore, C. A., Iqbal, R. I. S. and Alzaydi, A. A. (1979) Methane migration around sanitary landfills, *ASCE, Journal of the Geotechnical Engineering Division*, **105,** No. GT2, 131–144

Noble, G. (1976) *Sanitary Landfill Design Handbook*, Technomic, Westport, Connecticut

O'Leary, P. and Tansel, B. (1986) Land disposal of solid wastes, *Waste Age*, March, 68–78

Schroeder, P. R., Morgan, J. M., Walski, T. M. and Gibson, A. C. (1984) *The Hydrologic Evaluation of Landfill Performance (HELP) Model*; Vol 1, *User's Guide for Revision 1*, EPA/530-SW-84-009, Municipal Environmental Laboratory, USEPA, Cincinnati, Ohio

Sherard, J. L., Dunnigan, L. P. and Talbot, J. R. (1984) Basic properties of sand and gravel filters, *Journal of Geotechnical Engineering*, ASCE, **110**, no. 6

Sowers, G. F. (1968) Foundation problems in sanitary landfills, *ASCE, Journal of the Sanitary Engineering Division*, **94,** No. AS1, 103–116

Subtitle D Study (1986) Phase 1 Report, EPA/530-SW-86-054, USEPA

Tchobanoglous, G., Theis, H. and Eliassen, R. (1977) *Solid Wastes*, McGraw-Hill, New York, pp. 328–331

Thornthwaite, C. W. and Mather, J. R. (1957) Instructions and tables for computing potential evapotranspiration and the water balance, Publications in Climatology, Lab of Climatology, Drexel Institute of Technology, Centerton, New Jersey

USEPA (1975) *An Evaluation of Landfill Gas Migration and a Prototype Gas Migration Barrier*, Winston-Salem, Department of Public Works, EPA/530/SW-79d

USEPA (1980) *Classifying Solid Waste Disposal Facilities*, SW-828

USEPA (1983) *Landfill and Surface Impoundment Performance Evaluation*, SW-869

Weast, R. C. and Melvin, J. (1980) *Handbook of Chemistry and Physics*, CRC Press, Boca Raton, Florida

York, D., Lesser, N., Bellatty, T., Irsai, E. and Patel, A. (1977) Terminal development on a refuse fill site, *Proc Conf. on Geotechnical Practice for Disposal of Solid Waste Material*, ASCE, Ann Arbor, Michigan pp. 810–830

Zerlaut, G. A. and Stake, A. M. (1975) Chemical aspects of plastic waste management. In *Recycling and Disposal of Solid Wastes*, Ed. Yen, T. F., Ann Arbor Science, Ann Arbor, Michigan

Chapter 8

Soil structure

Soil structure represents the relative arrangement of soil particles in a soil mass. The engineering properties of a soil mass are affected by its composition, stress history and structure. Some of the factors contributing to the soil structure are particle shape, particle and pore size distribution, mass forces, interparticle forces, cementation, presence of organics, diagenetic bonds, electrolyte concentration, dielectric constant, etc.

This chapter describes the physicochemical aspects of soil structure, forms of organics found at waste sites, their interaction with soils, and how certain index properties of soils may be used to study the effect of organics on the engineering properties of soils.

8.1 Bonds

Two types of bonds that are important in the study of clay minerals are *primary bonds*, which hold the atoms together, and *secondary bonds*, which hold together water molecules of adjacent sheets of a crystalline lattice and affect its mineral characteristics.

8.1.1 Primary bonds

The primary bonds consist of ionic bonds and covalent bonds. *Ionic bonds* are formed when elements of completely different electronegativity release or gain electrons in the outer shell of their atoms. Sodium chloride is an example of such bonds. *Covalent bonds* are formed when two atoms share the valence electrons in their outer shells. The electrons occupy a molecular orbit formed by overlapping atomic orbits. This bond may be formed between elements sharing pairs of electrons of identical electronegativity, or of somewhat different electronegativity. A carbon–hydrogen bond, which is most common in organic compounds, is a covalent bond as there is not much difference between the electronegativity of the two elements. If the electronegative difference is between 0.5 and 1.7, and the centres of bonded ions in the molecule do not coincide, these bonds are known as *polar covalent bonds*. Such molecules have a positive and a negative charge and are known as polar molecules or *dipoles*. Water is an example of dipoles. However, if the arrangement within the molecule is symmetric, then polar covalent bonds yield non-polar molecules. Carbon tetracholoride is one such example. Ionic and covalent bonds are considered strong and are not affected by the stresses commonly encountered in geotechnical engineering practice.

8.1.2 Secondary bonds

The force of attraction between polar molecules or dipoles is called *dipole force*. If the positively charged end of the participating dipole is hydrogen then the attracting force is termed a *hydrogen bond*. A hydrogen bond is stronger than an ordinary dipole bond. Properties of clay minerals and their interaction with water are significantly influenced by the hydrogen bond.

Van der Waals forces are the attractive forces caused by interaction between electrical fields of neighbouring molecules. They are affected by the interparticle spacing and the dielectric constant of the medium separating the clay particles. The attraction between the dipoles is also a form of van der Waals force. These bonds are relatively weak and unlike polar bonds they are non-directional.

8.2 Fine-grained soils

Fine-grained soils consist of silts and clays. Chemical analysis indicates that clay minerals are crystalline in structure and consist of four elements: oxygen, silicon, aluminium and iron. They have an equivalent diameter of approximately 2 μm, a large surface area, and are plastic in nature. They are generally plate-like in shape with a few clay minerals having tubular form. The plates are usually very thin compared with the other two dimensions. Because of their large surface area a high percentage of constituting molecules are distributed on the surface. The *specific surface*, defined as the surface area per unit weight, is much larger than that of silt and sand, which is less than 1 m^2/g (Table 8.1). The interparticle forces which determine the soil fabric or geometric arrangement of the particles are much larger than the Newtonian forces.

If clay particles fracture, other electrical forces may develop at the edges, exposing the internal ions, which are usually positive in their charge. The edges may attract dipoles, anions or negatively charged faces of other clay particles. The van der Waals

Table 8.1 General characteristics of clay minerals with inorganic salts

Mineral	*CEC* (mEq/100 g)	*Relative density*	*Specific surface* (m^2/g)	*Liquid limit*	*Plastic limit*	*Expansion index*
Kaolinite	3–15	2.60–2.68	10–20	30–60	25–35	
Sodium				53	21	0.20
Calcium				38	11	0.06
Illite	10–40	2.60–3.00	65–100	60–120	35–60	
Sodium				61	34	0.15
Calcium				90	40	0.21
Montmorillonite (smectite group)	80–150	2.35–2.70	700–840	100–900	50–100	
Sodium				700	97	2.5
Calcium				177	63	0.80

Sources: Grim, 1968; Lambe and Whitman, 1969.

forces and the electrical forces constitute a net particle force. The interparticle forces are greatly affected by the externally applied stresses.

8.2.1 Clay minerals

The crystalline structure of a clay is composed of two basic building blocks, a silicon tetrahedron (SiO_2), where a silicon atom is surrounded by four oxygen atoms, and an octahedron consisting of aluminium [gibsite, $Al_2(OH)_6$] or magnesium [brucite, $Mg_3(OH)_6$] surrounded by oxygen atoms or hydroxyl groups (Figure 8.1). Different clay minerals are formed as the sheets of the basic units are stacked on top of each other with different ions bonding them together. Different clay minerals are formed if during their formation the normal locations of aluminium, magnesium or silicon ions are occupied partially or wholly by other ions. This substitution of one kind of ion with the others, without change in crystal structure, is known as *isomorphous substitution*. The isomorphous substitution causes a charge deficiency which is balanced by the absorption of external cations. For example the replacement of trivalent Al^{3+} with bivalent Mg^{2+} results in an access of negatively charged oxygen anions. The ability of a mineral to adsorb external cations is known as its *base exchange* or *cation exchange capacity* (CEC). Other factors which contribute toward the cation exchange capacity are broken bonds at particle edges, electrolyte concentration, temperature, organic contents, and replacement of hydrogen from hydroxyl. The CEC of a soil is expressed in milliequivalents per 100 g of dry soil (mEq/100 g). The externally adsorbed cations can be replaced by other cations. This is known as the *ion exchange*. The replacing power for some of the commonly occurring cations in soil minerals is, from low to high:

$$Na^+ < K^+ < Mg^{2+} < Ca^{2+} < Al^{3+} < Fe^{3+}$$

The properties of clay minerals can be altered by exchanging organic or inorganic ions, or by water adsorbed on the particle surface.

Although there are many types of clay minerals with various substituted and exchangable ions, they are generally characterized into three major groups. Some of their index properties are given in Table 8.1. Note that the values reported in the literature vary widely even for the same exchangeable cation.

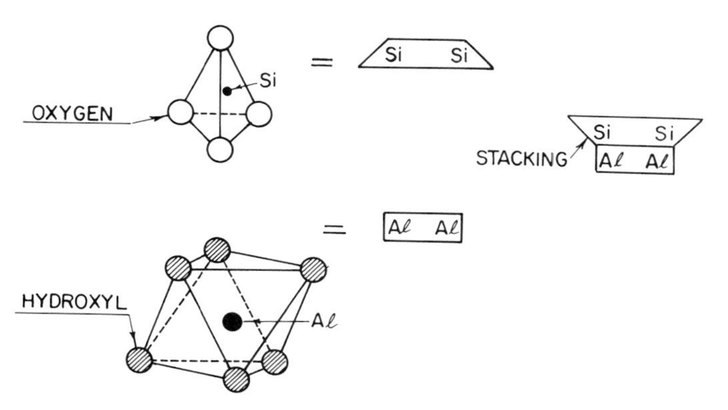

Figure 8.1 Silica tetrahedron and alumina octahedron (Reproduced from Yong and Warkinten, 1966, by permission)

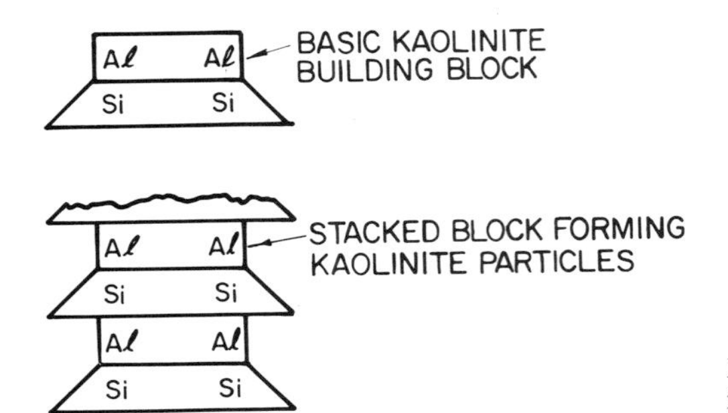

Figure 8.2 Structure of kaolinite (Reproduced from Yong and Warkinten, 1966, by permission)

8.2.2 Kaolinite

The kaolinite crystal layers consist of silica tetrahedra and alumina octahedral sheets held together by the strong hydrogen bond between oxygen and hydroxyl groups (Figure 8.2). Since it contains one sheet each of silica and alumina it is called a one-to-one (1 : 1) mineral. These sheets can extend greatly in two directions and typically such crystals are 70–100 layers thick. The kaolinite platelets are about 0.05 μm thick, have a diameter to thickness ratio of about 20 and are usually hexagonal in shape. Kaolinite has a low swelling potential and liquid limit. Halloysite is also a 1 : 1 mineral which is related to kaolinite except that it is tubular in form.

8.2.3 Illite

The illite crystals consist of alumina sheets between two silica sheets (two-to-one mineral, 2 : 1), which are stacked with potassium ions occupying the space between the adjacent sheets of the primary mineral and bind them together (Figure 8.3). There is considerable isomorphous substitution of silica in illite, which is partly balanced by interlayer potassium. The illite particles are about 0.01 μm thick with a diameter-to-thickness ratio of about 50. Illite has a moderate swelling potential and a liquid limit higher than that of kaolinite.

8.2.4 Chlorite

Chlorite is a 2 : 1 : 1 mineral with an alumina sheet between two silica sheets and also either a brucite or a gibsite sheet. It is commonly found in clay soils. There may be considerable isomorphous substitution.

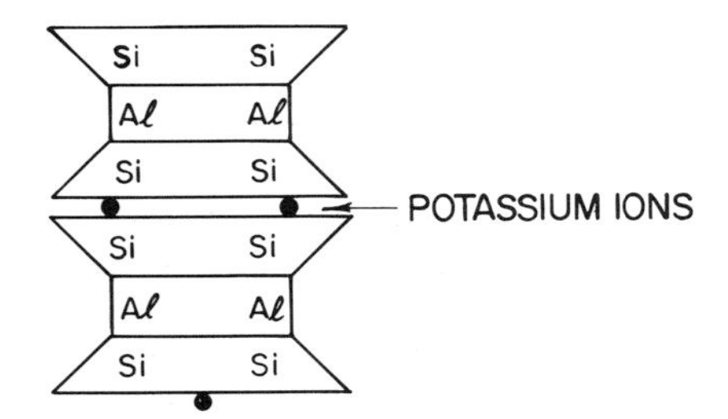

Figure 8.3 Structure of illite (Reproduced from Yong and Warkinten, 1966, by permission)

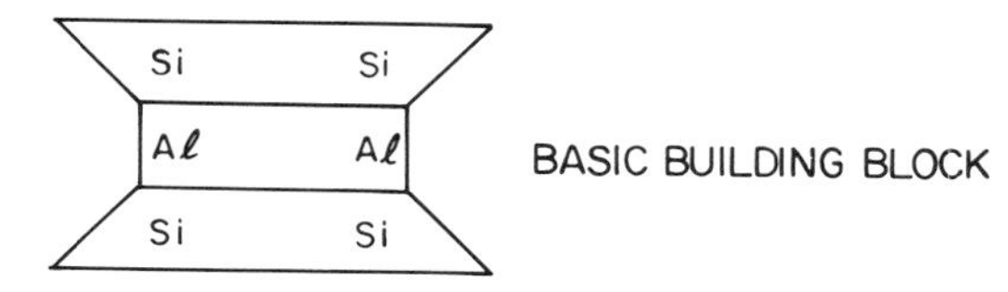

WATER MOLECULES

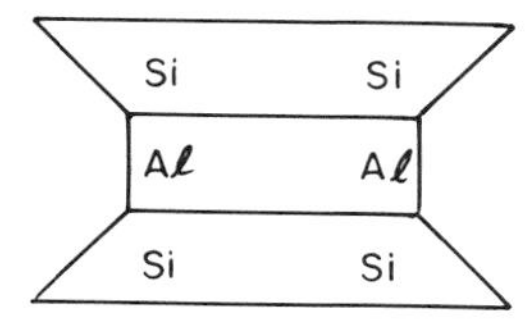

Figure 8.4 Structure of montmorillonite (Reproduced from Yong and Warkinten, 1966, by permission)

8.2.5 Smectite

The basic building sheets for smectite are the same as for illite (2:1), except that there is no potassium ion present. The bond holding the sheets is between oxygen ions facing each other. It is a very weak bond and is easily broken by water entering between the layers (Figure 8.4). Besides water, other polar or cationic organic fluids may also enter clay layers. There is extensive substitution of silica and alumina, resulting in considerable charge deficiency.

In montmorillonite, which is a member of the smectite group, one out of every six trivalent aluminium ions is replaced with a bivalent magnesium ion. The resulting charge deficiency is balanced by an externally absorbed sodium ion. Montmorillonite has a particle thickness of about 0.0001 μm with a diameter-to-thickness ratio of about 400. The water molecules enter easily between the layers, which expand considerably, yielding much smaller particles with a very large specific surface. Swelling potential, activity and liquid limit are the highest in this group of clays. Sodium montmorillonite is used as drilling mud and in slurry wall construction.

8.2.6 Clay–water interaction

When a clay particle is placed in water, a high concentration of cations occurs near its surface. This is due to:

1. The release of adsorbed cations from the clay surface.
2. The ionized water molecules interacting with the surface oxygen of clay particles.
3. The release of hydrogen from ionization of hydroxyl groups.
4. Dissolving of precipitated salts, if present.

When the pH value is high ($pH = 1/\log_{10} H^+$) the hydroxyl group has a greater tendency to form hydrogen ions. The released cations try to move away from the clay particle in order to equalize the charge distribution throughout the suspension. Attraction from the negatively charged surface of the particle restricts this movement, resulting in a much greater concentration of cations near the clay surface with only a few anions present. The negative charges on the clay surface and the balancing cations surrounding it are together called the *diffuse double layer* (Figure 8.5). The ions in the double layer can move in and out of it. The thickness of the double layer

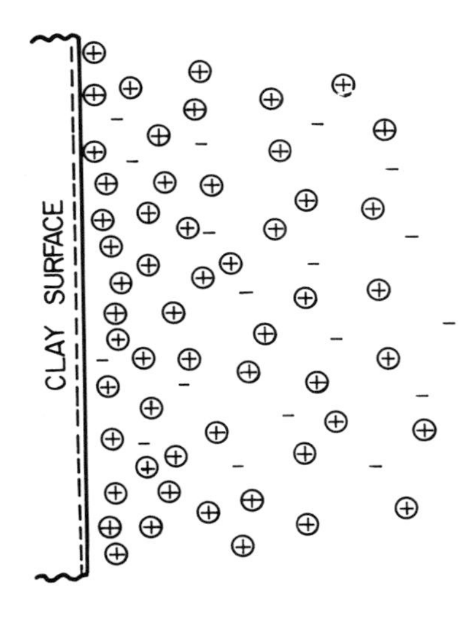

Figure 8.5 Clay–water system (Reproduced from Mitchell, 1976, by permission of John Wiley & Sons Inc.)

depends on the cation valence, electrolyte concentration, temperature, and dielectric constant (the measure of ease with which molecules can be polarized and orientated in an electric field) of the medium.

The theoretical expression for distribution of ions adjacent to negatively charged surfaces is due to Gouy-Chapman. The thickness of the double layer is given by Equation 8.1:

$$t_{\mathrm{dl}} = [\varepsilon \boldsymbol{k} T \div 8\pi n e^2 v^2]^{1/2} \qquad (8.1)$$

where ε is the dielectric constant of the medium, $\boldsymbol{k}$ is Boltzman's constant (1.38×10^{-16} erg/K), T is the absolute temperature (K), n the electrolyte concentration (ions/cm^3), e the unit electronic charge (16.0×10^{-20} C or 4.8×10^{-10} e.s.u.), and v is the cation valence.

The thickness of the double layer varies directly as the square root of temperature and the dielectric constant, and inversely as the valence and the square root of electrolyte concentration. As the temperature increases the dielectric constant decreases. For water the product of dielectric constant and temperature decreases from 2.4×10^{-4} to 2.20×10^{-4} as the temperature increases from 0 to 60°C. Thus with rising temperature the thickness of the double layer will actually decrease rather than increase. The thickness of the double layer will also increase as the cation valence and dielectric constant decrease (Mitchell, 1976).

8.2.7 Clay structure

The geometrical arrangement of particles in a clay mass is the net result of the repulsive forces between the double layers and the attractive van der Waals forces, the electrical force and the bonding due to organic and inorganic materials present on particle surfaces.

In pure water or in an electrolyte of low concentration of monovalent ions, the double layer is thicker and the net force between the particles is repulsive. If the pH is high the broken edges of the clay particles ionize negatively giving further impetus to the repulsive forces. The repulsive forces dominate and the clay platelets align themselves in parallel orientation forming what is known as a *dispersed* structure. Soils with dispersed structure exhibit greater swelling and lower permeability.

When the thickness of the double layer is reduced the clay particles can come closer to each other before the double layers interact. With the interparticle distance reduced, the van der Waals attractive forces become the determining factor for soil structure. If, in addition, the pH is low, the edges ionize positively, thus encouraging edge-to-face contacts. The clay particles flock to each other in a random fashion yielding what is known as *flocculated* structure. These are the two basic structures. The actual structure may be very complex. For further details see Mitchell (1976).

The engineering properties and behaviour of a clay soil are determined by its mineral constituents, type of adsorbed ion, environmental conditions during deposition or placement, stress history, nature of organic or inorganic chemicals percolating or seepage through it, etc.

8.3 Organic compounds in industrial waste

A very few of the organic compounds available to date have been investigated for their effect on properties of soils. The only practical way to pursue this type of study is to look at each class of organic chemicals rather than each chemical itself. For details on various classes of organic compounds see Hart and Schuetz (1972) and Madsen and Mitchell (1987). Only a brief description of various classes of these compounds is given here.

The simplest form of organic compounds are the *saturated hydrocarbons* which are called *alkanes*. The alkanes contain only hydrogen and carbon in a covalent bond. Methane, CH_4, is their simplest form (—C—). Increasing numbers of carbon atoms form chains ($\equiv$C—C$\equiv$) (e.g. propane, $CH_3CH_2CH_3$) or rings yielding other compounds (e.g. cyclopropane, $(CH_2)_3$). Fragments of saturated hydrocarbons (*alkyl* groups) are obtained by removing one of the hydrogen atoms from the parent compound. This class of compounds is relatively inert, practically insoluble in water, and shows no reaction with dilute acids or bases. When they react with other compounds hydrogen atoms are replaced (e.g. dichloromethane, CH_2Cl_2; ethyl bromide, CH_3CH_2Br).

Unsaturated hydrocarbons contain carbon–carbon double bonds (=C=C=; e.g. ethene CH_2=CH_2) in the *alkene* group, and carbon–carbon triple bonds (—C$\equiv$C—; e.g. acetylene, CH$\equiv$CH) in the *alkyne* group. They are highly reactive and form many of the compounds that pose an environmental hazard. They react by addition to the double bond (both alkenes and alkynes) or by addition of hydrogen (alkynes only). The unsaturated hydrocarbons built around the *benzene ring*

 C_6H_6

with carbon occupying each of the corners of a hexagon are called aromatic compounds or *arenes*. The carbon–carbon bond is neither single nor double but a hybrid. These compounds exhibit low reactivity. In their reaction with other compounds one or more of the hydrogens are replaced.

Alcohols are derived when a hydroxyl (—OH) group replaces a hydrogen atom in saturated or unsaturated hydrocarbons (e.g. ethanol, CH_3CH_2—OH), and *phenols* are obtained if a hydroxyl (—OH) group replaces a hydrogen atom in an aromatic ring

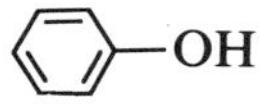

⌬—OH

(e.g. *p*-chlorophenol, ClC_6H_4—OH).

Alcohols and phenols may be considered an organic equivalent of water with one of the hydrogen atoms replaced by an organic group. Alcohols with short carbon chains (methyl, CH_3—OH; ethyl, CH_3CH_2—OH) have high water solubility. The solubility decreases with increasing carbon chain length. Alcohols with more than one hydroxyl group per molecule are called *polyhydric alcohols*. An example is ethylene glycol (CH_2OHCH_2OH) which is a permanent antifreeze.

Ethers (—O—) are isomers of alcohol. They have low water solubility, but they are good solvents for many organic compounds. Their reactivity is low. An example of an ether is diethyl ether, CH_3CH_2—O—CH_2CH_3.

Aldehydes and *ketones* contain *carbonyl* groups:

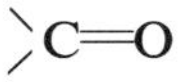

$$>C=O$$

and are highly reactive. In aldehydes one of the two groups attached to the carbonyl is hydrogen:

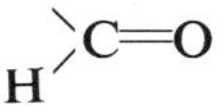

$$\overset{\diagdown}{\underset{H\diagup}{}}C=O$$

whereas, in ketones, both groups attached to carbonyl groups are organics:

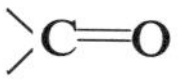

$$>C=O$$

Formaldehyde (HCOH) and acetone (CH_3COCH_3) belong to these groups.

Carboxylic acids (*organic acids*) contain a carboxyl group:

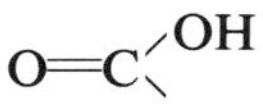

$$O=C\overset{\diagup OH}{\underset{\diagdown}{}}$$

and are acidic because of their ability to donate a proton to more basic substances. These acids are weaker than inorganic acids. Those with fewer carbon atoms or lower molecular weight are soluble in water, and the solubility decreases with increasing molecular weight. Acetic acid (CH_3COOH) is an example of a carboxylic acid.

Amines (—NH_2) or *organic bases* have structure akin to the inorganic base ammonia. They are called primary, secondary or tertiary amines as one, two or three of the hydrogen atoms of ammonia are replaced by organic groups. The replacing organic group may be *aliphatic* (compounds related to saturated or unsaturated open-chain or cyclic compounds, but containing no benzene group), aromatic, or a combination of the two. *Aromatic amines*:

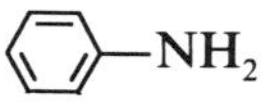

⌬—NH_2

often have *aniline* in their names.

Some of the properties of organic compounds are given in Table 8.2.

Table 8.2 Properties of organic liquids and water (20°C)

Description	*Unit weight* (g/cm³)	*Water solubility* (g/l)	*Dielectric constant*	*Dipole moment* (debye)
Water	0.98	∞	80.4	1.83
Hydrocarbons and related compounds				
Heptane	0.68	<0.03	2.0	0
Benzene	0.88	0.7	2.28	0
Xylene	0.88	<0.3	2.4	0
Trichloroethylene	1.46	1.0	3.4	0.9
Tetrachloromethane	1.59	0.8	2.2	0
Nitrobenzene	1.20	2	35.7	4.22
Alcohols and phenols				
Methanol	0.79	∞	31.2	1.66
Ethenol	0.79	∞	25.0	1.69
Phenol	1.06	∞	13.13	1.45
Aldehydes and ketones				
Acetone	0.79	∞	21.4	2.74
Acids				
Acetic acid	1.05	∞	6.2	1.04
Base				
Aniline	1.02	36	6.9	1.55

8.4 Soil organic interaction

The interaction between pore fluid and clay minerals is due to absorption, cation exchange and through intrusion of water and other molecules in clay mineral sheets. When water-soluble organics invade a clay–water system, with time, water and other cations in the diffused double layer may be replaced by the organics. The bond strength of an organic molecule is quite different from that of water. The dielectric constant of the organics is much less than that of water. This causes certain changes in the fabric of the soil. Since strength, deformation and permeability properties are influenced by the soil fabric, the behaviour of a soil will change with the changes in soil fabric. The organic molecules can enter the interlayer spaces, causing the clay layers to swell. This will have the effect of reducing both shear strength and permeability.

Organics with low water solubility, such as hydrocarbons and related compounds, are not able to displace the pore water and will have little effect on soil properties. However, if these fluids are forced through a soil under pressure, they may cause discontinuities within the soil mass. In such cases soil properties will be affected. When these compounds are mixed with dry clay minerals, the clay behaves like a non-plastic material. This is attributed to the low dielectric constant and lack of polar bonds.

The organic acids can dissolve clay minerals owing to the protons replacing the adsorbed cations on the clay surface, and possible replacement of aluminium, magnesium and silicon with hydrogen. The rates of dissolution increase with increasing acid concentration, clay surface area, time of exposure, amount of magnesium oxide in the clay mineral, and decreasing particle size. If the organic acids are not present in the given waste, they may be produced by anaerobic decomposition. Because of the large number of variables involved, it is difficult to predict the long-term behaviour of clay liners and containment structures.

Organic bases accept protons and become positively charged cations. They may change the chemistry of clay by replacing the adsorbed water, by interaction with adsorbed cation, by exchanging interlayer cations, or by interaction with surface oxygen of the clay mineral. If in the process the thickness of the double layer is reduced the attractive forces may become predominant. This may result in greater shear strength and permeability.

8.4.1 Atterberg limits

Atterberg limits are used as an index of a soil's consistency (Chapter 4). For inorganic compounds the Atterberg limits are shown in Table 8.1. For a given mineral the liquid limit has a larger range of values than the plastic limit. For highly swelling minerals like montmorillonite, adsorption of higher valency cations results in a considerably lower value of the liquid limit.

The effect of organic compounds on Atterberg limits is shown in Table 8.3. Note that concentrated organic liquids have much more effect on the Atterberg limits than their aqueous solutions even at high concentrations. With hydrocarbons, alcohols and ketones the soils cease to exhibit plastic properties. As seen from Table 8.2 the dielectric constants of these chemicals are much less than that of water. The

Table 8.3 Atterberg limits of some clay minerals with organics

Material	*Liquid limit*	*Plastic limit*	*Plasticity index*
Kaolinite			
Water	58	34	24
20% methanol	60	32	28
60% methanol	59	32	27
Pure methanol	74	45	29
Georgia kaolinite			
Water	66	34	32
Pure acetone	58	NP	0
Illite–chlorite			
Water	33	18	15
53 mg/l heptane	36	16	20
Pure heptane	24	NP	0
Illite–chlorite			
Water	33	18	15
20% acetic acid	30	15	15
60% acetic acid	30	20	10
Pure acetic acid	37	28	11
Sodium-montmorillonite			
Water	425	58	367
Pure acetone	66	NP	0

NP, non-plastic.
Sources: Acar and Ghosh, 1986; Bowders and Daniel, 1987; Daniel and Liljestrand, 1984.

hydrocarbons are essentially non-polar but polar moments for the other compounds are close to that of water. Considering these factors it appears that the dielectric constant plays a greater role in determining soil–liquid interaction. However, in aqueous solution the effect on plastic properties is not discernible.

Acetic acid shows a considerable increase in the Atterberg limits, but change in the plasticity index is minimal. The higher Atterberg limits are probably due to an increase in basal spacings caused by acetic acid.

It must be noted that the effect of organic compounds on a soil cannot be judged from the Atterberg limits. As described in the following section, free swelling appears to correlate better with the soil properties than the Atterberg limits.

8.4.2 Free expansion

When a dry soil is soaked in a liquid then the soil–liquid interaction influences the water absorption or swelling of the soil. The change in the volume of the soil (height) as a percentage of its original dry volume (height) is defined as the free expansion, E_w, and is given by the equation (Head, 1981):

$$E_w = (V_f - V_i) \times 100/V_i \tag{8.2}$$

where V_i is the initial dry volume of the soil, and V_f the final wet volume of the soil.

To correlate the free expansion of different chemicals an expansion index, E_i, is defined as follows:

$$E_i = E_{wc}/E_{ww} \tag{8.3}$$

where E_{wc} is the soil free expansion with the chemical, and E_{ww} is the free expansion with demineralized water.

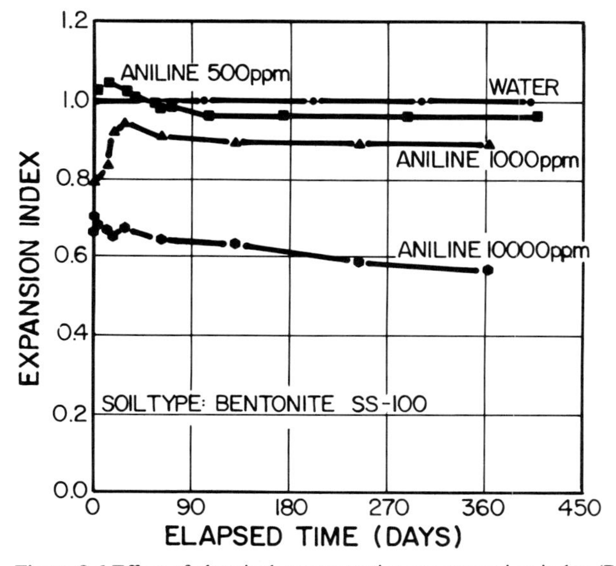

Figure 8.6 Effect of chemical concentration on expansion index (Reproduced from Khera *et al.*, 1987, by permission)

Expansion indices for a 'contaminant resistant' SS-100 with aniline of three different concentrations are shown in Figure 8.6. Its value decreases with increasing chemical concentration. At lower concentrations the value of E_i increases with time. At higher concentrations the effect is reversed. Tests with other organic and inorganic chemicals show that the expansion index is affected by the nature of the chemical (Khera *et al.*, 1987).

The values of Atterberg limits did not show any trend with changing chemical concentrations but the expansion index is sensitive to chemical type and concentration even at low strength. An expansion test is easier to perform and requires less skill than the Atterberg limit test, even though it takes longer to complete. Although the behaviour of a soil suspension is not the same as its *in situ* response, E_i can provide a qualitative measure of the influence of a permeant on the soil structure. The equilibrium period for an expansion test could serve as a guide to the selection of the proper time for obtaining more reliable values of hydraulic conductivity and other engineering properties of soils.

Summary

The engineering properties of a soil mass are affected by its composition, fabric and stress history. Clay minerals have a crystalline structure with the silicon tetrahedron and aluminium octahedron as their basic building blocks. The ability of a clay mineral to adsorb external cations is known as its cation exchange capacity (CEC). When a clay particle is placed in water a high concentration of cations occurs near its negatively charged surface, giving rise to the so-called diffuse double layer. The thickness of the double layer varies directly as the temperature and the dielectric constant, and inversely as the valence and the square root of electrolyte concentration, and has strong influence on the permeability of the soil. Water-soluble organics may replace the cations in the diffuse double layer. The differences in the bond strength of organic molecules and their lower dielectric constant cause changes in the fabric and affect its strength, deformation and permeability. The extent of these changes may be determined from the results of expansion tests and Atterberg limit tests. The former yield more reliable tests and are easier to perform.

Notation

CEC cation exchange capacity (mEq/g)
e charge (16.0×10^{-20} C or 4.8×10^{-10} e.s.u.)
E_i expansion index
E_{wc} soil free expansion with the chemical (L)
E_{ww} free expansion with water (L)
I_p plasticity index
k Boltzman's constant (1.38×10^{-16} erg/K)
n electrolyte concentration (ions/cm^3)
T absolute temperature (K)
v cation valence
V_f final wet volume of the soil (L^3)
V_i initial dry volume of the soil (L^3)
ε dielectric constant of the medium

References

Acar, Y. B. and Ghosh, A. (1986) Role of activity in hydraulic conductivity of compacted soils permeated with acetone, *Int. Symp. on Environmental Geotechnology*, Vol. 1, pp. 403–412

Bowders, J. J. and Daniel, D. D. (1987) Hydraulic conductivity of compacted clay to dilute organic chemicals, *ASCE, Journal of Geotechnical Engineering*, **113,** No. 12, 1432–1448

Daniel, D. E. and Liljestrand, H. M. (1984) *Effects of Landfill Leachates on Natural Liner Systems*, Report GR83-6, Geotechnical Engineering Center, University of Texas, Austin, Texas

Grim, R. E. (1968) *Clay Mineralogy*, 2nd Edn, McGraw-Hill, New York

Hart, H. and Schuetz, R. D. (1972) *Organic Chemistry: A Short Course*, 4th Edn, Houghton Mifflin, Boston, Massachusetts

Head, K. H. (1981) *Manual of Soil Laboratory Testing*, Vols. I and II, John Wiley & Sons, New York

Khera, R. P., Wu, Y. H. and Umer, M. K. (1987) Durability of slurry cut-off walls around the hazardous waste sites, *Proc. 2nd Int. Conf. on New Frontiers for Hazardous Waste Management*, EPA/600/9-87/018F, pp. 433–440

Lambe, T. W. and Whitman, R. V. (1969) *Soil Mechanics*, John Wiley & Sons, New York

Madsen, F. T. and Mitchell, J. K. (1987) *Chemical Effects on Clay Hydraulic Conductivity and their Determination*, Open File Report, Environmental Institute for Waste Management Studies, University of Alabama, Tuscaloosa

Mitchell, J. K. (1976) *Fundamentals of Soil Behavior*, John Wiley & Sons, New York

Chapter 9

Hydraulic properties

The movement of water through saturated soils was studied by Darcy, who showed that for laminar flow the rate of flow or discharge, q, was proportional to the hydraulic gradient, i.

$$q = -kiA \tag{9.1}$$

where k is the hydraulic conductivity and A the total cross-sectional area of soil perpendicular to the direction of flow.

$$i = h/L$$

where h is the total head loss (pressure head + elevation head), and L is the distance in which the head loss occurs.

The factors that affect k are the nature of the permeant, its density, viscosity and temperature; soil type, structure, particle shape, size and size distribution; pore size and pore size distribution, fissures, joints, stratification and other discontinuities; etc.

The effect of the density and viscosity of the permeant can be accounted for by considering *intrinsic permeability*, which is given by:

$$K = k\mu/\rho g \tag{9.2}$$

where μ is the dynamic viscosity of the fluid (M/LT), ρ is its density (M/L^3), and g is the acceleration due to gravity (L/T^2).

9.1 Laboratory measurements

Measurements of hydraulic conductivity can be made in the field and in the laboratory. Field measurements may be more appropriate but are costly and time consuming, and therefore less frequently used. Laboratory tests are generally easier to perform but there are certain difficulties associated with them. If a soil is inhomogeneous, stratified, fissured or jointed, the size of the sample may not be large enough to be representative of the soil in the field. Soil may be disturbed during sampling, or it may not even be feasible to obtain a sample in materials at waste sites. If the hydraulic conductivity is to be determined for a compacted sample and the specimen is prepared by laboratory compaction, the laboratory compacted specimen may not be representative of the material in the field. Anisotropy in the material's hydraulic conductivity will yield erroneous values if the direction of the flow in the laboratory does not correspond to the field flow direction.

Although several standard methods are available for determining hydraulic conductivity of soils, there is no standardized method for soils interfacing with liquid waste materials. Other parameters such as state of stress, hydraulic gradient, degree of saturation, etc., influence the outcome but are not standardized. A few of the commonly used methods are described here. For further details see Olson and Daniel (1981).

9.2 Permeant

In soil mechanics distilled water is generally used for hydraulic conductivity determination. However, the tests have shown considerable decrease in permeability values with distilled water when compared with those where pore fluid is used as the permeant. In an undisturbed clay containing illite clay mineral, leaching by distilled water resulted in the removal of a much higher proportion of monovalent cations than divalent cations (Figure 9.1). The monovalent cations continued to leach out while the extraction of the divalent cations stopped (Yong, 1986). The ease with which the monovalent ions can be leached out indicates that the soils that derive their low permeability from the abundance of monovalent ions, such as sodium bentonite, are most susceptible to permeability changes.

To determine the effect of leachate or other waste liquids on the permeability of a soil the baseline hydraulic conductivity is often obtained with 0.01N $CaSO_4$ or with tap water. It is reasoned that 0.01N $CaSO_4$ is representative of the salt concentration found in soils and that the distilled water can leach out some of the soil constituents. For the purpose of standardization 0.005N $CaSO_4$ has been suggested in the draft proposals of the ASTM.

Typical values of hydraulic conductivity for some natural soils are given in Table 9.1.

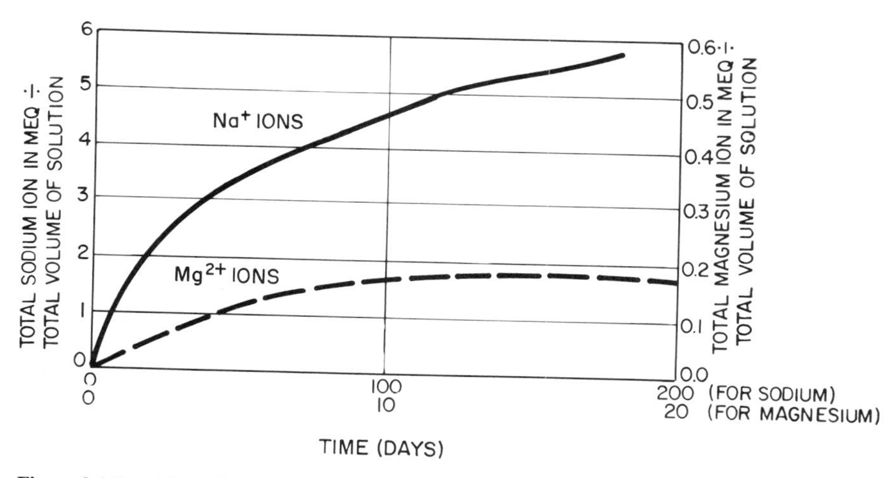

Figure 9.1 Leaching of salts from soils with distilled water (Reproduced from Young, 1986)

Table 9.1 Hydraulic conductivity of soils and rocks

Soil or rock formation	*Range of k* (cm/s)
Gravel	1–5
Clean sand	10^{-3}–10^{-2}
Clean sand and gravel mixtures	10^{-3}–10^{-1}
Medium to coarse sand	10^{-2}–10^{-1}
Very fine to fine sand	10^{-4}–10^{-3}
Silty sand	10^{-5}–10^{-2}
Glacial till	10^{-10}–10^{-4}
Homogeneous clays (unweathered)	10^{-9}–10^{-7}
Shale	10^{-11}–10^{-7}
Sandstone	10^{-8}–10^{-4}
Limestone	10^{-7}–10^{-4}
Fractured rocks	10^{-6}–10^{-2}

Sources: Freeze and Cherry, 1979; Peck *et al.*, 1974.

9.3 Permeability tests

Several methods are available for determining the permeability. A few of the common types are described in the following section, along with their applications, advantages and limitations.

9.3.1 Compaction mould cell

The standard compaction mould (ASTM D698-78) with a diameter of 101.6 mm (4 in) and a height of 116.5 mm (4.5 in) serves as a permeameter for compacted soils. Its base is modified to accept a porous disc as shown in Figure 9.2. The chamber above the compacted soil is filled with the desired fluid. A constant head is applied at the upper end of the specimen. The liquid is allowed to permeate through the soil until a constant flow rate is indicated. The hydraulic conductivity of the soil is determined from the total amount of flow, Q, in a given time t. Based on Equation 9.1 the hydraulic conductivity is given by:

$$k = QL/hAt \tag{9.3}$$

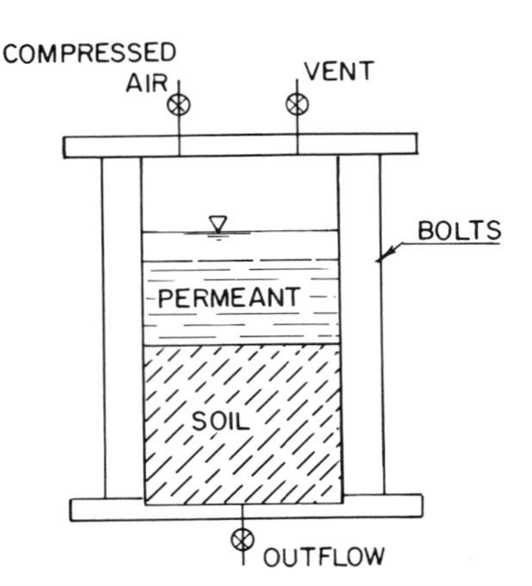

Figure 9.2 Compaction mould rigid wall permeameter (Reproduced from Brown and Anderson, 1983)

Although this test is straightforward it has no provision for applying vertical stress to the test specimen. If the soil shrinks because of interaction with the permeant, flow through shrinkage cracks cannot be controlled. Large voids may develop a soil–cell wall interface causing leakage. There is no satisfactory method available to ensure saturation of the soil sample.

A smaller compaction mould with a 33.3 mm diameter and 72.7 mm height has also been used as a permeability cell (Davis *et al.*, 1985). Soil is compacted by a tamper producing kneading action similar to that of a sheepsfoot roller. Because of its small size this cell is not desirable.

9.3.2 Double ring cell

A double-ring rigid wall permeameter was reported by Anderson *et al.* (1984) to be effective in overcoming the leakage problem of the compaction mould cell. In this permeameter a ring is built into the base plate so that the area within the ring is half the total area of the base. The flow adjacent to the side walls is isolated from the flow-through area enclosed by the ring. Leakage is indicated if the flow rates from the central core and the outer annular region are different. Like the compaction permeameter, no provision is made for the application of the confining pressure.

9.3.3 Oedometer cell permeameter

A fixed ring consolidation permeameter cell is shown in Figure 9.3. The soil specimen is placed in a ring with a diameter of 40–100 mm and a height of up to 100 mm. After mounting the specimen the consolidometer is placed in a loading frame and the desired axial load is applied. The air is flushed out from the bottom. One of the two outlets at the bottom is closed and a head is applied through the other outlet.

The head is applied using a stand pipe of a small diameter, a. A constant head is maintained at the exit. The hydraulic conductivity is determined from Equation 9.4 which is derived from Equation 9.1.

$$k=(aL/At)\times(\ln h_1/h_2) \tag{9.4}$$

where t is the total time of flow, h_1 the head at the beginning of the flow, and h_2 the head at the end of the flow period.

For low-permeability soils considerable time may be necessary to complete a test. This may introduce errors from leakage and evaporation. Additional pressure, dp, is applied to the stand pipe to reduce the test duration. The hydraulic conductivity is then given by:

$$k=(aL/At)\times\ln\frac{h_1+\mathrm{d}p/\gamma_w}{h_2+\mathrm{d}p/\gamma_w} \tag{9.5}$$

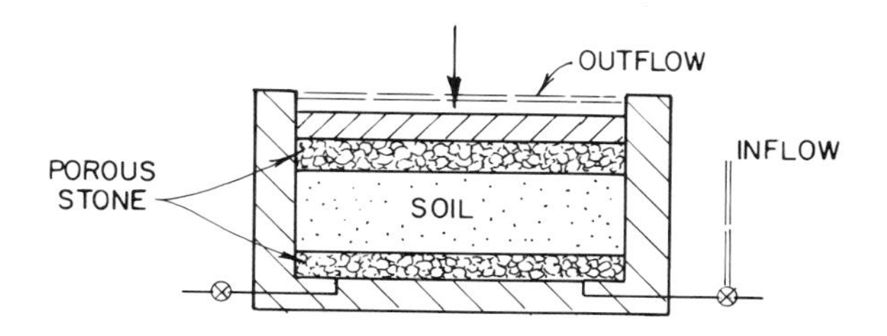

Figure 9.3 Consolidation ring permeameter (Reproduced from Olson and Daniel, 1981)

The hydraulic conductivity can also be computed from the rate of settlement using one-dimensional consolidation theory (Terzaghi, 1943).

This cell allows simulation of the field stresses, and reduces shrinkage crack formation and sidewall leakage. The hydraulic conductivity values can be determined at different void ratios. Because of the small thickness of the specimen a test is completed in a short time. Laboratory prepared or undisturbed specimens can be used. It is, however, difficult to perform tests at small consolidation pressure.

9.3.4 Flexible wall cell

A flexible permeability cell is similar to a triaxial cell (Figure 9.4). The specimen diameter may range from 38 to 152 mm. The height of the specimen is kept small, about 0.5–1 mm diameter or even smaller, to reduce the time of testing. The specimen is enclosed in flexible latex membranes with porous discs at each end. The membranes are sealed to the bottom pedestal and the top cap. To protect the membranes from disintegration by corrosive permeants, a Teflon tape (1 mil thick) is wrapped around the specimen before enclosing it in the membrane (Daniel *et al.*, 1984; Khera *et al.*, 1986). Alternatively a flexible butyl rubber or neoprene membrane may be used. Usually two drains are provided at each end to facilitate flushing the entrapped air from the system.

The pressure system commonly used in research and engineering practice (Edil and Erickson, 1985) utilizes compressed air. For toxic or corrosive permeants specially designed chambers are used to isolate the sensitive equipment and the personnel from potential hazards. One such pressure system is shown in Figure 9.5. Volume change and pressure measurements can be made for the pressure chamber at each end of the specimen.

Before performing a permeability test the soil specimen is saturated using back pressure. The value of the pore pressure parameter, B, is computed from Equation

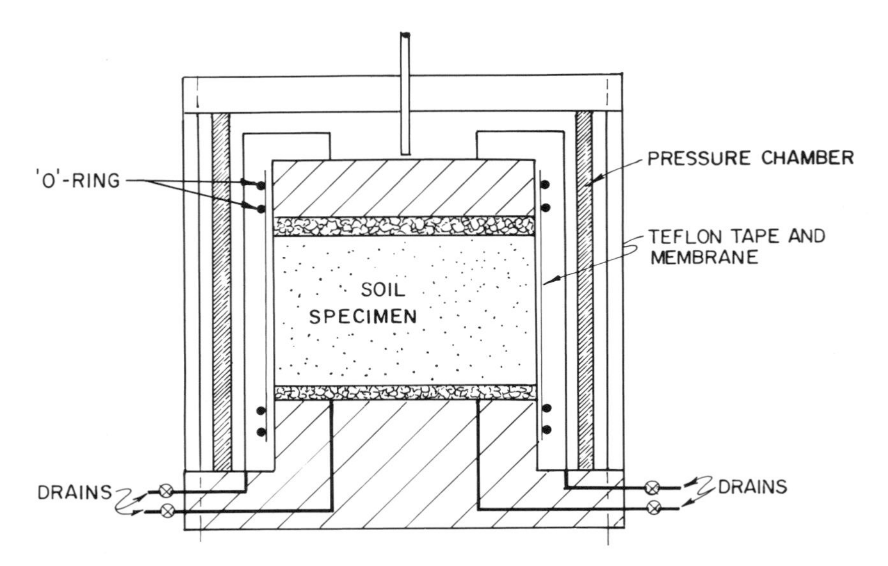

Figure 9.4 Flexible wall permeameter (Reproduced from Daniel *et al.*, 1984)

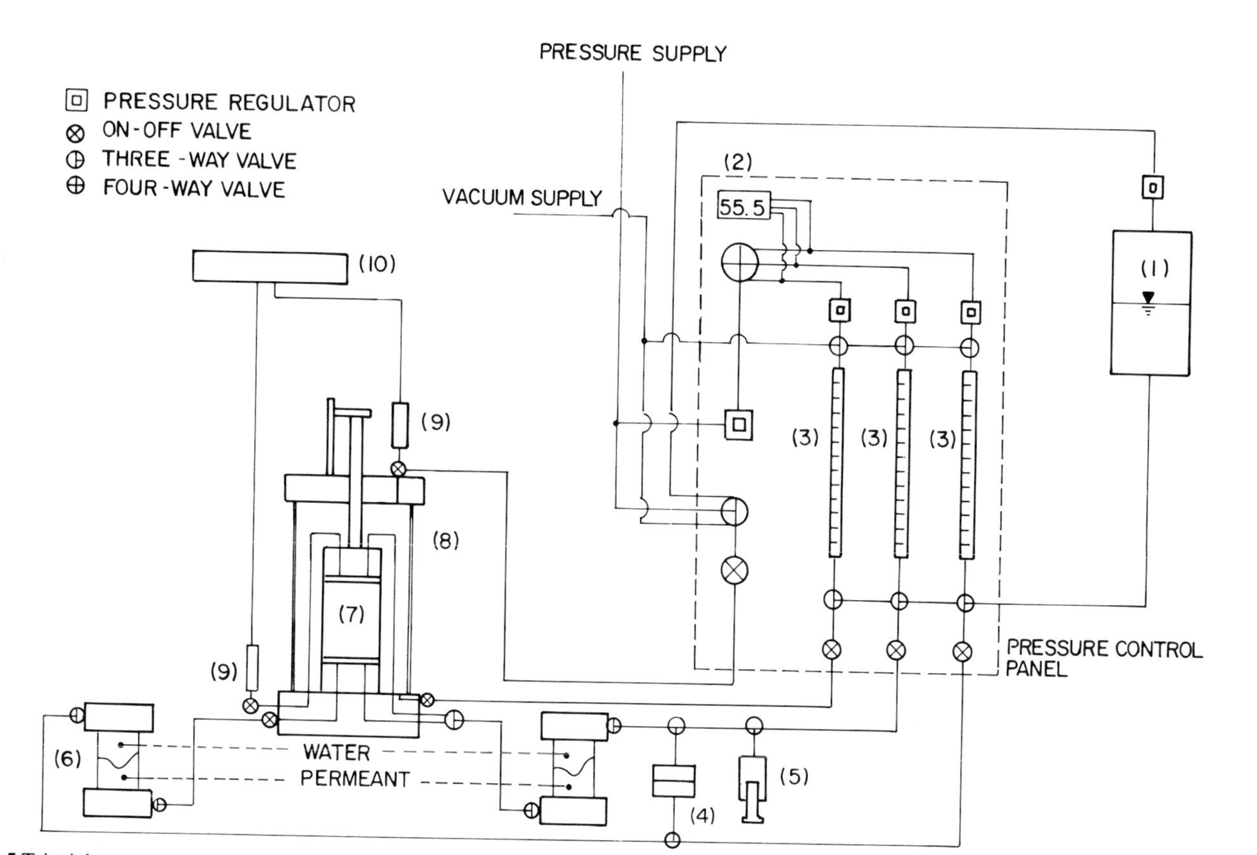

Figure 9.5 Triaxial pressure system and flow pump: Trautwein pressure system. 1, water reservoir; 2, pressure indicator; 3, volume change burettes; 4, differential transducer; 5, flow pump; 6, permeant interface device; 7, soil specimen; 8, flexible wall permeameter; 9, pressure transducer; 10, data logger. The differential transducer is connected to the data logger.

5.25. The process is repeated until a B value of 0.95 or greater is achieved. Dunn and Mitchell (1984) applied a small vacuum to the specimen followed by a back pressure while maintaining the flow through the specimen under an imposed hydraulic gradient. The value of B was greater than 0.95 when back pressure was between 300 and 350 kPa. In the absence of the flow, B was less than 0.9 at a back pressure of 450 kPa.

The saturation back pressure should be maintained during permeability tests. If it is released the dissolved gases will come out of solution and the result may be a more than 50% decrease in measured permeability (Zimmie *et al.*, 1981).

To reduce the time of testing, hydraulic gradients as high as 200 are used. Chamber pressure and the upstream fluid pressure are raised so that the effective stress on the upstream end of the specimen is equal to the consolidation pressure. This results in higher effective stress on the downstream end of the specimen, causing it to consolidate further. This yields a lower value of permeability. An alternative approach is to lower the effective stress on the upstream side and raise the effective stress on the downstream side by an equal amount; that is, the average change in effective stress is zero.

Example 9.1

A specimen for permeability test has been consolidated under a chamber pressure of 500 kPa and a back pressure of 300 kPa. If for the desired hydraulic gradient a head difference of 100 kPa is desired between the upstream and downstream ends of the specimen, determine the chamber pressure and pore pressure if the hydraulic gradient is obtained by (a) maintaining effective stress on the upstream end, (b) maintaining an average change in effective stress equal to zero.

Solution

Stresses (kPa) at upstream end:

	σ_3	u	σ_3'	du	dσ_3
Initial	500	300	200		
Final (a)	600	400	200	100	0
(b)	500	350	150	+50	−50

Stresses at downstream end:

	σ_3	u	σ_3'	du	dσ_3
Initial	500	300	200	0	
Final (a)	600	300	300	0	100
(b)	500	250	250	−50	+50

In the flexible wall permeameter there is no leakage between the specimen and the confining membrane. However, there may be leakage through the membranes and through the seals at the top and bottom of the specimen. For long duration tests and where the permeability of a soil is low, corrections for leakage are recommended. If specimen height changes are to be measured and field anisotropy stress conditions are to be simulated a triaxial cell is used as permeameter.

For a rigid wall permeameter the application of back pressure may even cause several fold increase in hydraulic conductivity (Edil and Erickson, 1985).

9.3.5 Flow pump method

In the flow pump method the permeant is injected or withdrawn from the specimen at a constant rate of flow. The induced head difference across the length of the specimen is monitored with a differential transducer (Olsen *et al.*, 1985). The arrangement of the flow pump is shown in Figure 9.5. This method allows the evaluation of the osmotic component of flow resulting from electrical, chemical and thermal gradients (Figure 9.6). The time needed to reach the steady-state condition was less than 1 min for a sand and little over 200 min for a silty clay.

The advantages of this method are that direct flow measurements are not necessary, permeability measurements are obtained in a much shorter period of time, the applied hydraulic gradient is small and compares well with the field values, and errors due to seepage-induced consolidation can be recognized. However, if the testing procedure requires that a certain number of pore volumes be passed through the test specimen, then the time of testing will increase substantially because of the low applied gradient. An alternative would be to prepare the test specimen with the given chemicals.

9.3.6 API test

This American Petroleum Institute (API) test is most commonly used for testing the hydraulic conductivity of bentonite slurry. The apparatus consists of a fixed ring with a porous stone and outflow at the base as shown in Figure 9.7. The permeant is subjected to an air pressure.The test can be completed in a few hours. Hadge and Barvenik (1985) adopted it for quality control of backfill at a slurry wall construction site. The agreement between API and flexible wall permeabilities was within ±66%. If this test is adopted at other sites its accuracy must be verified against results from a flexible wall permeameter.

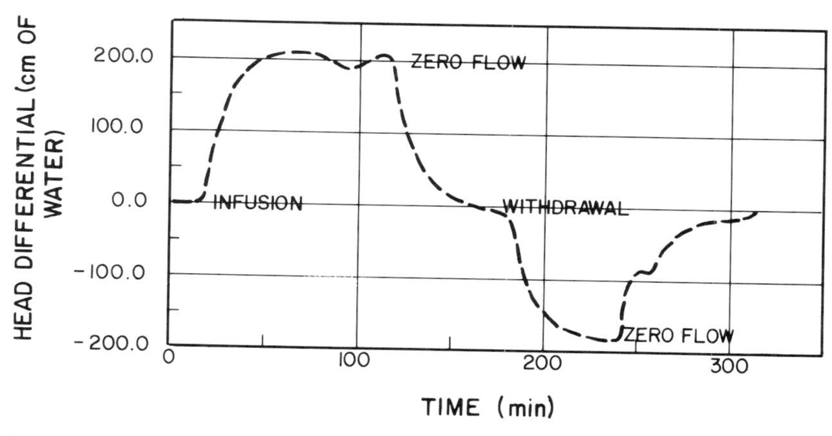

Figure 9.6 Head difference from flow pump measurements (Reproduced from Khera *et al.*, 1988)

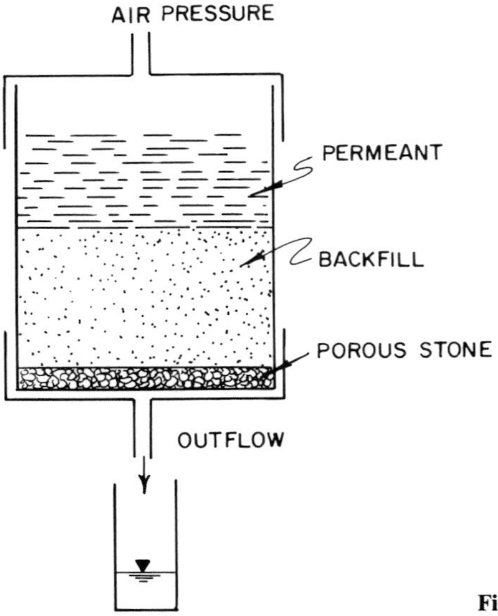

AIR PRESSURE

PERMEANT

BACKFILL

POROUS STONE

OUTFLOW

Figure 9.7 Schematic of an API filter press test for slurry

9.4 Factors affecting permeability

A few of the significant factors that have an effect on soil permeability are discussed in the following sections.

9.4.1 Fine-grained components

Addition of a small percentage of fines in a coarse-grained soil can decrease its permeability by several orders of magnitude. The plastic fines are much more effective in reducing permeability than the non-plastic fines (D'Appolonia, 1980).

9.4.2 Compaction effect

In a compacted soil the permeability depends on the method of compaction and the water content. Soils compacted wet of optimum have a permeability of two to three orders of magnitude lower than samples compacted dry of optimum. Kneading compaction yields lower permeability than static compaction (Mitchell *et al.*, 1965). Static compaction is suggested for laboratory specimens over kneading or impact compaction when these tests are to be used for field control. This will yield higher but more realistic values of hydraulic conductivity (Dunn and Mitchell, 1984). Further discussion of this will be found in Chapter 10.

9.4.3 Sample saturation

In unsaturated soil, voids occupied by air or other gases are not available to permeant to flow through. Therefore, in the absence of complete saturation of the sample, its measured permeability will be lower. Mitchell *et al.* (1965) reported an increase in permeability by a factor of four to five as the degree of saturation increased from 85 to 98%. For lower degrees of saturation the difference will be even less.

Various methods have been used for specimen saturation. Often compacted samples are allowed to soak from the bottom to achieve saturation. Sometimes a vacuum is applied while permitting inflow at the bottom. These methods do not always yield complete saturation. In a flexible wall cell the use of back pressure yields the best results (see Chapter 5).

9.4.4 Hydraulic gradient effect

The hydraulic gradients that exist across compacted soil liners seldom reach 20. Federal regulations limit the height of the leachate to 300 mm (1 ft) in a landfill. Since a soil liner is almost never less than 3 ft (914 mm) in thickness the hydraulic gradient will be less than one.

Hydraulic gradients from 1 to 500 have been reported for laboratory tests. The higher gradients require the application of larger effective stresses at the downstream end of the specimen. These stresses cause changes in the specimen volume and the structure of the void system. In materials of softer consistency or low density, permeability decreases (Dunn and Mitchell, 1984; Edil and Erickson, 1985; Korfiatis *et al.*, 1986) with increasing gradient (Figure 9.8). The impact on stiffer materials is less. High gradients may also cause migration of soil particles resulting in either clogging of the larger pores or increase in pore size and pore volume due to erosion. The permeability values could thus be either lower or higher and are not always reliable.

At low hydraulic gradients some organics may not permeate a water-saturated soil, but at higher gradients macropores and other discontinuities develop in the soil, resulting in greater permeability. These conditions may occur in the rigid wall permeameter. In the flexible wall permeameter the increases in permeability are small.

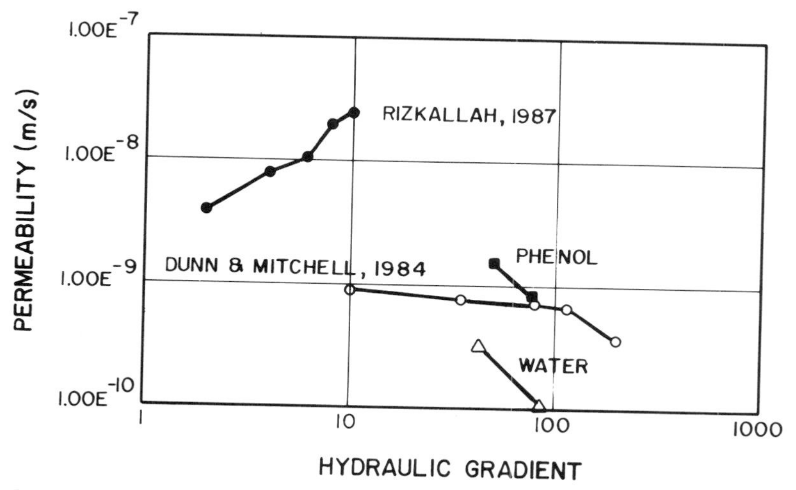

Figure 9.8 Effect of hydraulic gradient and consolidation pressure on permeability

The permeability may be underestimated in a flexible wall permeameter as the cracks will close if a high confining pressure is applied. On the other hand, if the confining stresses are small the effect of increasing the gradient may be reversed as shown in Figure 9.8 (Rizkallah, 1987). For the tests shown the confining pressures were 40 and 80 kPa and the original specimens had shrinkage fissures. The application of increasing gradients seemingly increased the size of the cracks similar to those reported for rigid wall permeameters, resulting in higher permeability. This phenomenon is consistent with the concept of hydraulic fracture. The critical value for the ratio of change in pore pressure to the effective stress (du/σ') is 0.5 to 1.0 (Bjerrum *et al.*, 1972).

It is recommended that a low hydraulic gradient be used for the determination of hydraulic conductivity with due regard to the effective confining stress. The flow pump method of permeability measurement may be a better choice than the other methods. However, if a higher gradient must be used, in the case of dense soils, its detrimental effect may be somewhat reduced if the upstream back pressure is increased and the downstream back pressure is decreased by half the amount needed for the desired hydraulic gradient (see Example 9.1). The need for proper hydraulic gradient cannot be overemphasized.

9.5 Physicochemical changes

The nature of permeating chemicals can change the permeability characteristics of clay soils by alterations in soil fabric, increase or decrease in the diffused double layer thickness, solution of soil components by strong acids and bases, blockage of soil voids due to precipitation, and growth of microorganisms. An extensive summary and discussion of the effect of permeants on the hydraulic conductivity of various naturally occurring and commercially available clay soils may be found in EPA (1986).

9.6 Inorganic compounds

Inorganic compounds may influence the engineering properties of soils with clay components. A change in chemical or electrolyte concentration of the pore fluid in a soil has an effect on the hydraulic conductivity. Soils with greater swelling potential, such as bentonite, show larger changes in permeability with the exchange of the adsorbed cations. The monovalent sodium ion with a bigger hydrated radius causes a large interlayer spacing and more dispersion of the clay particles than divalent calcium ions, which have a smaller hydrated radius. The affinity of clay minerals for cations increases as the ion valence increases. However, the high concentration of lower valence ions in a solution can negate the greater replacing power of higher ions. The exchanging of cations affects the thickness of the double layer. As a result the soil structure may alter, its volume may change, it may develop cracks and macropores, etc.

Alther *et al.* (1985) reported an increase in hydraulic conductivity of more than one order of magnitude for a bentonite as the concentrations of chlorides of potassium, sodium, magnesium and calcium were increased (Figure 9.9). For a divalent cation the permeability appears to stabilize after a small increase in salt concentration. It is

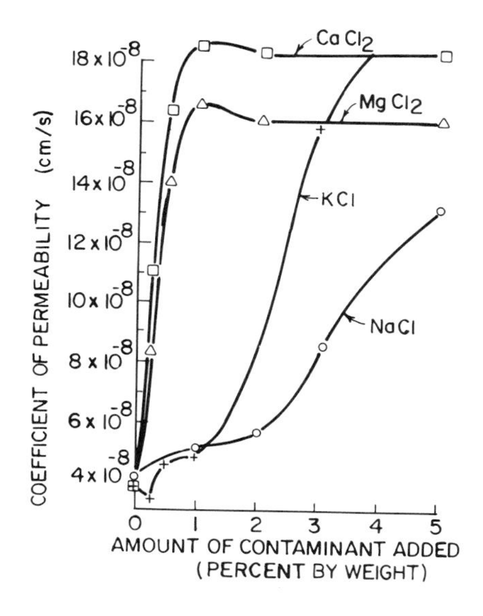

Figure 9.9 Effect of inorganic salts on permeability of bentonite (Reproduced from Alther *et al.*, 1985, by permission)

evident that the hydraulic conductivity of a soil is influenced much more by the introduction or removal of monovalent ions than divalent ions.

Yong (1986) reported that in an undisturbed clay distilled water removed much larger proportions of monovalent cations (up to 30 times) than divalent cations. Soils that realize their low permeability from the abundance of monovalent ions are most likely to exhibit changes in permeability when these ions are replaced by higher valence ions. Since sodium montmorillonite derives its low permeability characteristics from the presence of monovalent sodium ions it falls into this category of materials.

The change in the permeability of illite and kaolinite was less than that of smectite. This is attributed to their much lower cation exchange capacity.

9.6.1 Inorganic acids

Increasing concentration of inorganic acids triggers flocculation, inhibits swelling, and increases dissolution of clay minerals. Kaolinite shows the least solubility and smectite the most. Concentrated acids have caused increases of several orders of magnitude in the permeability of clay soils, essentially by dissolution. With dilute acids there may be an initial tendency toward decrease in permeability due to clogging of the pores by precipitated salts. In some instances this is followed by an increase in permeability as the precipitates leach out.

Dilute hydrochloric acid with pH values of 1, 3, and 5 had little effect on the hydraulic conductivity (Lentz *et al.*, 1985).

9.6.2 Inorganic bases

Bases tend to cause dispersion of soils and may even dissolve soil particles. Sodium hydroxide in tap water with a pH of 9 and 11 had no effect on the permeability of compacted clays.

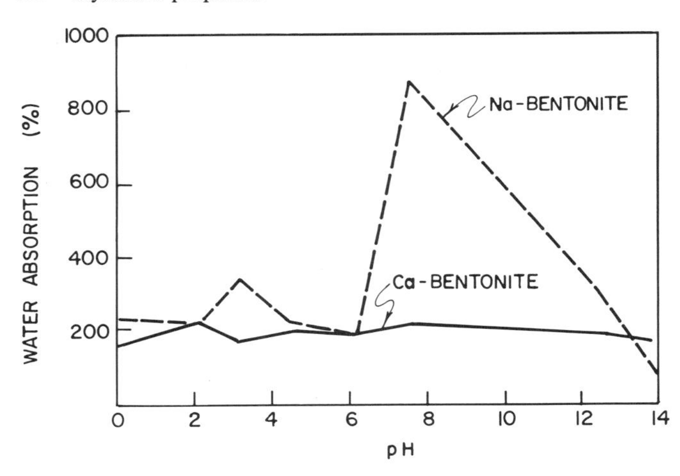

Figure 9.10 Influence of pH on the water absorption capacity of sodium and calcium bentonite (After Hermanns *et al.*, 1987)

A pH of 13 caused a decrease in the hydraulic conductivity by a factor of about 2.5–13. The decrease in permeability was small for the monovalent ions and was attributed to the precipitation of salts from the tap water. The larger decreases, observed for magnesium montmorillonite, were believed to be from the replacement of divalent magnesium ions with monovalent sodium ions (Lentz *et al.*, 1985).

The effect on water absorption capacity of sodium and calcium bentonites is shown in Figure 9.10. A lower permeability is to be expected for pH between 8 and 11 with sodium bentonite because of its high swelling potential in this range of pH. Furthermore, the changes in pH drastically reduce the water absorption capacity of sodium montmorillonite. This will result in increased permeability. Sodium bentonite possesses high water absorption capacity. In slurry wall construction it is recommended that the pH of slurry be maintained around eight (Xanthakos, 1979). At this pH the water absorption capacity is the highest (Figure 9.10). However, as the pH value rises or falls the water absorption capacity decreases significantly. The reduced water absorption capacity indicates a decrease in double layer thickness and, therefore, an increase in soil permeability (Mitchell, 1976). Since the pH in the waste environment can vary over a wide range, the possibilities of permeability changes in backfills containing sodium bentonite are high.

Data on soil–bentonite backfill and slurry wall filter cake, where an API type of filter press was used, showed an increase in permeability by a factor of 5–10 with a 5% solution of sodium hydroxide (D'Appolonia, 1980). The differences in soils, moisture content, density, and the permeameter used in the two investigations may be responsible for the divergent results.

9.7 Organic compounds

Test data with organic compounds showing large changes in hydraulic conductivity were reported by Anderson (1982) (Figure 9.11). His study drew the attention of many researchers to this area. Since then a large body of data has become available on the permeability of compacted soils. The results are, however, conflicting. With the same chemicals, hydraulic conductivity has been reported to increase, decrease, or show no change. Some of this behaviour has been attributed to the interaction between soil and the chemical; the other to the appearance of cracks, large pores, and channels in the soil specimen. And finally the testing procedures and the equipment used influence the measured value of permeability.

When a pore fluid has a low dielectric constant, it will have a tendency to curtail soil swelling and, therefore, an increase in its permeability. Figure 9.12 shows the

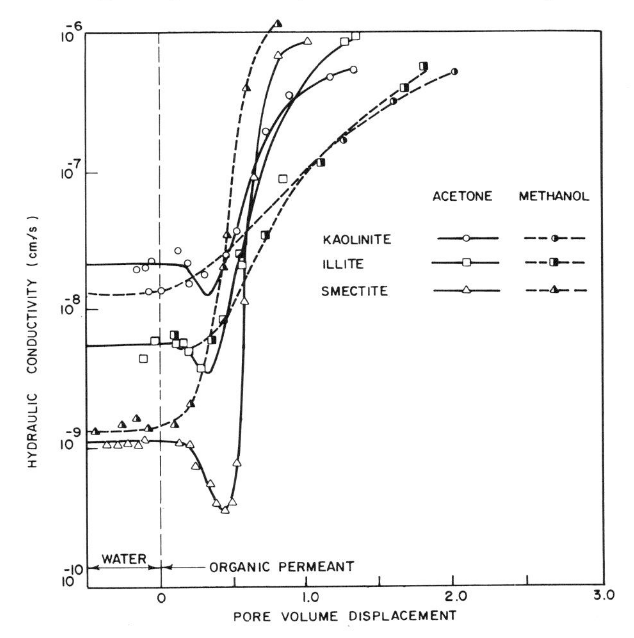

Figure 9.11 Hydraulic conductivity with organic permeants in a rigid wall permeameter (Reproduced from Anderson, 1982, by permission)

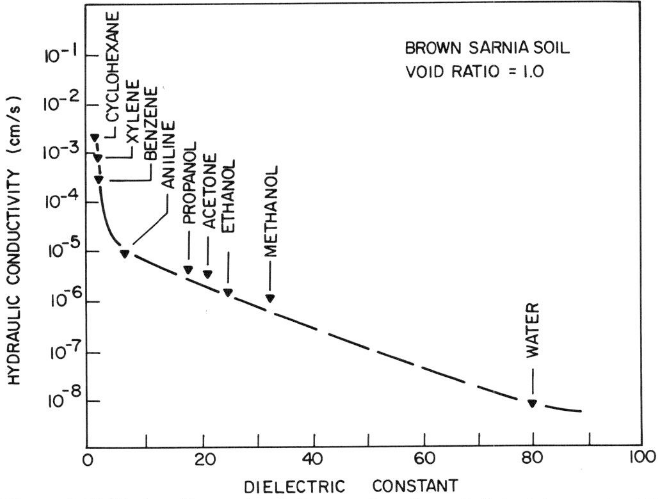

Figure 9.12 Hydraulic conductivity as a function of dielectric constant (Reproduced from Fernandez and Quigley, 1985, by permission)

relationship between hydraulic conductivity and the dielectric constant for an illite-rich lean clay. The lowest permeability is obtained for water, which is polar and has the highest dielectric constant (80). Alcohols and acetone, which are also polar, with a dielectric constant ranging from 20 to 35, show intermediate values. Non-polar hydrocarbons and related compounds with dielectric constant of less than 3 have the highest hydraulic conductivity.

9.7.1 Hydrocarbons

Hydrocarbons (saturated, unsaturated, halogenated and aromatic compounds) usually have a very low water solubility, a negligible dipolar moment, and a low dielectric constant. Some of the properties of organic compounds and their effect on the permeability of soils are shown in Table 9.2. In most instances where soils were compacted and then tested for permeability with pure compounds using rigid wall permeameters, the hydraulic conductivity increased by several orders of magnitude. In such cases cracks and other discontinuities developed within the soil and flow occurred through these discontinuities. In flexible wall permeameters the same compounds showed a decrease in hydraulic conductivity. Less than 10% of the water in the soil pores was displaced by the compound (Fernandez and Quigly, 1985). The low permeability in a flexible wall permeameter is caused by restraining of cracks and macropores. Furthermore the flow occurs only through a small proportion of the specimen.

Complete replacement of pore water was obtained by sequential permeation of the soil, first with alcohol (ethanol), then with a hydrocarbon (benzene). This resulted in a permeability increase of four orders of magnitude (Figure 9.13). When the process of sequential permeation was reversed and the pore hydrocarbon was replaced by water the permeability decreased by the same order of magnitude.

The changes in the hydraulic conductivity are negligible for any concentration within the solubility limits of the compound (Table 9.2).

9.7.2 Alcohols, phenols and ketones

The high water solubility of acetone and phenol allows them to replace pore water easily. In a rigid wall permeameter the hydraulic conductivity may show increases of as much as three orders of magnitude. These compounds may initially show decreases in permeability but this is followed by an increase of two to three orders of magnitude as more of the pore water is replaced. The permeability change shows a decrease if water is used again as the permeant. There is not much change in permeability if the chemical concentrations are less than 75%, or if a flexible wall permeameter is used irrespective of the concentration (Table 9.2).

9.7.3 Organic acids

The only data available among the organic acids are for acetic acid. Both increases and decreases in hydraulic conductivity have been reported for pure and dilute states. The presence of any salts in the soil, which the acid can react with, may result in an initial decrease in permeability due to clogging of the pores by the precipitates. Long duration tests are necessary to establish if eventually the precipitates will leach out. If this occurs it may lead to larger pore sizes, piping, and consequently higher permeability. The clay particles themselves may also be dissolved by the acid. This will have similar effects.

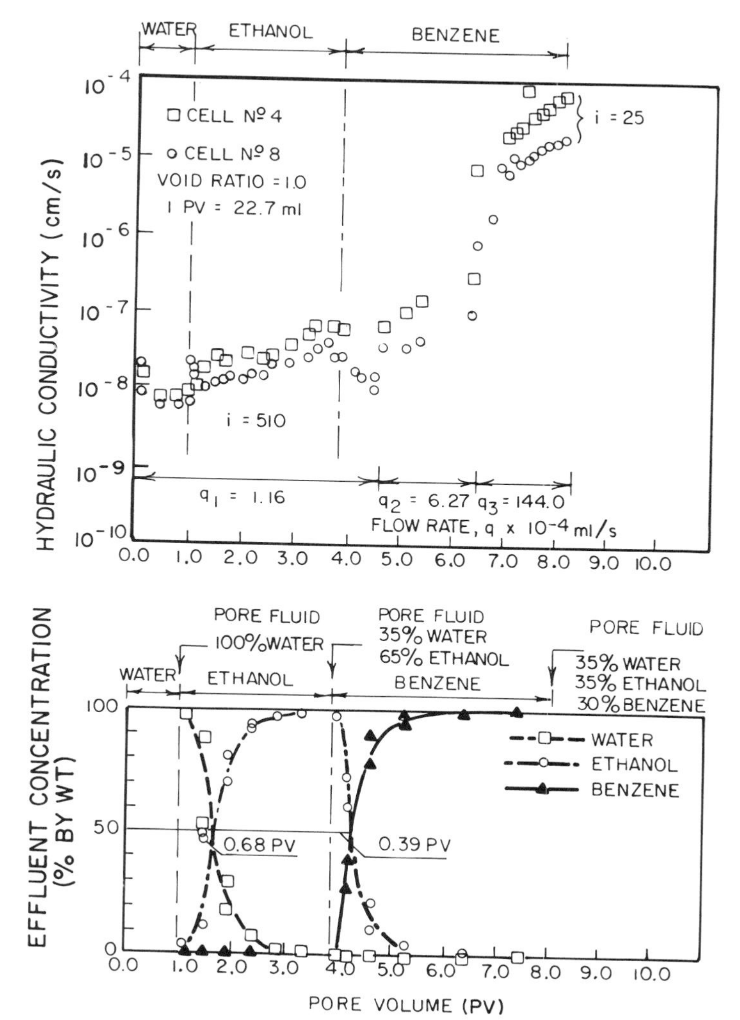

Figure 9.13 Effect of pore fluid on permeability (Reproduced from Fernandez and Quigley, 1985, by permission)

9.7.4 Organic bases

The only organic base for which test data are available is aniline. Large permeability increases occur with pure chemical and not with dilute solutions. Changes occurring in the soil structure may be reversed by permeation with water. There is not much migration or dissolution of particles.

9.7.5 Mixed chemicals

There are few data available on the effect of mixed organics on sodium bentonite. Hermanns *et al.* (1987) reported that after 2 months of permeation with a reconstituted leachate the hydraulic conductivity of the mix containing sodium bentonite was

Table 9.2 Organic compounds and their effect on hydraulic conductivity

Class of compound	*Compound*	*Soil type*	*Soil state*	*Concentration*	*Permeameter type*	*Permeability with water (cm/s)*	*Permeability with compound (cm/s)*	*k of chemical/ k of water*	*Remarks*	*Source*
Hydrocarbon and related compounds	Heptane	Kaolinite	Compacted	Pure	FW	7.7×10^{-8}	2.3×10^{-9}	0.03		Bowders and Daniel (1987)
					RW	1×10^{-7}	2.5×10^{-5}	250		Bowders and Daniel (1987)
		Illite–chlorite			RW	1.5×10^{-8}	1×10^{-5}	660		Bowders and Daniel (1987)
				53 mg/l	FW	5×10^{-9}	5×10^{-9}	1		Bowders and Daniel (1987)
					RW	1.3×10^{-8}	1.3×10^{-8}	1		Bowders and Daniel (1987)
	Benzene	Natural soil	Compacted	Pure	RW	1×10^{-8}	1×10^{-8}	1		Anderson (1982)
	Xylene	Non-calcareous smectitic	Compacted	99%	RW	1×10^{-9}	5×10^{-7}	500		
		Bentonite–clay–soil mix	Compacted (field)	Paint	Field perc.	1×10^{-8}	1×10^{-6}	100		Brown *et al.* (1983)
	Tetra-chloro-methane	Sand–bentonite mix	Consolidated	Pure	FW	4×10^{-9}	6×10^{-7}	150		Evans *et al.* (1985)
				720 mg/l	FW	4×10^{-8}	4×10^{-8}	1		
			Slurry							
	Nitro-benzene	Georgia kaolinite	Compacted	Pure	FW	5.2×10^{-8}	1.5×10^{-10}	0.003		Acar *et al.*(1985)
				0.10%	FW			0.68		
	Chlorinated hydro-carbons	Soil–bentonite mix	Compacted (kneading)	Pure	RW	2.7×10^{-6}	2.4×10^{-5}	60		Davis *et al.* (1985)
		Contaminated soil	Undisturbed	Pure	RW	3.3×10^{-7}	3.1×10^{-5}	94		Davis *et al.* (1985)
Alcohols and phenols	Methanol	Kaolinite	Compacted	Pure	FW	5.9×10^{-8}	1×10^{-7}	1.7		Bowders and Daniel (1987)
					RW	1×10^{-7}	7.5×10^{-7}	7.5		
		Lufkin clay	Compacted	Pure	FW	8×10^{-9}	3×10^{-8}	3.8		Foreman and Daniel (1986)
					RW	1×10^{-9}	1×10^{-6}	1000		Brown and Anderson (1983)

		Smectite soil–bentonite mix	Slurry	Pure	DR	4×10^{-8}	2×10^{-5}	500		Anderson *et al.* (1985)
	Phenol	Brunswick soil	Compacted (static)	4.7 g/l	FW	6×10^{-8}	4.5×10^{-8}	0.75		Korfiatis *et al.* (1986)
		Woodbury clay		4.7 g/l	FW	2×10^{-8}	1.6×10^{-8}	0.8		Korfiatis *et al.* (1986)
		Ca-bentonite–cement mix 200/200 kg/m^3	Slurry	350 mg/l	FW	5.2×10^{-8}	2.8×10^{-8}	0.54	25 days value	Hermanns *et al.* (1987)
		Na-bentonite–cement mix 200/200 kg/m^3	Slurry	350mg/l	FW	1×10^{-6}	1×10^{-6}	1	25 days value	Hermanns *et al.* (1987)
Ketones	Acetone	Non-calcareous smectite	Compacted	99% pure	RW	1×10^{-9}	9×10^{-7}	900		Anderson (1982)
		Kaolinite–mont-morillonite–sand mix	Compacted	Pure	FW, C	3×10^{-7}	4×10^{-7}	1.33	$A = 0.36$	Acar and D'Hollosy (1987)
			Compacted	Pure	FW, C	3×10^{-7}	8×10^{-7}	2.7	$A = 1.39$	Acar and D'Hollosy (1987)
Organic acids	Acetic acid	Sand–bentonite mix	Consolidated slurry	Dilute, pH 1	FW	6×10^{-8}	2×10^{-7}	3.33		Evans *et al.* (1985)
		Non-calcareous smectitic	Compacted	99% pure	RW	1.1×10^{-9}	8×10^{-9}	7.3		Anderson (1982)
		Micaceous clay	Compacted (field)	92.7 + 4% benzene phenol + unknown	Field perc.	1×10^{-7}	1×10^{-7}	1		Brown *et al.* (1983)
Leachate		Na-bentonite–cement, 35 : 200	Slurry		FW	1×10^{-8}	1×10^{-6}	100	80 day value	Hermanns *et al.* (1987)
		Ca-bentonite–cement, 200 : 200			FW	9×10^{-10}	1×10^{-10}	0.11	80 day value	Hermanns *et al.* (1987)
		Na-bentonite–cement, 45 : 200	Slurry		FW	8×10^{-8}	2×10^{-6}	25	40 day value	Naußbaumer (1987)
		Ca-bentonite–cement, 200 : 175			FW	1×10^{-9}	1.05×10^{-8}	9.5	40 day value	Naußbaumer (1987)

several times larger than that containing calcium bentonite. After 30 days of immersion in the same leachate, the samples with sodium bentonite disintegrated and lost 90% of the material, while the samples with calcium bentonite showed a 5% increase in its weight, which was attributed to the absorption of leachate. Nussbaumer (1987) also reported similar results based on his test data (Table 9.2). Ryan (1987) noted that a leachate from a sanitary landfill, which contained several organic chemicals (phenolics, phenol, acetone, benzene, toluene, etc.), with none having concentrations more than 75 ppm, was incompatible with different bentonites and yielded unacceptable values of permeability. The aforementioned findings indicate that, if the individual chemicals do not interact with sodium bentonite, there is no guarantee that they will not affect it adversely when more than one of them are present in the aqueous permeant.

9.8 Microorganisms

Microorganisms have the beneficial effect of attenuation. They may produce certain organic acids which can react with trace contaminants, oxidize or reduce certain metals and chemicals, and deposit them in the soil voids thereby retarding the rate of flow. If the development of microorganisms in the field is not a possibility they should not be allowed to develop in the laboratory tests either. A disinfectant may be used for this purpose.

9.9 Volume change

Changes in soil volume occur as a result of changes in the clay structure. The volume changes exhibited by the smectite group are much greater than those for the kaolinite or illite group. These volume changes may not be visually observable. Its effect on the soil mass may be the development of cracks, fissures, joints, change in pore space distribution, swelling, etc. After these changes, the clay may have a greater or lesser permeability than before.

Organic cations may destroy the swelling capacity of smectite. Grim (1968) reported a decrease in the water absorption capacity of smectite due to coating of interlayer spaces by the organics. The water adsorption capacity decreased with increasing size of the organic ion. Volume change may also result from changes in effective stress. When a hydraulic gradient is applied the effective stresses change. In a constant head permeameter once the soil has reached equilibrium under a given hydraulic gradient there is no further change in the effective stresses. In the falling head permeameter, however, effective stress changes with head variations.

9.10 Expansion index and permeability

The advantage of the expansion index over the liquid limit was pointed out in an earlier section. Observe in Figure 9.14 that the expansion indices show a linear relationship with hydraulic conductivity which increases as the expansion index decreases. These results are in agreement with the Gouy-Chapman theory, which depicts contraction of the double layer as the primary reason for the higher magnitude of permeability in a clay mass.

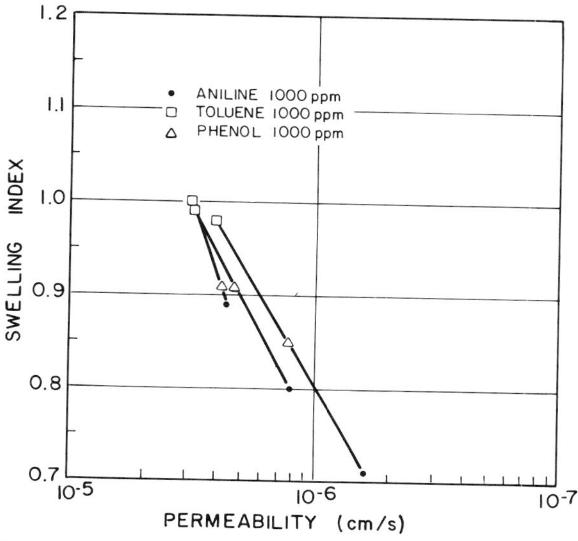

Figure 9.14 Expansion index and permeability for soil with 10% chemically treated bentonite (Reproduced from Khera *et al.*, 1987, by permission)

9.11 Soil fluid compatibility

It is recommended that the effect of a given permeant on the permeability of the soil be investigated. If actual samples of the permeant are not available, similar samples from other sites, or specially prepared samples with the expected constituents, can be used.

In determining soil–permeant compatibility a test must be continued until the permeability has essentially become constant. This may be assumed to occur when the slope of a regression line for permeability *versus* the time plot does not vary appreciably from zero at the 95% confidence level. The passage of unit pore volume of liquid should be considered as minimum for compatibility studies (Peirce, 1984). If the effluent shows turbidity a greater pore volume may have to be displaced so that the effect of dissolution can be fully evaluated. For bentonite 20–50 pore volume displacements have been suggested (Hughes, 1977).

An alternative approach is to pretreat the bentonite–soil mixture with the waste liquid and then perform the permeability test (Hermanns *et al.*, 1987; Khera *et al.*, 1987). Currently, this is practised for bentonite admix lining for brine ponds.

9.12 Use of laboratory tests

1. Use the results of laboratory tests for comparing the performance of different types of materials which are possible candidates for use in liners.
2. Do not use field moisture content and energy input values that are less than the laboratory values.
3. If a liner is to be put to service before it is completely hydrated, allow only partial hydration of the laboratory specimens.

4. When estimating the field permeability from laboratory tests, use the largest laboratory values. For other test conditions, such as magnitude of hydraulic gradient, type of permeameter, etc., follow the recommendations described in the previous sections.
5. Use carefully obtained undisturbed tube samples and block samples for permeability checking.

9.13 Field permeability

Laboratory permeability values are often lower than those obtained in the field. This results in considerable underestimation of the leakage in the field. For natural soils the ratio of field permeability to laboratory permeability has been reported to be as high as 46 000, with the vast majority of them larger than one except some which were less than one (Olson and Daniel, 1981).

The coefficients of permeability for laboratory and field compacted soils are given in Table 9.3. Permeability of the laboratory prepared samples and the undisturbed samples is several orders of magnitude smaller than that in the field, if the liner thickness is small. However, for thicker liner there is a good agreement between the laboratory, the undisturbed samples, and the field values.

Some of the factors contributing to higher permeability are:

1. The inhomogeneous nature of the compacted soil. Small zones of high permeability govern the overall permeability.
2. Disparity in soil structure between the field and the laboratory samples due to differences in the method of compaction.
3. Formation of desiccation cracks, if compacted soil is not protected.
4. Disturbance of the soil samples obtained from the field, including smearing of the surfaces.
5. Considerably higher confining pressures used in the laboratory than that existing in the field.
6. Use of much higher gradients in the laboratory than in the field.

Methods for rectifying some of these difficulties are described in other chapters.

Table 9.3 Effect of test method on hydraulic conductivity

Test type	k (cm/s)		
	Soil 1	*Soil 2*	*Soil 3*
Laboratory compacted sample	1×10^{-8}	2×10^{-9}	1×10^{-9}
Field permeability test	5×10^{-6} [a,b]	3×10^{-6} [a,b]	
Full-scale test	9×10^{-6} [a]	4×10^{-6} [a]	2×10^{-8} [c,d] 3×10^{-9}
Laboratory tests on hand-carved and tube samples	1×10^{-8} [a]	1×10^{-8} [a]	1.1×10^{-8} [c] -5×10^{-9}

[a] 6 in (150 mm) liner thickness, [b] single and double ring infiltrometers 12–44 in diameter, [c] 3 ft 11 in (1.2 m) liner thickness, [d] 15 m square lysimeters.
Sources: Day and Daniel, 1985; Reades *et al.*, 1987.

9.14 Hydraulic properties of municipal refuse

Laboratory measurement of hydraulic conductivity of municipal refuse is not a routine step in an investigation programme. Estimates are usually made based on local experiences and published data. Field pumping tests may be necessary for a proper determination of the hydraulic properties of refuse. Laboratory data on baled refuse (Fang, 1983) indicate hydraulic conductivity of 2 ft/day (7×10^{-4} cm/s) for dense specimens (unit weight of 71 lb/ft^3) to 15×10^{-3} cm/s (42.6 ft/day) for loose specimens (unit weight of 35.8 lb/ft^3). Intermediate values of 5×10^{-3} cm/s (13.6 ft/day) and 3.5×10^{-3} cm/s (10 ft/day) were measured for unit weights of 49 lb/ft^3 and 52.2 lb/ft^3, respectively. Laboratory data on shredded refuse (Fungaroli and Steiner, 1979) suggest hydraulic conductivity in the range of 10^{-2} to 10^{-4} cm/s. In the absence of field measurement a value of 10^{-3} cm/s is appropriate (Oweis *et al.*, 1990).

Leachate build-up occurs in landfills with no underdrain system and where the hydraulic conductivity of the underlying formation is substantially smaller than that of refuse. For refuse on an impervious base, the leachate build-up can be estimated using Equation 7.9.

The leachate build-up may be measured in the field and knowing the value of percolation (using for example the water balance method or actual data), the hydraulic conductivity of refuse, k_r, can be determined from Equation 7.9. This technique was applied to a landfill in the Hackensack Meadows in Northern New Jersey (Oweis and Khera, 1986). The back calculated k_r value was 2.6×10^{-3} cm/s. A field pumping test at the same landfill yielded a k_r value of about 10^{-3} cm/s. The critical parameter in removing leachate by pumping from refuse appears to be the specific yield or effective porosity (ratio of the volume of drainable leachate to the total volume of the refuse). The pumping test referred to indicated a projected specific yield of about 10%, which is at the low end of the range for fine sands. Figure 9.15 shows a distance–drawdown curve for the pumping test. The pumping rate was 12 gal/min (45 l/min).

The falling head test can be used for estimating the hydraulic conductivity of refuse. The values derived are, however, largely affected by the local composition. A high concentration of plastics yields apparently lower hydraulic conductivity.

Because of the heterogeneous nature of refuse and its random channels, values of hydraulic conductivity are more difficult to determine than in soils composed of aggregate. The common assumption, for example, is that leachate is generated only after refuse reaches its field capacity. In some cases, however, leachate is generated before the landfill as a whole reaches its field capacity, and not all the leachate flows to a well designed surface drainage system. Leachate has been observed to spring from the sides of a landfill despite the presence of a functional drain at the toe. The hydraulic conductivity produced by side channelling or 'fingering' is commonly known as 'secondary permeability'.

The moisture content at field capacity is less than the saturation water content. Therefore, initially moisture transport occurs under unsaturated conditions. Above the field capacity the hydraulic conductivity is of the same order of magnitude as at saturation.

9.15 Hydraulic properties of mineral waste

The coefficient of permeability of coarse coal waste can vary over a wide range depending on the variability in particle size distribution and unit weight *in situ*.

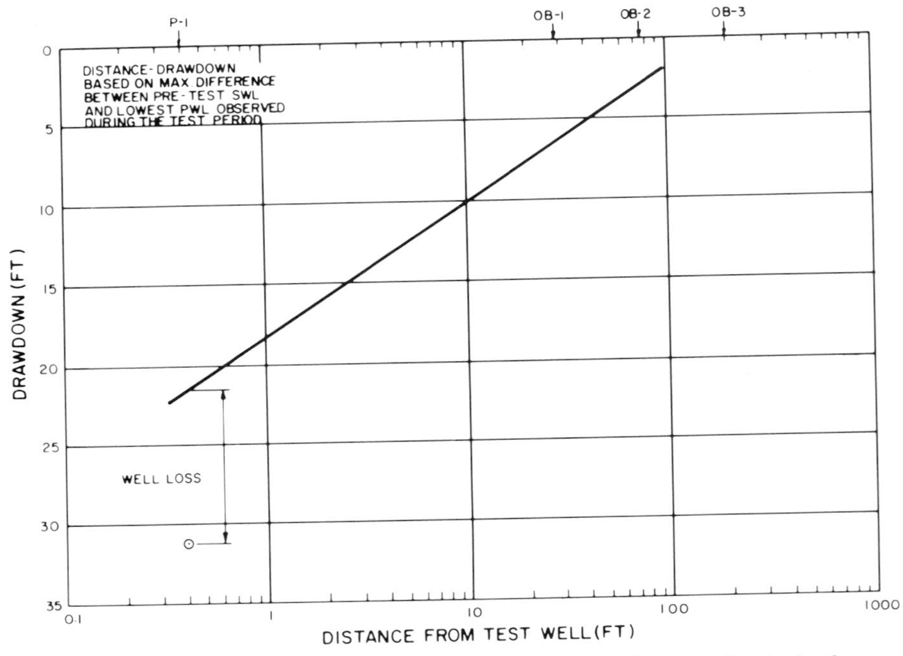

Figure 9.15 Distance–drawdown data. SWL, static water level; PWL, pumped water level.

Reported laboratory values have ranged between 10^{-3} and 10^{-6} cm/s. For fresh uncompacted waste it may be as high as 10^{-1} cm/s and for well compacted waste which has undergone weathering it may be as low as 10^{-8} cm/s. The hydraulic conductivity for fine coal waste (tailings) has been reported between 10^{-3} and 10^{-7} cm/s (Holubec, 1976).

The coefficient of permeability of fly ash ranges between 10^{-4} and 10^{-7} cm/s. McLaren and DiGioia (1987) reported a mean value for hydraulic conductivity of class F fly ash to be 1.32×10^{-5} cm/s and that of class C fly ash as 1.13×10^{-5} cm/s. The value of permeability decreases with time and increasing density. The effect of compacted unit weight on permeability is shown in Figure 9.16. If lime or cement is used as a stabilizing agent, the effect of curing time on permeability should also be investigated as permeability has been reported to be affected by curing time (Bowders and Daniel, 1987). The hydraulic conductivity values for bottom ash are much higher and range between 5×10^{-3} and 10^{-1} cm/s. At disposal sites of fly ash, bottom ash is frequently used as drainage material.

The hydraulic conductivity of FGD sludges varies between 5×10^{-3} and 5×10^{-6} cm/s. Permeability of 10^{-9} cm/s was reported for sludge from a double alkali scrubbing system containing 30 and 50% fly ash (Krizek *et al.*, 1976). The lower values are more common to sulphite sludges and the higher to sulphate-rich sludges. When the sludge is mixed with fly ash its hydraulic conductivity increases but with increased degree of compaction the hydraulic conductivity decreases. A few selected data points plotted in Figure 9.16 show a general decreasing trend in permeability with increasing unit weight (Krizek *et al.*, 1987). With the addition of lime and passage of time there may be some tendency toward a decrease in permeability.

Pulp and paper mill sludges have a permeability range of 10^{-7} to 10^{-8} cm/s. The addition of 10% fly ash increased its permeability by one to two orders of magnitude at low pressure but with increasing effective stress the permeability decreased rapidly (Andersland and Mathew, 1973).

Because of the great variability in waste materials composition and other factors such as the density, moulding moisture content, method of disposal, age, etc., it is imperative that site-specific laboratory and field tests be conducted for proper evaluation of permeability of waste materials.

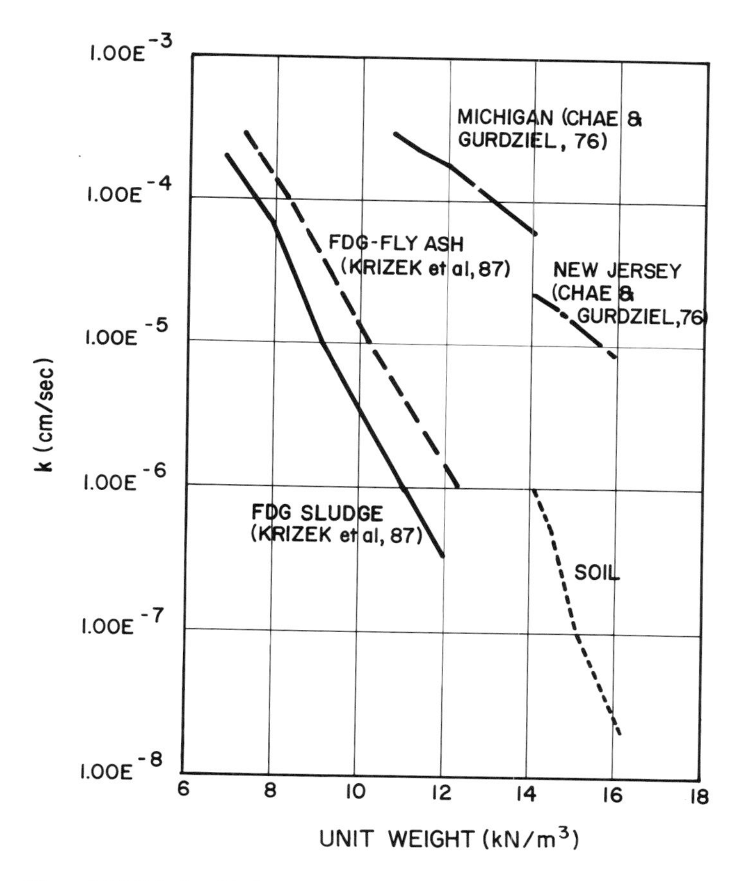

Figure 9.16 Relationship between permeability and unit weight.

Summary

Field methods of measuring permeability are more appropriate for inhomogeneous deposits. The laboratory methods consist of a fixed wall permeameter, double wall permeameter, flexible wall permeameter, consolidation ring permeameter, and constant rate of flow permeameter. Although there is no standardized testing method, the flexible wall method is most commonly used. Back pressure is used to ensure complete saturation. The hydraulic gradient and the effective stress may have a significant effect on the test results.

Sodium bentonite is susceptible to changes when interfacing with waste chemicals. In most instances the aqueous solutions have less adverse effect on soil–bentonite admixtures than the concentrated chemicals. Where several organics were mixed the permeability was higher even at low concentrations. Water, which has the highest dielectric constant, yields the lowest values for permeability and the permeability increases with decreasing dielectric constant. The results reported in the literature are often contradictory and no theory is available which can predict the response of a soil to various chemicals. Soil expansion or sedimentation tests may be used for assessing the soil–waste compatibility using the actual leachate or laboratory constituted leachate. Where permeability tests are performed, a base line value is determined with 0.01N $CaSO_4$ solution or tap water.

Permeability values for various waste types are given. Because of the large range of these values site-specific tests will often be necessary.

Notation

A	total cross-sectional area of soil
dp	change in pressure
du	change in pore pressure
$d\sigma_3$	change in chamber pressure
h	total head loss
h_1	head at the beginning of flow
h_2	head at the end of flow period
i	hydraulic gradient
K	intrinsic permeability
k	coefficient of permeability or hydraulic conductivity
L	distance in which the head loss occurs
ρ	density
q	rate of flow
Q	total amount of flow
t	total time of flow
u	pore pressure
γ_w	unit weight of water
σ_3	total chamber pressure
σ_3'	effective consolidation stress
μ	dynamic viscosity

References

Acar, Y. B. and D'Hollosy, E. (1987) Assessment of pore fluid effects using flexible wall and consolidation permeameters, *Geotechnical Practice for Waste Disposal '87*, Proc. Spec. Conf. Geot. Special Publ. No. 13, Ed. Woods, R. D., pp. 231–245

Acar, Y. B., Hamidon, A., Field, S. D. and Scott, L. (1985) The effect of organic fluids on hydraulic conductivity of compacted kaolinite, *Hydraulic Barriers in Soil and Rock*, Eds Johnson, A. I., Frobel, R. K., Cavalli W. J. and Pettersson, C. B. ASTM, STP 874, pp. 171–187

Alther, G., Evans, J. C., Fang, H. and Witmer, K. (1985) Influence of inorganic permeants upon the permeability of bentonite. In *Hydraulic Barriers in Soil and Rock*, Eds Johnson, A. I., Frobel, R. K., Cavalli, N. J. and Pettersson, C. B. ASTM, STP 874, pp. 64–74

Anderson, D. (1982) Does landfill leachate make clay liners more permeable?, *ASCE, Civil Engineering*, September, 66–68

Bjerrum, L. (1972) *Embankments on Soft Grounds*, Proc. of the Specialty Conf., Performance of Earth and Earth-Supported Structures, Purdue University, June 1972, Vol. 2, pp. 1–45

Bowders, J. J. and Daniel, D. D. (1987) Hydraulic conductivity of compacted clay to dilute organic chemicals, *ASCE, Journal of Geotechnical Engineering*, **113,** No. 12, 1432–1448

Brown, K. W. and Anderson, D. C. (1983) *Project Summary: Effect of Organic Solvents on the Permeability of Clay Soils*, EPA-600/S2-83-061

Brown, K. W., Thomas, J. C. and Green, J. W. (1983) The influence of selected organic liquids on the permeability of clay liners, *Land Dsposal of Hazardous Waste: Proc. 9th Annual Research Symposium*, Fort Mitchell, Kansas, EPA-600/9-83-018

Chae, Y. S. and Gurdziel, T. J. (1976) New Jersey fly ash as structural fill, *New Horizons in Construction Materials*, Vol. 1, Ed. Fang, H. Y., Envo Publishing Co. Inc., pp. 1–13

D'Appolonia, D. J. (1980) Soil bentonite slurry trench cutoffs, *ASCE, Journal of the Geotechnical Engineering Division*, **106,** No. GT4, 399–417

Daniel, D. E., Trautwein, S. J., Boynton, S. S. and Foremen, D. E. (1984) Permeability testing with flexible-wall permeameters, *ASTM, Geotechnical Testing Journal*, **7,** No. 3, 113–122

Davis, K. E., Herring, M. C. and Hosea, J. T. (1985) *Slurry Wall Materials Evaluation to Prevent Groundwater Contamination from Organic Constituents*, EPA/600/9-85/025, pp. 289–302

Day, S. R. and Daniel, D. E. (1985) Hydraulic conductivity of two prototype clay liners, *ASCE, Journal of Geotechnical Engineering*, **111,** No. 8, 957–970

Dunn, R. J. and Mitchell, J. K. (1984) Fluid conductivity testing of fine-grained soils, *ASCE, Journal of Geotechnical Engineering*, **110,** No. 10, 1648–1665

Edil, T. B. and Erickson, A. E. (1985) Procedure and equipment factors affecting permeability testing of a bentonite–sand liner material, *Hydraulic Barriers in Soil and Rock*, Eds Johnson, A. I., Frobel, R. K., Cavalli, N. J. and Pettersson, C. B. ASTM, STP 874, pp. 155–170

EPA/530-SW-86-007 (1986) *Design, Construction, and Evaluation of Clay Liners for Waste Management Facilities*, 663pp

Evans, J. C., Kugelman, I. J. and Fang, H. Y. (1985) Organic fluid effects on strength, deformation and permeability of soil–bentonite slurry walls, *Toxic and Hazardous Wastes, Proc. 7th Mid-Atlantic Industrial Waste Conf.*, Ed. Kugelman, I. J., Lehigh University, pp. 275–291

Fang, H. Y. (1983) *Physical Properties of Compacted Disposal Materials*, Unpublished report

Fernandez, F. and Quigley, R. M. (1985) Hydraulic conductivity of natural clays permeated with simple liquid hydrocarbons, *Canadian Geotechnical Journal*, **22,** No. 2, 205–214

Foreman, D. E. and Daniel, D. E. (1986) Permeation of compacted clay with organic chemicals, *ASCE, Journal of Geotechnical Engineering*, **112,** No. 7, 669–681

Freeze, R. A. and Cherry, J. A. (1979) *Groundwater*, Prentice-Hall, Englewood Cliffs, New Jersey

Fungaroli, A. A. and Steiner, R. L. (1979) *Investigations of Sanitary Landfill Behavior*, Final Report, USEPA 600/2-79-053, Vol. 1, p. 314

Grim, R. E. (1968) *Clay Mineralogy*, 2nd Edn, McGraw-Hill, New York

Hadge, W. and Barvenik, M. J. (1985) Upgrading soil bentonite cutoff wall technology for containment of hazardous waste, presented at *Geotechnical Aspects of Waste Management*, Met Sec, ASCE, December 1985

Hermanns, R., Meseck, H. and Reuter, E. (1987) Sind Dichtwandmassen beständig gegenüber den Sickerwässern aus Altlasten? Mitteilung des Instituts für Grundbau und Bodenmechanik, TU Braunschweig, Heft Nr. 23, Ed. Meseck, H., Dichtwände und Dichtsohlen, Braunschweig, Federal Republic of Germany, pp. 113–154

Holubec, I. (1976) Geotechnical aspects of coal waste embankments, *Canadian Geotechnical Journal*, **13,** No. 1, 27–39

Hughes, J. (1977) A method for the evaluation of bentonites as soil sealants for the control of highly contaminated industrial wastes, *Purdue Industrial Waste Conference*, pp. 814–819

Khera, R. P., Wu, Y. H. and Umer, M. K. (1986) *Durability of Slurry Cut-Off Walls around the Hazardous Waste Sites*, Progress Report, NSF Industry/University Cooperative Research Center, NJIT, Newark, New Jersey

Khera, R. P., Wu, Y. H. and Umer, M. K. (1987) Durability of slurry cut-off walls around the hazardous waste sites, *Proc. 2nd Int. Conf. on New Frontiers for Hazardous Waste Management*, August 1987, EPA/600/9-87/018F, pp. 433–440

Khera, R. P., Thilliyar, M. and Moradia, H. (1988) Cracking of backfill materials in soil bentonite walls, *Report SITE-13, NSF Industry/University Cooperative Research Center*, New Jersey Commission on Science and Technology, Advanced Technology Center, NJIT, Newark, New Jersey

Korfiatis, G. P., Demetracopoulos, A. C. and Schuring, J. R. (1986) Laboratory testing for permeability and dispersivity of cohesive soils, *International Symposium on Environmental Geotechnology*, Vol. 1, Ed. Fang, H. Y., pp. 363–369

Krizek, R. J., Giger, M. W. and Legatski, L. K. (1976) Engineering properties of sulfur dioxide scrubber sludge with fly ash. In *New Horizons in Construction Materials*, Vol. 1, Ed. Fang, H. Y., Envo Publishing Co. Inc., pp. 67–81

Lentz, R. W., Horst, W. D. and Uppot, J. O. (1985) The permeability of clays to acidic and caustic permeants, *Hydraulic Barriers in Soil and Rock*, Eds Johnson, A. I., Frobel, R. K., Cavalli, N. J. and Pettersson, C. B., ASTM, STP 874 pp. 127–139

McLaren, R. J. and DiGioia, A. M. (1987) The typical engineering properties of fly ash, *Geotechnical Practice for Waste Disposal '87*, Proc. Spec. Conf. Geot. Special Publ. No. 13, Ed. Woods, R. D.

Mitchell, J. K. (1976) *Fundamentals of Soil Behavior*, John Wiley & Sons, New York

Mitchell, J. K. and Madsen, F. T. (1987) Chemical effects on clay hydraulic conductivity, *Geotechnical Practice for Waste Disposal '87*, Proc. Spec. Conf. Geot. Special Publ. No. 13, Ed. Woods, R. D., pp. 87–116

Mitchell, J. K., Hooper, D. N. and Campanella, R. G. (1965) Permeability of compacted clay, *ASCE, Journal of Soil Mechanics and Foundations Division*, **91**, No. S4, Part 1, 41–66

Nussbaumer, M. (1987) Beispiele für die Herstellung von Dichtwänden im Schlitzwandverfahren, *Mitteilung des Instituts für Grundbau und Bodenmechanik*, TU Braunschweig, Heft Nr. 23, Ed. Meseck, H., Dichtwände und Dichtsohlen, Braunschweig, Federal Republic of Germany, pp. 21–34

Olson, R. E. and Daniel, D. E. (1981) Measurement of the hydraulic conductivity of fine-grained soils, *ASTM, Permeability and Groundwater Contaminant Transport*, STP 746, 18–64

Olsen, H. W., Morin, R. H. and Nichols, R. W. (1985) Flow pump application in triaxial testing, *Advanced Triaxial Testing of Soil and Rock*, Eds Donaghe, R. T., Chaney, R. C. and Silver, M. L. ASTM, STP 977, pp. 68–81

Oweis, I. A. and Khera, R. (1986) Criteria for geotechnical construction on sanitary landfills, *Int. Symp. on Environmental Geotechnology*, Vol. 1, Ed. Fang, H. Y., pp. 205–222

Oweis, I. S., Smith, D., Ellwood, B. and Greene, D. (1990) Hydraulic properties of refuse, *Journal of the Geotechnical Engineering Division, ASCE*, April

Peck, R. B., Hanson, W. E. and Thornburn, T. H. (1974) *Foundation Engineering*, 2nd Edn, John Wiley & Sons, New York

Peirce, J. J. (1984) *Effects of Inorganic Leachates on Clay Permeability*, Final Report, USEPA, Contract No. 68-03-3149, 24-1, Department of Civil and Environmental Engineering, Duke University

Reades, D. W., Poland, R. J., Kelly, G. and King, S. (1987) Discussion: hydraulic conductivity of two prototype clay liners, *ASCE, Journal of Geotechnical Engineering*, **113,** No. 7, 809–813

Rizkallah, V. (1987) Geotechnical properties of polluted dredged material, *Geotechnical Practice for Waste Disposal '87*, Proc. Spec. Conf. Geot. Special Publ. No. 13, Ed. Woods, R. D., pp. 759–771

Ryan, C. R. (1987) Vertical barriers in soil for pollution containment, *Geotechnical Practice for Waste Disposal '87*, Proc. Special Conf. Geot. Special Publ. No. 13, Ed. Woods, R. D., pp. 182–204

Terzaghi, K. (1943) *Theoretical Soil Mechanics*, John Wiley & Sons, New York

Vallee, R. P. and Andersland, O. B. (1974) Field consolidation of high ash paper mill sludge, *ASCE, Journal of the Geotechnical Engineering Division*, **100,** No. GT3, 309–328

Yong, R. N. (1986) Selected leaching effects on some mechanical properties of sensitive clay, *Int. Symp. on Environmental Geotechnology*, Vol. 1, pp. 349–362

Zimmie,T. F., Doynow, J. S. and Wardell, J. T. (1981) Permeability testing of soils for hazardous waste disposal sites, *Proc. of the Tenth ICSMFE*, Stockholm, Vol. 2, pp. 403–406

Chapter 10

Ground modification and compaction

The conditions at a given site may not always provide proper support for the foundations. It may, however, be possible to improve the properties of the subsurface materials. This is done by compaction, dynamic compaction, preloading or precompression, drainage, reinforcing, use of additives, grouting, replacement of the poor materials, etc. The method selected would depend on the type of materials to be treated, the available time for construction, the tolerable settlement of the structure, etc. Some of the methods for stabilization of weak materials are discussed here. For additional discussion see Mitchell (1981).

10.1 Compaction

Compaction of soils or waste materials increases their density and strength, and reduces their permeability. The loose materials are placed in lifts varying in thickness from a few inches to a maximum of about 3 ft in the case of garbage.

For design and control of construction both laboratory and field tests are conducted. For heterogeneous materials, such as those found in a municipal landfill, reliable laboratory tests are difficult to design and conduct. Although field tests are expensive and time consuming, they yield more reliable data.

10.1.1 Coarse-grained soils

For clean coarse-grained soils with less than 12% fines the extent of compaction is measured in terms of its relative density:

$$D_r = \frac{100(e_{max} - e)}{e_{max} - e_{min}} \tag{10.1}$$

where e_{max} is the void ratio of the soil in its loosest state, e_{min} the void ratio of the soil in its densest state, and e the actual void ratio.

In terms of unit weights the relative density is given by:

$$D_r = \frac{\gamma_{max}(\gamma - \gamma_{min})}{\gamma(\gamma_{max} - \gamma_{min})} \tag{10.2}$$

where γ_{max} is the unit weight of soil in its densest state, γ_{min} the unit weight of the soil in its loosest state, and γ the actual unit weight of the soil.

For details of the testing method see ASTM D 2049.

10.1.2 Mixed soils

In the standard compaction test (ASTM D-698) the soil is compacted in a mould of volume 1/30 ft^3 in three layers with each layer receiving 25 blows of a 5.5 lb hammer falling 12 in. This method is preferred for control of compaction for clay liners. An alternative method (ASTM D-1557) uses five layers, 25 blows, a 10 lb hammer and an 18 in drop. The degree of compaction is affected by type of soil, energy input, and water content.

To determine the compaction characteristics of a given soil, it is compacted at various water contents using the same compactive effort. As shown in Figure 10.1, the typical plot of dry density and water content is a paraboloid. Other shapes of curves are also reported (Winterkorn and Fang, 1975). Water content corresponding to the maximum dry density is known as the optimum moisture content. Both the optimum moisture content and the maximum dry density are dependent on the energy input. The zero-air-void curve (Figure 10.1a) represents the upper limit of the soil density at any given moisture content and serves as a check on the test data.

An increase in energy input increases the maximum dry density but decreases the optimum moisture content. The relative increase in dry density is greater for soils of higher plasticity (Monahan, 1986). Thus poor control in compaction operation during the construction of a liner, which of design must contain materials of high plasticity, may result in considerable variation in its permeability.

10.1.3 Compaction effect on permeability

Effect of compaction moisture content on soil permeability is shown in Figure 10.1b. An increase in the compaction moisture content results in a decrease in the soil permeability. This decrease is attributed to the soil structure. Both the moisture content and the method of compaction affect the soil structure. The tendency toward more dispersed structure increases with increasing moisture content and increasing soil shearing during compaction. The kneading action of a plunger type of compactor or a sheepsfoot roller results in greater soil shearing. For water content greater than the optimum (Figure 10.1b) the kneading compaction causes a further decrease in permeability. The permeability of a soil compacted wet of optimum may be more than three orders of magnitude lower than that compacted dry of optimum (Mitchell, 1976). Soils with high plasticity have a thicker double layer, possess greater dispersivity, and exhibit lower permeability.

10.1.4 Compaction effect on shear strength

A soil compacted dry of optimum fails at lower strain and exhibits higher shear strength than if compacted wet of optimum. If during subsequent saturation the soil is permitted to swell then the strain at failure is large regardless of the compaction moisture content. However, upon saturation the strength of a soil compacted dry of optimum may be lower than that of a soil compacted wet of optimum.

In a compacted earthen liner, there is little restraining pressure before placement of the waste, thus allowing the soil to swell as it becomes saturated. The soil must be compacted wet of optimum firstly to obtain lower permeability and secondly to reduce the detrimental effect on soil strength from swelling during saturation. Also, higher compaction moisture content allows the soil to undergo larger deformation without rupturing. If the foundation soils of a liner undergo differential settlements,

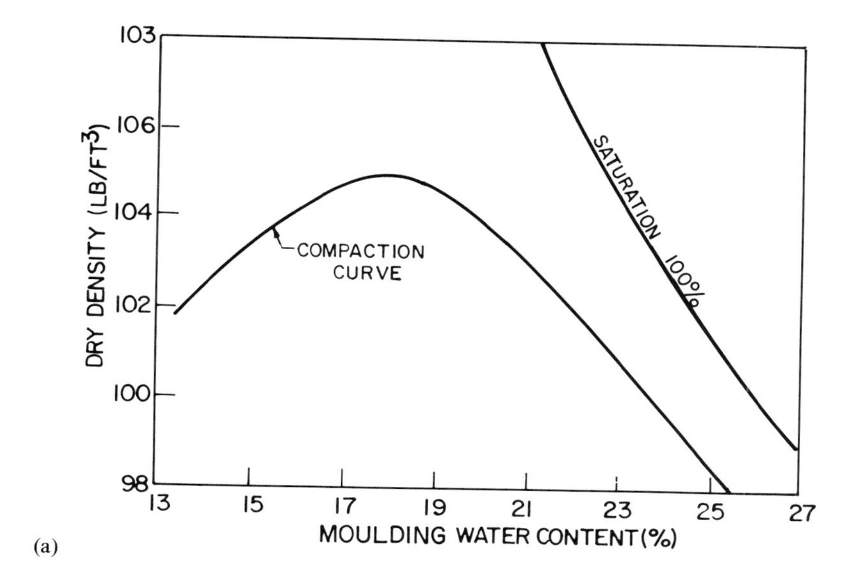

(a)

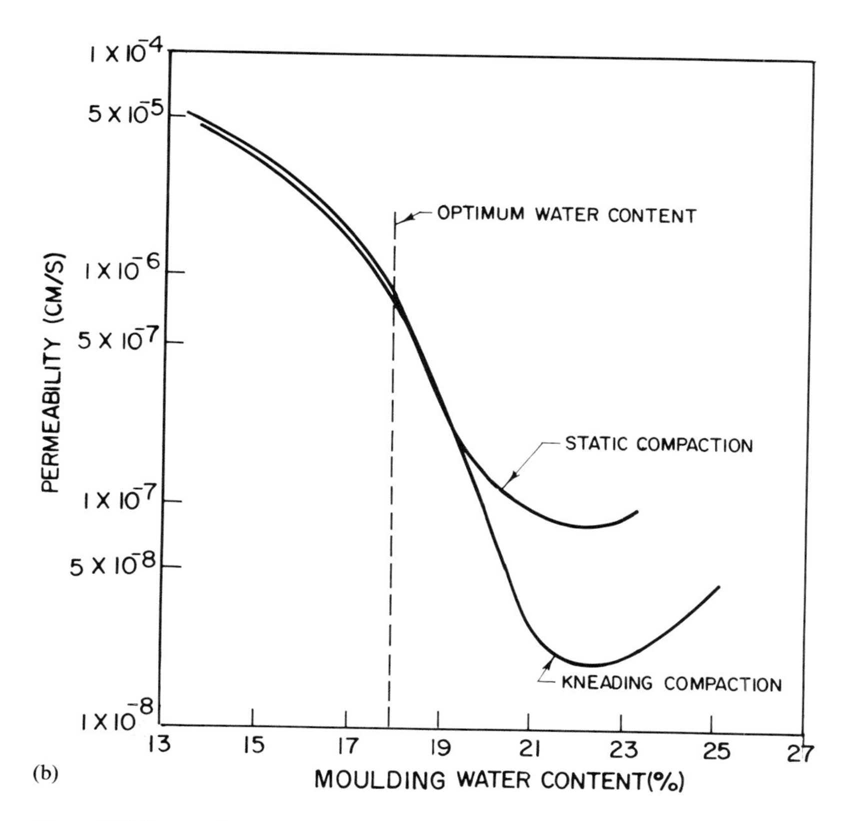

(b)

Figure 10.1 Compaction curve and effect on permeability (Reproduced from Mitchell, 1976, by permission of John Wiley & Sons Inc.)

damage to the liner will be less if it was compacted wet of optimum. Moisture content about 2–3% above the optimum value is maintained during field compaction.

10.1.5 Compaction equipment

Many types of field equipment are available for compaction (Table 10.1). In selecting the equipment consideration should be given to the soil type, compaction moisture conditions, and the intended use of the compacted soil.

10.1.6 Compaction of waste products

Compaction has proved to be an effective means of reducing the volume of waste, increasing the life of disposal sites, and improving its engineering properties. Compaction curves for some mineral and industrial wastes are given in Figure 10.2. For a given material there would be a range of values rather than a single curve. The curves shown for fly ash indicate a variation in the maximum dry unit weight and optimum moisture content to be more than 30% and 100%, respectively.

10.1.6.1 Ashes

For fly ash the maximum dry unit weight ranges between 74.4 and 127.5 lb/ft^3 (11.7 and 20 kN/m^3) and the optimum moisture content 9–50%. The average dry unit weight and optimum moisture content for class F fly ash are about 82.8 lb/ft^3 (13 kN/m^3) and 25%, respectively. The corresponding values for class C fly ash are 94.2 lb/ft^3 (14.8 kN/m^3) and 20%. In compaction tests on fresh fly ash utilize a new sample for each test as the recompacted fly ash behaves quite differently. The dry unit weight increases with increased relative density of the fly ash. The presence of unburned carbon reduces its relative density, unit weight and strength.

Usually the higher unit weights are associated with lower optimum moisture content. Some compaction characteristics of fly ash are given in Table 10.2.

Heavy pneumatic rollers are effective in compacting 0.3 m thick lifts with about six passes per lift at a water content below the optimum value. For fresh fly ash compaction must be carried out immediately after placement because of the rapid commencement of pozzolanic action. Borex has been used as a retardant to mitigate the difficulty associated with rapid hydration (Manz and Manz, 1985). Considerable scatter in the field compaction data has been reported (Leonards and Bailey, 1982).

For bottom ash the maximum dry density may range between 68.8 and 117.8 lb/ft^3 (10.8–18.5 kN/m^3) with w_{opt} from 13 to 26%. The moisture density curve is relatively flat and in some instances the mositure content has no effect on the compacted unit weight, indicating a behaviour similar to that of coarse-grained soils. Maximum and minimum relative densities for boiler slag, which also behaves as granular material, are in the range of 91–110 lb/ft^3 (13.2–16 kN/m^3) and 71–88 lb/ft^3 (10.3–12.8 kN/m^3), respectively (DiGioia *et al.*, 1977).

For incinerator residue the maximum dry density ranges between 82.8 and 108.2 lb/ft^3 (13–17 kN/m^3) and the optimum water content is 15–26%.

10.1.6.2 Coal mine waste

The dry density of mine rock and coarse coal refuse from coal mining operations may range from 12.5 to 21 kN/m^3 and the optimum moisture content is 6–14%. Because of the varied nature of the coarse refuse materials, considerable variations will exist from

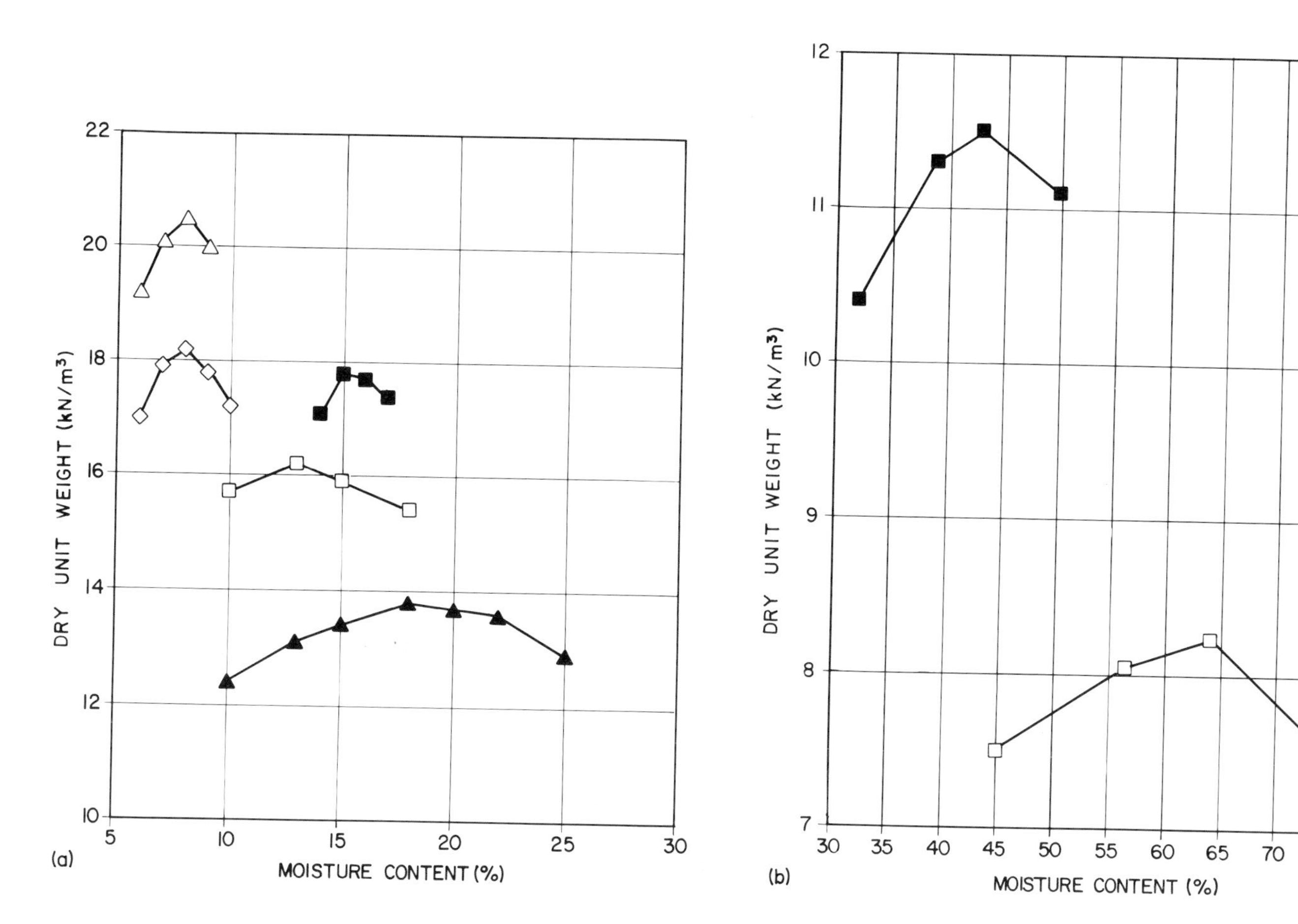

Figure 10.2 Compaction tests on waste: (a) ■, fly ash, western USA; □, fly ash, New Jersey; ◇, coal waste, eastern USA; △, coal waste, western USA; ▲, fly ash, Pennsylvania; (b) ■, FGD sludge; □, paper solid waste

Table 10.1 Compaction equipment and methods

Equipment type	Applicability	Requirements for compaction of 95–100% standard Proctor maximum density: Compacted life thickness (in)	Passes or coverages	Dimensions and weight of equipment	Possible variations in equipment
Sheepsfoot rollers	For fine-grained soils or coarse-grained soils with more than 20% passing No. 200 sieve. Not suitable for clean coarse-grained soils. Particularly appropriate for compaction of soil liners	6	4–6 passes for fine-grained soil 6–8 passes for coarse-grained soil	*Soil type* / *Foot contact area* (ft²) / *Foot contact pressures* (psi): Fine-grained soil PI > 30 / 5–12 / 250–500 Fine-grained soil PI < 30 / 7–14 / 200–400 Coarse-grained soil / 10–14 / 150–250 Efficient compaction of soils wet of optimum requires less contact pressure than the same soils at lower moisture contents	For highway and airfield work, articulated self-propelled rollers are commonly used. For smaller projects, towed 40–60 in drums are used. Foot contact pressure should be regulated to avoid shearing the soil on the third or fourth pass
Rubber tire roller	For clean, coarse-grained soils with 4–8% passing the No. 200 sieve	10	3–5 coverages	Tire inflation pressures of 35–130 psi for clean granular material or base course and subgrade compaction. Wheel load 18 000–25 000 lb	Wide variety of rubber tire compaction equipment available. For cohesive soils, light-wheel loads, such as provided by wobble-wheel equipment, may be substituted for heavy-wheel load if lift thickness is decreased. For granular soils, large-size tires are desirable to avoid shear and rutting
	For fly ash, fine-grained soils or well graded, coarse-grained soils with more than 8% passing the No. 200 sieve	6–8	4–6 coverages	Tire inflation pressures in excess of 65 psi for fine-grained soils of high plasticity. For uniform clean sands or silty fine sands, use large size tires with pressures of 40–50 psi	

Smooth wheel rollers	Appropriate for subgrade or base course compaction of well graded sand–gravel mixtures, coarse and mixed coal mine waste	8–12	4 coverages	Tandem type rollers for base course or subgrade compaction 10–15 ton weight, 300–500 lb/lineal in of width of rear roller	Three-wheel rollers obtainable in wide range of sizes. Two-wheel tandem rollers are available in the range of 1–20 ton weight. Three-axle tandem rollers are generally used in the range of 10–20 ton weight. Very heavy rollers are used for proof rolling of subgrade or base course
	May be used for fine-grained soils other than in earth dams. Not suitable for clean well graded sands or silty uniform sands	6–8	6 coverages	Three-wheel roller for compaction of fine-grained soil; weights from 5–6 ton for materials of low plasticity to 10 ton for materials of high plasticity	
Vibrating sheetsfoot rollers	For coarse-grained soils, sand–gravel mixtures	8–12	3–5	1–20 ton ballasted weight. Dynamic force up to 20 ton	May have either fixed or variable cyclic frequency
Vibrating smooth drum rollers	For coarse-grained soils, sand–gravel mixtures, rock fills	6–12 (soil) to 36 (rock)	3–5 4–6	1–20 ton ballasted weight. Dynamic force up to 20 ton	May have either fixed or variable cyclic frequency
Vibrating baseplate compactors	For coarse-grained soils with less than about 12% passing No. 200 sieve. Best suited for materials with 4–8% passing No. 200 sieve, placed thoroughly wet	8–10	3 coverages	Single pads or plates should weigh no less than 200 lb. May be used in tandem where working space is available. For clean coarse-grained soil, vibration frequency should be no less than 1600 cycles/min	Vibrating pads or plates are available, hand-propelled, single or in gangs, with width of coverage from $1\frac{1}{2}$ to 15 ft. Various types of vibrating-drum equipment should be considered for compaction in large areas
Crawler tractor	Best suited for coarse-grained soils with less than 4–8% passing No. 200 sieve, placed thoroughly wet	6–10	3–4 coverages	Vehicle with Standard tracks having contact pressure not less than 10 psi	Tractor weight up to 85 ton
Power tamper or rammer	For difficult access, trench backfill. Suitable for all inorganic soils	4–6 in for silt or clay, 6 in for coarse-grained soils	2 coverages	30 lb minimum weight. Considerable range is tolerable, depending on materials and conditions	Weights up to 250 lb, foot diameter 4–10 in

test to test. During handling and disposal, and from the mechanical action of compaction equipment, considerable particle breakdown occurs. In design, while selecting strength parameters, the effect of both physical breakdown and chemical weathering must be considered. Some compaction characteristics of coal waste are given in Table 10.3

Saxena *et al.* (1984) showed that the type and size of compaction equipment and the thickness of lift had considerable influence on the strength and permeability of compacted coal refuse. Smooth drum, vibratory, and vibratory sheepsfoot rollers were all found to be effective, with the smooth roller performing the best. Materials compacted with a smooth drum roller had the highest shear strength, lowest measured field permeability and a relatively impermeable surface. The recommended lift thickness is 0.3 m or less, and the compaction moisture content is near the optimum value.

As shown in Figure 10.2, coal waste shows a clearly defined maximum dry density and optimum moisture content. At most disposal sites, the moisture content is not controlled during compaction. In the absence of proper compaction, upon saturation, the materials become soft and compressible. Furthermore, coal mine waste contains a considerable proportion of combustible materials. To reduce the danger of rapid ignition and sustained combustion, and to obtain higher strength and lower permeability, proper management of compaction is essential. Also, from an environmental and economical view point it makes sense to control compaction as the water seeping through the waste is acidic, requiring costly treatment.

Selected mine stone waste has been used in the UK to construct liners and covers at disposal sites for municipal, commercial and inert wastes. A hydraulic conductivity of 10^{-5} cm/s is considered acceptable at such sites (Rainbow and Nutting, 1986).

10.1.6.3 Flue gas desulphurization sludge

The maximum dry density of FGD ranges from 70 to 100 lb/ft^3 (11–16 kN/m^3), and the optimum water content from 45 to 15%. The sulphite sludges are at the lower end of the density and the higher end of the optimum water content ranges. The reverse is true of gypsum or sulphate sludge.

Because of the high water content of the processed sludge the field dry density is less than the laboratory values and ranges between 5 and 10 kN/m^3 (31.8–63.7 lb/ft^3). At these water contents application of energy larger than 20–40% of the standard compaction value is non-productive. The laboratory tests show an increase in optimum moisture content and a decrease in dry density if dry fly ash, lime or soil is added to the sludge. In the field this advantage has not been realized fully because of the difficulty of obtaining a homogeneous mix (Krizek *et al.*, 1977; Ullrich and Hagerty, 1987).

10.1.6.4 Sanitary landfill

To minimize settlement-related problems in a landfill, articles such as tires and washing machines must be removed. The sorted materials are more easily and uniformly compacted, reducing the risk of larger sudden and long-term settlements. It is usually necessary to place a few inches to a few feet of thick granular fill on the refuse before compaction, or it may not be possible to operate the equipment. A thicker blanket serves as a rigid mat for the placement of the foundation and reduces the amount of differential settlement.

Table 10.2 Compaction of fly ash (modified compaction)

Source	*Optimum moisture* (%)	*Maximum dry*[a] *unit weight* (kN/m³)	*Relative density*
Arizona (low sulphur)	22.9	14.2[a]	2.35
Michigan	32–20	11.7–14.6	2.36–2.61
New Jersey	13.6	16.2	2.54
Pennsylvania	31–19	12.0–14.0	2.29–2.59
West Virginia Shelby tube samples	33–14	11.2–17.5	2.20–2.50
	—	14.9–16.0	2.26–2.37
Wisconsin	28–19	12.2–14.9	2.33–2.44
Wyoming (low sulphur)	9	18.5	2.75

[a]Standard compaction.
Sources: Chae and Gurdziel, 1976; Moulton *et al.*, 1976; Parker *et al.*, 1977; Srinivasan *et al.*, 1977.

Table 10.3 Compaction of coal waste

Source	*Optimum moisture* (%)	*Maximum dry unit weight* (kN/m³)	*Relative density*
Coarse waste			
Appalachian	21–9	14.5–18.9	1.75–2.5
UK	16–4	14.8–20.4	1.8–2.7
Fine waste			
Appalachian	20–18	10.2–11.8	1.4–1.66

Sources: Chen *et al.*, 1976; Holubec, 1976; Saxena *et al.*, 1984; Usmen, 1986.

Volume reductions of 2–17% have been reported from compaction. Excess pore pressures develop in saturated landfill materials from the weight of the roller, making it difficult to achieve any degree of compaction. Most of the reduction in volume occurs in five passes. Figure 10.3 illustrates the effect of compaction effort.

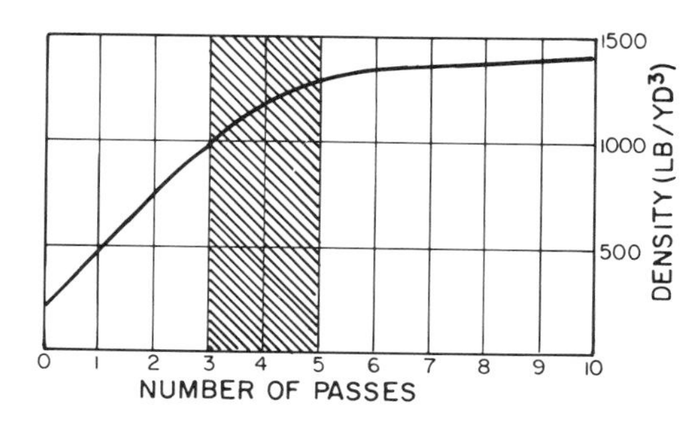

Figure 10.3 Effect of number of passes on density (Reproduced from Shoemaker, 1972, by permission)

10.1.7 Compaction of clay liners

10.1.7.1 Laboratory control

Design field density and compaction moisture content are based on laboratory tests. Since moisture content has a dominant effect on permeability, the standard compaction (ASTM D-698) test is more frequently used as the design basis since it yields an optimum moisture content higher than the modified compaction test (ASTM D-1557).

Although permeability is the governing criterion, in instances where strength and compressibility are of concern appropriate tests and testing methods must be used to evaluate the relevant parameters as described elsewhere.

10.1.7.2 Field requirements

Most specifications are written as performance specifications. The maximum dry density (usually 95% of the maximum dry density) and a hydraulic conductivity of 10^{-7} cm/s or less is generally specified.

A much higher degree of quality control is called for in the compaction of clay liners compared with the compaction requirements for routine foundations. A test fill may be used to select the soil, equipment and method of construction. In order to assure proper performance of a clay liner the following should be considered:

1. Inspect every aspect of the construction.
2. Ensure that the site provides a proper foundation for the waste and the liner and is not susceptible to failure or excessive differential settlement, which may be detrimental to the integrity of the liner. This may require replacement of pockets of soft materials and pervious materials with compacted fill, and sealing of cracks and joints, and any other correction measures that may be warranted.
3. Check the borrow area to ensure that there is an adequate supply of the material selected for the liner. If the variability of the available material is not within the acceptable range, other sources may have to be investigated.
4. Protect the materials stored at the site so that the fines are not washed out and the loss of moisture is minimum. The moisture loss is especially important for soils that exhibit reduced Atterberg limits on drying.
5. Use 2–3% higher moisture content than the laboratory optimum value. If water must be added it is usually sprinkled at the borrow area. When needed, additional water may be sprinkled during spreading and mixing operations. Before placement the soil must be thoroughly mixed and allowed sufficient time for uniform distribution of moisture. If the soil is too wet it is dried by aeration and drying in the sun. Compaction may not be permitted when the soil is frozen or after periods of heavy rain.
6. Maintain the lift thickness closely to design values. Tie in the lifts together by scarifying the underlying lifts before placing the next lift. Rollers with long tampers produce a better connection between successive layers.
7. If liner installation is to be done in segments, bevel or step the older segment to join them properly with the new ones. Where compaction is to be carried out on the side walls, more care is required in obtaining the proper connection at the joints.
8. Use the specified amount of compaction effort along with the specified type of roller. For thicker liners (>2 ft), frequently the type of roller used is determined by the ease of its availability. Liners less than 3 ft thick are not recommended.

9. Break down large clods of clay sufficiently. Depending on the type of equipment and the size of the lift the acceptable clod size may be 1–2 in for hand compactors or one-quarter to one-third the thickness of the lift for rollers.
10. If adding bentonite to the native soil, its powder form may produce a more homogeneous mix than the granular form. Mixing may be done in the field or in a cental plant, the latter being preferable. When a central plant is used water is added after soil and bentonite have been mixed for a couple of minutes. Mixing is continued for another 10 min or more to produce a homogeneous material.
11. Do not permit the compacted fill to dry. Proper protection against desiccation may be provided by covering the fill with soil, flexible membrane, or other suitable material.
12. Perform field tests to ensure that the desired density values have been achieved. It may be necessary to perform field permeability tests. This may be done by using a single ring or double ring infiltrometer.

The number of verification tests to be performed could be assessed based on probabilistic techniques (Spigolon and Kelly, 1984; Harrop-Williams, 1987). The tests include laboratory and field density tests, laboratory hydraulic conductivity tests on undisturbed samples, field permeability tests, and certain index properties tests. A key to these procedures is the estimation of the coefficient of variation for the property under study. Consider, for example, that it is desired to determine the number of field density tests for each lift of liner material placed daily. The size of the lift (the block) is chosen such that the testing does not impede construction, and the locations to be tested are selected at random. The average number of tests is determined based on (Spigolon and Kelly, 1984):

$$n_u = (tv'/E)^2 \tag{10.3}$$

where n_u is the number of units in the sample, t is the probability factor from the t-distribution tables based on confidence level and sample size (a tabulation for large samples (size >30) is given in Table 10.4), E is the allowable sampling error of the expected mean, and v' is the coefficient of variation (standard deviation/mean), the known or estimated value of the block.

A block is an isolated quantity of the same composition and produced by essentially the same process. Its size is much greater than that of a sample (covered by Equation 10.3) taken from it. In practice, the coefficient of variation, v', is established based on experience with similar installation of the same materials. If v' is not known use a small sample to estimate its value, then determine the value of n_u from Equation 10.3. If the sample used was less than n_u, additional samples are called for. A range of

Table 10.4 *t*-factors for large samples

t-*factor*	*Probability of exceeding* E *significance*	
	One-sided tale	*Two-sided tale*
3	0.0013	0.0026
2.575	0.005	0.01
2.32	0.010	0.020
2.0	0.023	0.046
1.96	0.025	0.050
1.645	0.05	0.10

Table 10.5 Coefficient of variation for soil tests

Test type (%)	*Range of v′ value (%)*	*Suggested*
Angle of friction		
Sand, fly ash	5–10	10
Coefficient of consolidation	25–100	50
Maximum dry unit weight		
Soils	1–7	5
Fly ash	10–17	15
Liquid limit	2–48	10
Optimum moisture content	11–40	
Clay soils		20
Sandy gravelly soils, fly ashes		40
Plastic limit	9–29	30
Plasticity index	7–79	
Clay soils		30
Sandy gravelly soils		70
Slurry wall		
Slurry density	1–7	5
Backfill moisture content	17–20	20
Backfill, percentage fines	15–17	20
Relative density	5–25	10
Standard penetration test	27–85	30
Unconfined compressive strength	6–100	40

Sources: Lee *et al.*, 1983; Bergstrom *et al.*, 1987; McLaren and DiGioia, 1987.

values of v' for various tests is given in Table 10.5. Most of the spread in the values of v' is a result of different methods of testing, differences in testing procedures, deviation from the specified testing procedure, different testing apparatus, different operators, etc. For better results strict control of these parameters cannot be overemphasized (Lumb, 1974).

Consider for example that v' for percentage compaction is 3% and it is desired that the error of the average of the values tested not vary by more than 2% of the mean for the block (say a 100 × 200 ft area) then the number of tests for 99% confidence (a probability of 10 in 1000 that the error will exceed E) is

$$n_u = (2.32 \times 3/2)^2 = 12$$

The coefficient of variation for the hydraulic conductivity could be large (50% or even larger). Assuming E to be 15% of the mean and for 95% confidence (a probability of 50 in 1000 that E will exceed 15%),

$$n_u = (1.645 \times 50/15)^2 = 30$$

These assessments are sensitive to the value estimated for the coefficient of variation and other statistical parameters. The number of tests for quality control is usually regulated or based on experience with successful installations. The New Jersey Department of Environmental Protection requires hydraulic conductivity testing on undisturbed core samples of the clay liner from locations at 200 ft grid. A field infiltration test using a double ring infiltrometer is required for each 10 acres. Field density and moisture content are required at 50 ft grid. Clay stockpiles should be tested (one per 16 000 yd^3) for a particular site for particle size distribution, plasticity and hydraulic conductivity. These requirements, especially for field density, are somewhat excessive but perhaps provide the basis for establishing the statistical basis for less stringent requirements at later stages in the project.

Further details on compaction equipment, quality control and quality assurance may be found in EPA (1986) and Hilf (1975).

10.2 Dynamic compaction

The basic concepts of dynamic compaction (also known as dynamic consolidation, impact densification, or heavy tamping), as it is used today, were presented by Ménard and Broise (1975). The method consists of dropping heavy blocks of steel, concrete, or thick shells filled with concrete or sand weighing 5–20 ton, from heights of up to 30 m (100 ft). Drop weights as much as 180 ton have been used. The imprints of these blocks may be square, circular or octagonal. The blocks are lifted by cranes with a capacity up to 25 ton (23 tonne) and 100 ft (30 m) drop, or by tripods which have a higher load capacity and a greater drop height. Tripods permit a free drop which is more efficient than a crane drop. The applied energy in most instances is between 60 and 300 ft kip/ft^2 (100–400 tonne m/m^2).

The materials treated by dynamic compaction exhibit a higher bearing capacity and lower post-construction settlement. It is most suited for loose coarse-grained soils, rubble fills, and non-hazardous landfills. Materials below the water table cannot be treated very efficiently.

10.2.1 Drop weight treatment

The area to be treated is divided into grid patterns with each grid point receiving several blows in a given pass. Several passes may be necessary to obtain the desired results. Field studies are conducted for grid spacings, energy level of blows, number of blows per pass, number of passes, etc. To evaluate the extent of ground improvements, measurements are taken for drop weight imprints, heave, average settlement, changes in soil properties, pore water pressure, ground vibration, etc.

The initial grid spacing is generally greater than the thickness of the layer to be improved and the energy per blow is higher so that the compaction energy will influence the entire depth of the stratum. The high energy drops may form 1–2 m (3–6 ft) deep imprints. The most increase in crater depth occurs in the first half a dozen drops. The energy input per unit area determines the total amount of settlement. At the end of each pass the imprints are filled and the ground is levelled using the material heaved around them. The amount of compression at the end of each pass is determined from topographic maps prepared after each pass. The final pass is of low energy and is called 'ironing'. For the ironing pass a weight with a square footprint is used. The top 0–1 m (0–3 ft) of the materials may still not be compacted properly even with the ironing pass and may thus require the use of standard compaction equipment for densification. At sites where the upper materials are very soft, a working mat 1–2 m (3–6 ft) thick consisting of granular materials is provided to facilitate treatment (Charles *et al.*, 1981).

10.2.2 Depth of influence

The maximum depth to which a soil is influenced, D_{max} (m), is estimated by:

$$D_{max} = I(WH)^{0.5} \quad (10.4)$$

where I is the influence factor, W the weight of the dropping block in metric tonnes, and H the drop height in metres.

Ménard and Broise (1975) used 1 for I. Other suggested values are 0.5 (Leonards *et al.*, 1980), 0.65–0.80 (Lukas, 1980) and 0.3–0.8 (Mayne *et al.*, 1984). An I value of 0.5 appears to be most widely accepted at this time. Factors that influence I are soil type, initial soil density, depth to groundwater table, area of falling block, grid spacing, number and sequence of drops, number of passes, time elapsed between successive passes, method of determining ground improvement, definition of what denotes improvement, etc.

The ground improvement is determined by comparing the values before and after the treatment using cone penetration resistance, standard penetration resistance, vane shear strength, pressure meter modulus and limiting pressure, etc. In heterogeneous materials, such as municipal landfills, large-scale plate load tests may be more appropriate (Baker, 1982).

10.2.3 Construction vibrations

The impact from the weight causes transient vibrations in the ground. The guideline for estimating possible damage to adjacent structures is based on peak particle velocity, V_p, which depends on the site conditions and the scaled energy factor $(WH)^{0.5}/d$, where WH is the energy and d the horizontal distance from the impact point. Figure 10.4 shows the relationship between particle velocity and scaled energy. This may be used as a guide for determining the safe energy limits. Most data from sand, rubble and silty soils lie in a narrow range (Mayne *et al.*, 1984). A particle velocity of 50 mm/s (2 in/s) is considered the threshold of damage to residential structures (Wiss, 1981).

10.2.4 Dynamic compaction of municipal landfill

Charles *et al.* (1981) used dynamic compaction on a 15-year-old and up to 6 m (20 ft) thick municipal landfill. A 15-tonne weight with a base area of 4 m^2 was dropped ten times at each location from heights up to 20 m (66 ft). The primary grid spacing was 5 m (16 ft). There were one or two additional tamping stages at grid points offset from the original grid with the maximum energy input of 2600 kN m/m^2. High excess pore pressures were observed in some instances of dynamic compaction. The average settlement was 0.5 m (20 in). Under identical embankment loads the immediate settlement of the untreated refuse was three times larger than that of the treated refuse. About 5 years later the long-term settlement of the treated section was about 35 mm (1.4 in).

Welsh (1983) cites a roadway site with 20–40 ft (6–12 m) of refuse with 3 ft (1 m) cover. Stabilization was accomplished in three passes by using 18 ton weight, 92 ft drop, 10 drops per location, and at 15 ft (4.5 m) centres. Settlement of the treated area after 7 days was one-twentieth of the settlement of the untreated areas. D'Appolonia (1978) cites a case for a commercial structure in Rouen, France, where 20–30 ft of refuse were stabilized by two passes of dynamic compaction producing 2.3 ft (0.70 m) to over 4.9 ft (1.5 m) settlements. The modulus (E) for the treated area as obtained from the pressuremeter tests was at least double that of the untreated refuse.

Ménard (1984, unpublished) cites a case for a warehouse designed with floor loads of 400 lb/ft^2 and spread footing with 3 kip/ft^2. Refuse 20 ft to 55 ft thick was stabilized with three to six passes and with applied energy of 36–75 ton ft/ft^2. The pressuremeter

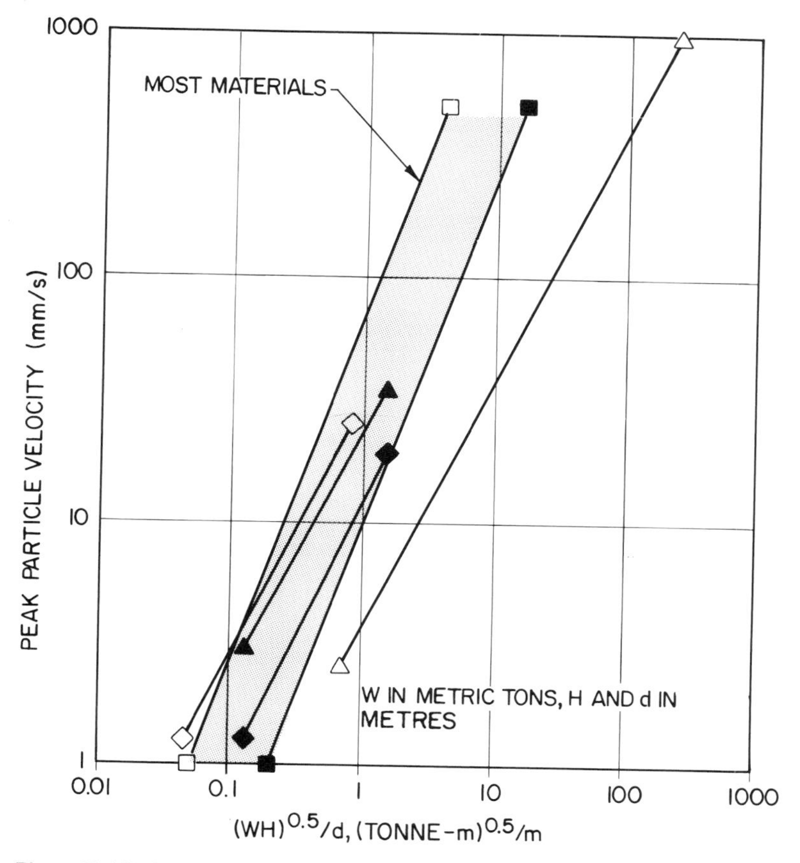

Figure 10.4 Scaled energy factor *versus* particle velocity: ■, most data; □, upper limit; ◇, rubble; △, wet sand; ◆, decomposed garbage; ▲, loose sand (Data from: ■, □, Mayne *et al.*, 1984; ◇, △, ◆, Lukas, 1980; ▲, Leonards *et al.*, 1980)

modulus E increased by a factor of 5 near the top and 2.3 at 21 ft depth. Load tests after treatment showed settlement of 0.4–0.5 in under 3 kip/ft^2 stress. For a railroad site where dynamic compaction was used to homogenize the fill up to six passes were necessary. Drop heights of 30–60 ft were used with weights of 9–15 ton.

Because of the substantial densification of refuse, the dynamic compaction would be expected to reduce the rate of decomposition and secondary compression due to the reduction of the surface area. It appears that proper tamping could perhaps reduce the primary settlement by 70% and the secondary settlement by 50%. The dynamic compaction process introduces high excess pore water pressure. To accelerate consolidation, use vertical drains in conjunction with dynamic compaction, especially in that zone of footings supporting the columns. In employing densification by dynamic compaction, the possibility of contaminants migrating to the surface during densification should be evaluated and environmental impact assessed.

10.3 Preloading

Precompression is an effective and economical means of improving weak and compressible materials. Precompression is commonly done through the use of earth fill or water as the surcharge. A drainage layer is placed on top of the compressible material before placing the earth fill on it. In water loading the area is surrounded by an embankment, the drainage blanket is covered with an impermeable membrane, which is then filled with water to provide the surcharge load. Under the surcharge load the underlying weak soils decrease in volume and gain in strength. Subsequently, when the actual loads are applied the resulting settlements are considerably smaller. Surcharging beyond the design load can further reduce post-construction settlement. Other methods of precompression include groundwater lowering, electro-osmosis, and application of vacuum beneath the ground surface covered with impermeable membranes (Holtz and Wagner, 1975).

10.3.1 Factors affecting settlement rate

In a soil not fully saturated, surcharging causes rapid settlement due to compression and dissolution of gases. As pore pressure decreases during consolidation the dissolved gases come out of the solution and expand. This reduces the rate of consolidation (Johnson, 1970a).

For loaded areas large in width compared with the thickness of the compressible layer, the strain distribution is fairly constant with depth. Otherwise, the strain decreases with depth. To determine the rate of consolidation more accurately, assume the soils to be composed of several layers, each with its own constant c_v value. Use Equations 5.18 and 5.19 to determine the equivalent thickness of the compressible layer. A decrease in strain with depth will increase the settlement rate.

For a relatively narrow surcharge load, the presence of thin seams of sand or silt will help drain the soil rapidly. However, at the edge of the loaded area high pore water pressure may develop in the permeable layers resulting in considerable decrease in their shear strength. To ensure safety against instability it may be necessary to provide berm or relieving wells at the toe of the slope. For a wide load, sand or silt layers may have little positive effect on the rate of settlement but the negative effect on stability may be present.

10.4 Surcharge design

In designing a surcharge system, the magnitude of the surcharge-induced stresses and their duration should be such that the post-surcharge settlements are within the tolerable limits of the proposed structures.

10.4.1 Compensation for primary consolidation

The relationship between surcharge load, settlements and time is shown in Figure 10.5. The average degree of consolidation is:

$$U_{f+s} = \delta_{cf}/\delta_{c(f+s)} \tag{10.5}$$

where δ_{cf} is the total primary consolidation settlement under the permanent load, and

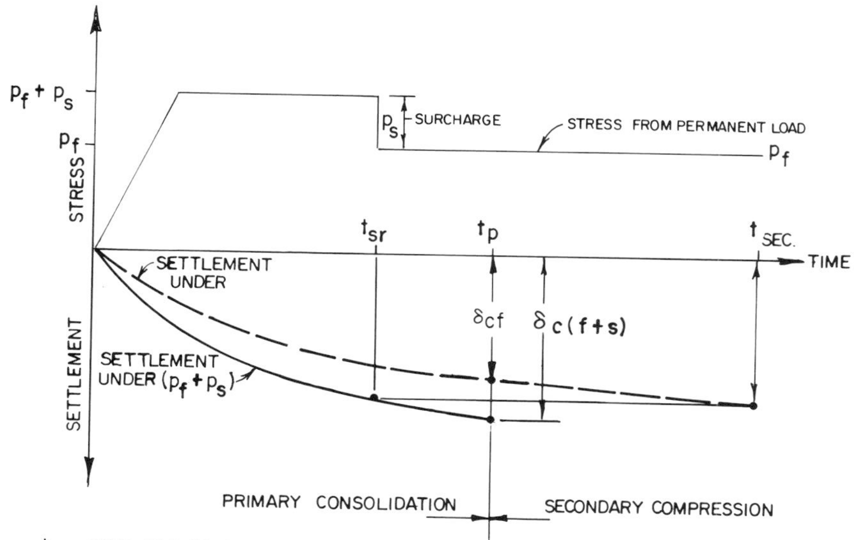

Figure 10.5 Settlement compounds under surcharge load (Reproduced from NAVFAC, 1982)

$\delta_{c(f+s)}$ is the total primary consolidation settlement under the permanent load plus surcharge load.

For normally consolidated soil of unit thickness (Equation 5.11c)

$$\delta_{cf} = \frac{C_c}{(1+e_0)} \log\left(\frac{\sigma'_{vo} + p_f}{\sigma'_{vo}}\right) \tag{10.6}$$

and

$$\delta_{c(f+s)} = \frac{C_c}{(1+e_0)} \log\left(\frac{\sigma'_{vo} + p_f + p_s}{\sigma'_{vo}}\right) \tag{10.7}$$

from which the degree of consolidation is given by:

$$U_{f+s} = \frac{\log(1 + p_f/\sigma'_{vo})}{\log\{1 + (p_f/\sigma'_{vo})[1 + (p_s/p_f)]\}} \tag{10.8}$$

The distribution of excess pore pressures through the depth of the compressible layer from the load is shown in Figure 10.6. At this time the consolidation ratio is given by (from Equation 5.16):

$$U_{z(f+s)} = 1 - u_e/u_i$$

$$= p_f/(p_f + p_s) \tag{10.9}$$

Assuming that for a given soil p_f/σ'_{vo} is 0.5, p_s/p_f is 1.0, and the computed value of U_{f+s} from Equation 10.8 is 0.59. From Figure 5.10 for U_{f+s} of 0.59 the consolidation ratio, U_z, is about 0.4. If p_s is removed at this time, the effective stress in the middle region would still be less than $(\sigma'_{vo}+p_f)$. The primary consolidation will continue after the removal of the surcharge (Figure 10.6 curve 1). To prevent further primary consolidation the surcharge should not be removed until after the excess pore pressure at this critical plane has dissipated. From Figure 5.9 it is seen that for $U_z=0.5$ the average degree of consolidation is about 0.64 and not 0.59.

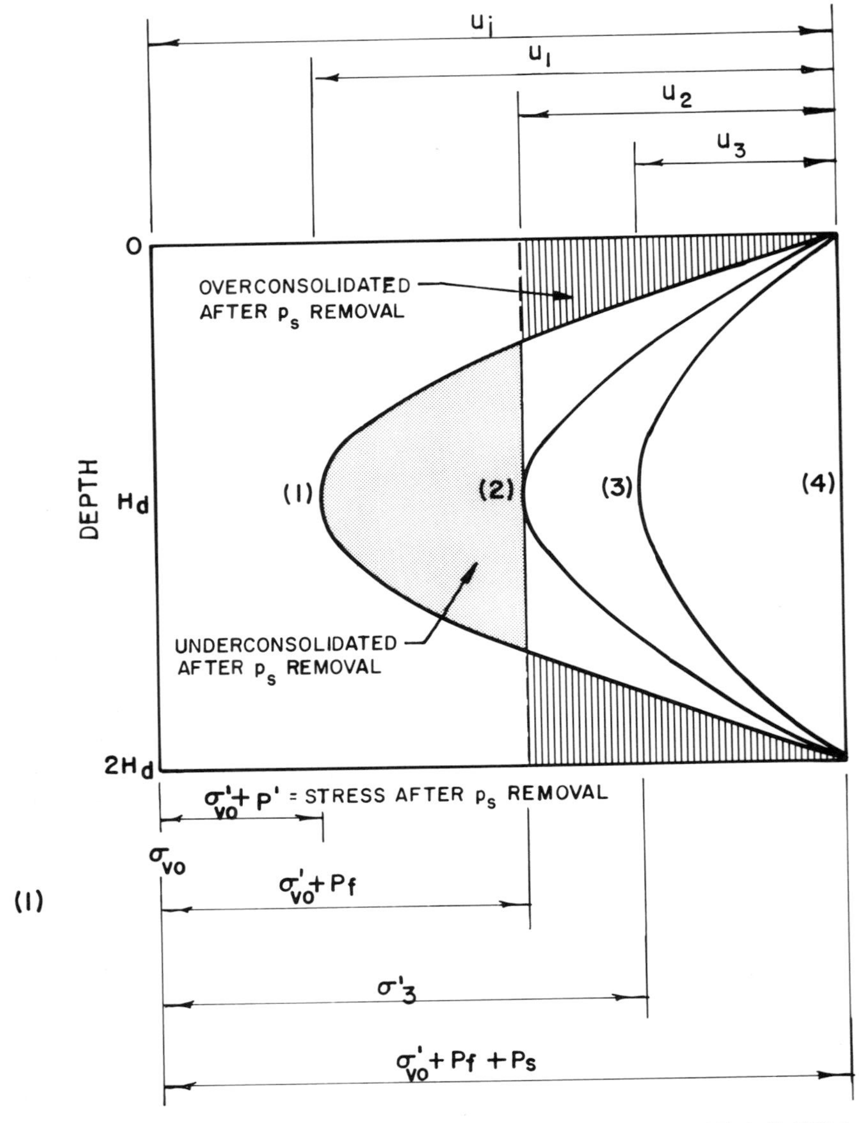

Figure 10.6 Excess pore pressure during surcharge stages (Reproduced from Mitchell, 1981, by permission)

10.4.2 Compensation for secondary compression

To compensate for primary and some of the secondary compression resulting from the permanent load, the surcharge removal should be at the time when the settlement has reached a value given by:

$$\delta_{(c,s)f} = \delta_{cf} + \delta_{sf} \tag{10.10}$$

where δ_{sf} is the computed settlement due to secondary compression under permanent load.

The consolidation ratio for this condition may be represented by curve 3 in Figure 10.6. At this stage the primary consolidation under permanent and surcharge load has not yet completed and there may be some excess pore water pressure in certain critical regions. However, this is of no consequence.

The average degree of consolidation required at the time of surcharge removal is:

$$U_{f+s} = \delta_{(c,s)f}/\delta_{c(f+s)} \tag{10.11}$$

$$= (\delta_{cf} + \delta_{sf})/\delta_{c(f+s)} \tag{10.12}$$

$$= \{\delta_{cf} + C_\alpha[1 - \delta_{cf}][\log(t_{sec}/t_p)]\}\delta_{c(f+s)} \tag{10.13}$$

where t_{sec} is the time for the completion of secondary compression under p_f, and t_p is the time for completion of primary consolidation under p_f.

From Equations 10.7, 10.8 and 10.13:

$$U_{f+s} = \frac{[1 - C_\alpha \log(t_{sec}/t_p)][\log(1 + p_f/\sigma'_{vo})] + (C_\alpha/C_c)(1 + e_0)\log(t_{sec}/t_p)}{\log[1 + p_f/\sigma'_{vo} \times (1 + p_s/p_f)]} \tag{10.14}$$

In this equation t_{sec}/t_p is usually taken as 10. The procedure for determining the value of p_s/p_f is the same as before.

10.4.3 Settlement after surcharge removal

Upon removal of the surcharge, the soil will show little or no secondary compression and may even exhibit some swelling. However, after some time the secondary compression does appear. As $(p_f + p_s)/\sigma'_{vo}$ increases C_α decreases and the time delay in reappearance of the secondary compression increases. Thereafter the secondary compression continues but at a lower rate depending upon the value of $(p_f + p_s)/\sigma'_{vo}$.

The amount of secondary compression may be determined (Johnson, 1970a) from:

$$\delta_{sec} = C_\alpha H_p \log(1 + dt/t_{sec}) \tag{10.15}$$

where dt is the time since the removal of surcharge load.

The length of time the surcharge load is left in place has an effect on the rate of secondary compression (Bjerrum, 1972). Figure 10.7 shows a series of pressure–void ratio curves each with a different duration of sustained load. If the precompression is carried out under the final load p_f, then the settlement from primary consolidation is represented by abc′f (Figure 10.7). Since some of the secondary compression occurs during the primary process, the actual settlement will occur along a curve abf (dashed curve). Surcharging under p_s for the same time period will cause total settlement along abd (shown dashed). Just before the removal of the surcharge the rate of secondary consolidation (point d) is about 0.5%/year. The reduction of stress from p_s to p_f moves

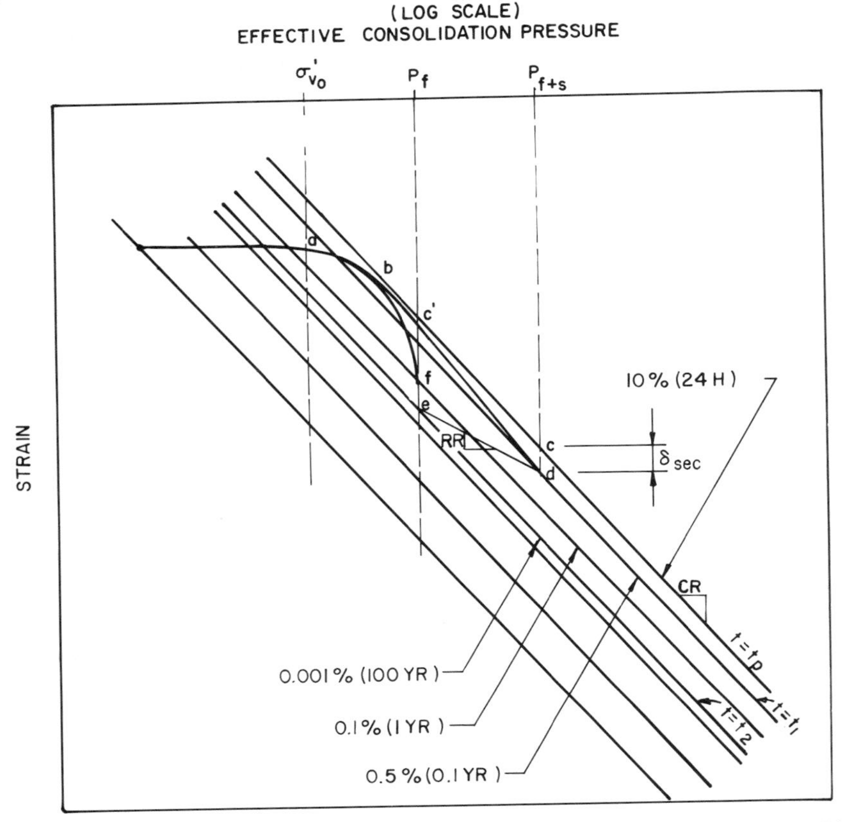

Figure 10.7 Surcharge design for reduction of secondary settlement (Reproduced from Bjerrum, 1972, by permission)

point d to point e. As seen in Figure 10.7 the corresponding rate of secondary consolidation is then only about 0.001%/year. In addition to the reduction of the rate of secondary compression, the surcharge produced 100 years of secondary compression (from point f to point e). Assuming the life of the structure is represented by point e, then in theory there should be no secondary compression. Such a conclusion is not valid for refuse since settlement would continue due to decomposition. For the surcharge to reduce the rate and magnitude of the secondary compression, it must remain in place for an adequate period of time so that the void ratio attained is lower than point f.

10.4.4 Construction control in preloading

A preloading job requires proper design, construction supervision, and field monitoring. Before placement of the surcharge, a layer of clean sand material (<3% passing a No. 200 sieve) is placed on the site. The drainage blanket may contain a perforated pipe to assist drainage of the sand. Where vertical drains are used there should be a proper hydraulic contact between the drainage blanket and the drains.

Since the soils being stabilized are of low shear strength, surcharge load is applied at a controlled rate allowing the soils to consolidate and possibly gain strength. The initial load increment is determined based on the field and laboratory test data for undrained conditions. The subsequent load increments are determined from consolidated undrained conditions. These values must be checked against the field piezometer and strength data. The strength data may be obtained from vane shear tests, cone penetration tests, pressuremeter tests, etc. The long-term factor of safety is based on consolidated drained conditions.

An appropriate safety factor should be used against shear failure. The stability analysis is based on an appropriate method such as a circular failure surface, sliding wedge or others (see Chapter 11). If the stability conditions cannot be satisfied, it may be necessary to provide berms requiring additional fill material.

10.4.5 Instrumentation

Field instruments are required in many cases to monitor settlements, rate of settlement, and lateral movement. Surface settlement plates and deeper settlement probes are installed so that both the surface and subsurface settlements can be determined. Groundwater profile and excess pore water pressure distribution are obtained from piezometers placed at selected depths (see Chapter 6 for details of instrumentation). Lateral movement and heave can be detected by using alignment stakes at varying distances from the toe of the embankment. Inclinometers at the edge of the surcharge can provide warning about potential stability problems. For a case history where inclinometers were used see Chapter 5.

10.5 Vertical drains

Vertical columns filled with sand have been successfully used to promote drainage and thus reduce the time for precompression (Figure 10.8). They can be installed by a displacement method where a hollow mandrel with a hinged lid at its base is driven into the ground. As the mandrel is extracted a mixture of sand and water is poured into it. The lid at the base opens and the sand forms a vertical drain. This method causes considerable remoulding of the soils and excess pore pressure. The non-displacement or 'jetted' drains, where high-pressure water jets are used to advance the pipe, do not have these shortcomings, but require large amounts of water.

In Europe, Japan and other countries geotextile drains have essentially replaced sand drains. In the USA their use started only in the last decade. The composite drains derive their strength from plastics and hydraulic conductivity from paper (more common) or polyester. Paper filters are of lower permeability than polyester filters. However, polyester filters have larger voids, which may become clogged by fine soil particles (Morrison, 1982). The fabric drains are the non-woven type, which are made from needle punched non-woven geotextiles. The thickness of these drains is between 2.5 and 5 mm and the width between 100 and 300 mm. Sandwich drains are made of coarse sand-filled geotextile casing of 60 mm diameter.

Synthetic drains when compared with sand drains are easier and faster to install, require generally no water during installation, and are much more economical. The drains are wound on a reel with the end attached to a lance which is advanced by vibratory hammers, rigs specifically constructed for such purpose, by hydraulic pressure, or by water jets if the soils are very sensitive.

Several factors must be considered in the design and installation of drain systems. Drains may be subjected to tensile forces during installation. The horizontal stresses from the soil will tend to compress the drain, dislodged soil particles will block the filters, and the remoulding of the soil during installation will reduce soil permeability. Each of these elements will have a tendency to reduce the rate of flow. Currently there are no methods available that can account for all these factors. Furthermore, there are many brands of wick drains, each using different materials and form of construction, adding to the number of variables influencing a drain system.

10.5.1 Design of vertical drains

The basic equation is based on Terzaghi's consolidation theory and since in the sand drains the flow is in a radial direction Equation 5.14 is written as:

$$\partial u_e/\partial t = c_h[\partial^2 u_e/\partial r^2 + (1/r)\partial u_e/\partial r] \tag{10.16}$$

where c_h is the coefficient of consolidation in the horizontal direction, r the radial distance from the centre of the drain, and u_e the excess pore water pressure.

It is assumed that each drain is a circular cylinder with a circular zone of influence from which it receives water. For a non-circular band drain an equivalent diameter is expressed as:

$$d_e = 2(B + t_d)/\pi \tag{10.17}$$

where d_e is the equivalent drain diameter (L), B the drain width (L), and t_d the drain thickness (L).

The equivalent diameter of the soil cylinder which is being drained is:

$D_e = 1.06L$ for triangular pattern

$D_e = 1.13L$ for square pattern

where L is the drain spacing (L).

For rigid surface load the vertical strain is equal and the solution is given by Equation 10.18a. Assuming no well resistance and disturbance during installation, the time of consolidation is given by Equation 10.18b:

$$U_h = 1 - e^{-(8T_h/\mu)} \tag{10.18a}$$

$$t = \frac{D_e^2 \mu}{8c_h} \ln \frac{1}{1 - U_h} \tag{10.18b}$$

where

$$T_h = c_h t / D_e^2$$

$$\mu = \frac{n^2}{n^2 - 1}\left[\ln(n) - 0.75 + \frac{1}{n^2}\left(1 - \frac{1}{4n^2 - 1}\right)\right] \tag{10.19}$$

$$\approx \ln(n) - 0.75 \qquad \text{for } n > 10$$

$$n = D_e/d_e$$

$$c_h = k_h/\Gamma_w m_v \quad (\mathrm{L^2/T})$$

and U_h is the average degree of consolidation considering horizontal flow only, k_h the coefficient of permeability in the horizontal direction (L/T), and m_v the coefficient of volume compressibility ($\mathrm{L^2/F}$).

For a flexible surface load (free strain) for $D_e/d_e > 5$ the same solution can be used as for equal vertical strain (Richart, 1959). If vertical and lateral drainage is to be considered their combined effect (three-dimensional consolidation effect) can be determined from the following:

$$U = 1 - (1 - U_v)(1 - U_h) \qquad (10.20)$$

where U_v is the average degree of consolidation considering vertical flow only.

In a drain with resistance to flow there would be a head loss, in which case μ is given by:

$$\mu = \ln(n) - 0.75 + \pi z(2l - z)\frac{k_h}{q_w} \qquad (10.21)$$

where l is the drain length for a typical drain closed at the bottom, z the distance from the open end of the drain, k_h the radial hydraulic conductivity of the soil to be drained, and q_w the discharge capacity of the drain.

For a typical drain 20 m deep if q_w/k_h is less than 3000 the drain resistance should be considered. The average degree of consolidation according to Equation 10.21 varies with depth and is largest near the open end of the drain. The average value occurs at various depths at different times, but ranges typically from $z = 0.3l$ to $z = 0.5l$. An average value of $0.4l$ could be used, hence

$$\mu = \ln(n) - 0.75 + 2l^2\frac{k_h}{q_w} \qquad (10.22)$$

The effect of disturbance is to increase the parameter μ which is given by:

$$\mu \approx \ln\left(\frac{n}{s}\right) + \frac{k_h}{k_d}\ln(s) - \tfrac{3}{4} + \pi z(2l - z)\left(\frac{k_h}{q_w}\right)\frac{1 - (k_h/k_d - 1)}{(k_h/k_d)(n/s)^2} \qquad (10.23)$$

where $s = d_d/d_e$, d_d is the diameter of the disturbed or smeared zone, and k_d is the radial hydraulic conductivity of the disturbed zone.

It is generally difficult to assess values for the parameters s and k_d. The disturbance effect is usually accounted for by using s values 1.5 to 3 (s is 1 for undisturbed soil), by reducing c_h or by reducing the drain diameter. The drain diameter is typically reduced by 20–30%. For further details see Hansbo (1979) and Van Zanten (1986).

Example 10.2

Determine the percentage of average consolidation in 1 year in a 10 m thick homogeneous clay layer with $c_h = 0.35\ \text{m}^2/\text{year}$ for a drain of thickness 3.5 mm and width 200 mm. The ratio of horizontal to vertical permeability is 2. The drains are arranged in a triangular pattern at 1 m centres. The clay is sandwiched between sand deposits.

Solution

$d_e = 2(200 + 3.5)/\pi = 129.5$ mm or 0.13 m

$D_e = 1.06 \times 1 = 1.06$ m

$n = 1.06/0.13 = 8.15$

From Equation 10.18,

$$\mu = \frac{(8.15)^2}{(8.15)^2 - 1}\left[\ln(8.15) - \tfrac{3}{4} + \frac{1}{(8.15)^2}\left(1 - \frac{1}{4(8.15)^2 - 1}\right)\right]$$

$$= 1.384$$

$$1 = \frac{(1.06)^2 \times 1.38}{8 \times 0.35} \ln \frac{1}{1 - U_h}$$

Therefore, $U_h = 0.836$,

$$c_v = c_h k_v / k_h$$

$$= 0.35/2 = 0.18 \text{ m}^2/\text{year}$$

From Equation 5.17,

$$T_v = 0.18 \times 1/5^2 = 0.007$$

From Table 5.1,

$$U_v = 0.1$$

From Equation 10.20,

$$U_t = 1 - (1 - 0.1)(1 - 0.836) = 0.852$$

Frequently, the engineers recommend 1.5–2 wick drains to replace one sand drain.

10.6 Settlement of municipal landfill

The settlement of a landfill stems from one or more of the following:

1. Settlement due to the reduction in void space and compression of loose materials from self-weight of the waste and the weight of the cover materials.
2. Occasional movement of smaller particles in larger voids resulting from collapse of larger bodies, seepage, abrupt drop in the water table, shock wave or vibration. This type of movement may cause unexpected depression on the surface.
3. Volume changes from biological decomposition and chemical reactions, which are accelerated at high moisture content, warmer temperature, poorer state of compaction and larger proportion of organic contents.
4. Dissolution from percolating water and leachate.
5. Settlement of the soft compressible soils underlying the landfill.

10.6.1 Landfill stabilization

The theories developed for fine-grained materials are not strictly applicable to landfill. The laboratory consolidation tests are not feasible because of the difficulty of obtaining reliable and representative test specimens. Field measurements are essential for developing methods for estimating the settlement of landfills. Chemical and biological decomposition influences secondary compression.

Designing and constructing foundations on old landfill sites require not only consideration of settlements but also finding solutions to problems such as venting of gases, corrosion of buried foundation elements, leachate control, and other construction difficulties.

The deformation under surcharge load occurs very rapidly and is treated like primary consolidation. Settlement under self-weight and long-term settlement from the applied loads are similar to secondary compression.

10.6.2 Settlement from self-weight

The settlement will occur during construction of the landfill. For a landfill of unit weight γ_l and height H_l the stress increment $\delta\sigma_v$ is taken as the average vertical effective stress or $\gamma_l H_l/2$. The settlement is determined from the method given in Section 5.4.2. The constrained modulus D is $(\partial\sigma_v/2)/0.435$ CR, or approximately $\gamma_l H_l/2$CR. To evaluate D, the compression ratio CR or compression index C_c may be estimated from Table 10.6. The settlement during construction is calculated using Equation 5.11c.

With increasing age and height of a landfill its rate of settlement from self-weight decreases. At greater depths anaerobic environment is likely to reduce the rate of decomposition. The rate of settlement may eventually become constant once a depth of about 100 ft (30 m) is reached (Yen and Scanlon, 1975).

10.6.3 Primary compression

Typical settlement–log time curves for a refuse landfill under a surcharge are shown in Figure 10.8 (Sheurs and Khera, 1980). Note that these curves look very much like those of peat or organic materials. Initially, the settlement occurs rapidly as in primary consolidation, and then it tapers off as in secondary compression.

Since the refuse is not always saturated, the early part of the settlement is denoted primary compression and not consolidation. Sowers (1973) stated that this settlement usually occurred in less than a month. The data in Figure 10.15 show that about 70–80% of the settlement took place within the first 3 months. The value c_V ranged between 0.15 and 5 ft^2/day.

The settlement from primary compression may be calculated using Equation 5.11. Various parameters for municipal landfill materials are given in Table 10.6.

10.6.4 Stress history effect

The stress history and the load increment ratio influence and settlements. For an interstate highway part of the old landfill was excavated before surcharging. Where the stress increase from surcharge load was less than the stress before the excavation the settlement was from 5 to 7%. Where the surcharge stress was over 40% above the pre-excavation stress the settlement ranged from 11.4 to 16.8% (Sheurs and Khera, 1980).

10.6.5 Secondary compression

The settlement of a landfill continues after the primary compression (Figure 10.12). The long-term settlement appears to be linear on a log time scale and can be determined by Equation 10.16 or as follows:

$$\delta_s = C_t H \log(t/t_p)/(1+e_0) \quad (10.24)$$

where C_t is the secondary compression index (change in void ratio per log cycle of time), t is the time at which the settlement value is required, and t_p is the time for completion of primary compression.

A range of C_α and C_t values are shown in Table 10.6. The average value of C_α may be taken as 0.2 with upper and lower limits of 0.32 and 0.13, respectively. In about 10 years' time C_α reaches a constant value of 0.01 to 0.02. However, if additional load is placed on the landfill the rate of secondary compression will increase (Bromwell, 1978).

Table 10.6 Deformation parameters for municipal landfills

Material	C_c	CR[a]	C_t	C_α
Peat	$0.75e_0$			
15-year-old landfill, Boston, Massachusetts		0.26		0.24
Lab test on simulated material		0.20		0.30
Old landfill, West Virginia		0.15		0.04
Low organic contents and conditions unfavourable to decomposition	$0.15e_0$	0.15	$0.03e_0$	0.024
High organic contents and conditions favourable to decomposition	$0.55e_0$	0.41	$0.09e_0$	0.072
Municipal waste, Melbourne, Australia	$0.1e_0$			0.06
15- to 20-year-old landfill, Michigan				0.02
10-year-old landfill, Elizabeth, New Jersey		0.08		0.02
		0.21		0.04
Landfill, Harrison, New Jersey	$0.25e_0$			
Recompacted municipal waste–soil mix				0.14–0.034

[a] $CR = C_c/(1 + e_0)$.
[b] Landfill compacted before preloading.
Sources: Bromwell, 1978; Dodt *et al.*, 1987; Keene, 1977; Moore and Pedler, 1977; Oweis and Khera, 1986; Sheurs and Khera, 1980; Sowers, 1973; Yen and Scanlon, 1975; York *et al.*, 1977.

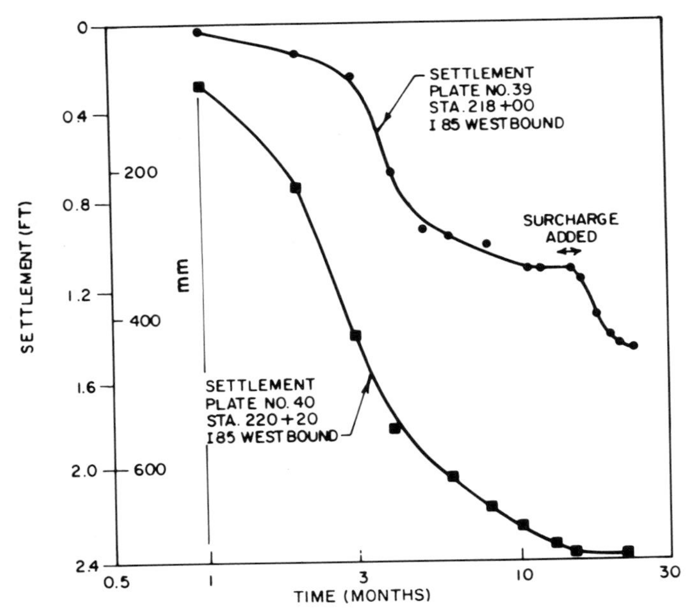

Figure 10.8 Typical time–settlement curve for refuse (Reproduced from Sheurs and Khera, 1980, by permission)

10.6.6 Surcharging of municipal waste materials

Experience with preloading of municipal waste has shown that the primary consolidation is completed within a month or two depending on the thickness of the landfill (Section 10.6.1). To reduce the amount of post-construction settlement, surcharge is left in place beyond the time needed for primary consolidation under this load. A considerable amount of secondary compression occurs during this period t_1 as represented by *cd* in Figure 10.7. If a structure is to be built on the treated landfill, the surcharge load and the load corresponding to the permanent load must be removed. When the load is removed the rebound occurs along *de* at a slope of RR. The construction of the structure imposes the permanent load. The settlement of the structure takes place along *de* and at any time t it is calculated from the following equation:

$$\delta = H\,\mathrm{RR}\log\left(\frac{\sigma'_{vo}+p_f}{\sigma'_{vo}}\right) + C_\alpha H\log(t/t_2) \qquad (10.25)$$

for $t > t_2$, and

$$\delta = H\,\mathrm{RR}\log\left(\frac{\sigma'_{vo}+p_f}{\sigma'_{vo}}\right) \qquad (10.26)$$

for $t \leqslant t_2$.

The value of t_2 is determined from the following equation:

$$\log(t_2/t_1) = [(\mathrm{CR}-\mathrm{RR})/C_\alpha]\{\log\frac{\sigma'_{vo}+p_f+p_s}{\sigma'_{vo}+p_f}\} \qquad (10.27)$$

where CR is the compression ratio, and RR the recompression ratio.

Example 10.3

A 20 ft thick layer of refuse is to be stabilized to support a warehouse having an average load of 500 lb/ft^2. For the refuse CR = 0.3, RR = 0.05, C_α = 0.05, and unit weight is 55 lb/ft^3. The site is to be graded, which requires adding 3 ft of sand with a unit weight of 120 lb/ft^3. Estimate the settlement after ten years if the preload of 500 lb/ft^2 and a surcharge of 500 lb/ft^2 are kept for a period of 1 year before construction.

$$\sigma'_{vo} = 3\times 120 + 10\times 55 = 910$$

From Equation 10.27,

$$\log(t_2/t_1) = \frac{0.3-0.05}{0.05} \times \log\left(\frac{910+500+500}{910+500}\right)$$

$$t_2/t_1 = 4.56$$

or $t_2 = 4.56$ years.

From Equation 10.25,

$$\delta = 20\times 0.05\log(910+500/500) + 0.05\times 20\log(10/4.56)$$

$$= 0.79\text{ ft}$$

10.7 Expressway in New Jersey

An experimental embankment was constructed for an expressway in the late 1970s in northern New Jersey. The landfill materials within the project area were in various degrees of decomposition with thin layers of soil cover. Information from various sources indicated the landfill age to range between 5 and 15 years.

The depth of landfill ranged between 16 and 28 ft (4.9 and 8.5 m). The underlying materials consisted of an average of 4 ft of gray-black clayey organic silt and about 3 ft of brown peat (locally referred to as the New Jersey Meadowmat), followed by a 20–25 ft (6–7.5 m) thick layer of well graded dense sand and gravel with less than 10% silt. The soils below this are varved clays.

The final pavement elevation of the eastbound roadway was established below the top of the landfill with the excavation extending to a point 5 ft (1.5 m) below the subgrade level. This required a removal of 5–21 ft (1.5–6.4 m) of the refuse. A section of the eastbound roadway is shown in Figure 10.9.

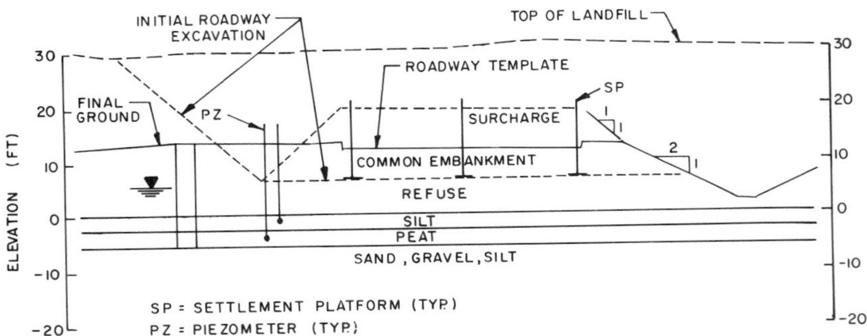

Figure 10.9 Section of eastbound roadway (Reproduced from Oweis and Khera, 1986, by permission)

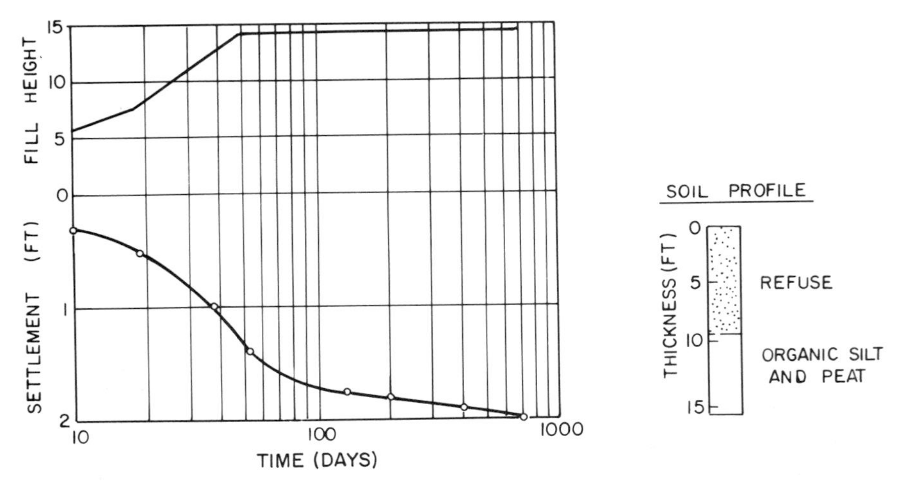

Figure 10.10 Eastbound roadway section on organic soils and refuse (Reproduced from Oweis and Khera, 1986, by permission)

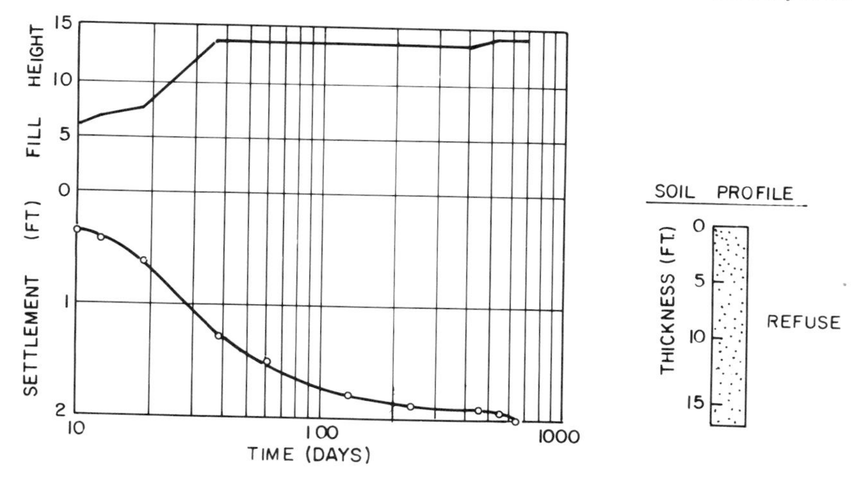

Figure 10.11 Eastbound roadway section on refuse (Reproduced from Oweis and Khera, 1986, by permission)

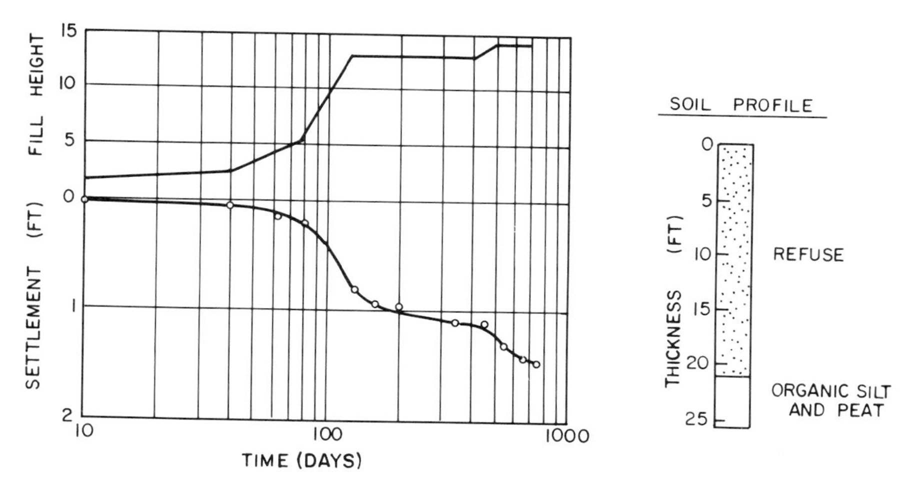

Figure 10.12 Westbound roadway section on organic soil and refuse (Reproduced from Oweis and Khera, 1986, by permission)

Along the westbound roadway the top of the existing refuse material was considerably lower (1–19 ft) than the design subgrade elevation. The materials excavated from the eastbound roadway were cleared of large objects and placed on the westbound roadway in about 2 ft (0.6 m) thick lifts and compacted using 3–5 passes of a 35 ton sheepsfoot compactor. The loading sequence, settlement observations, and the soil profile for two stations of the eastbound roadway are shown in Figures 10.10 and 10.11.

In one of the eastbound sections silt and peat make up almost two-thirds of the total compressible material (Figure 10.10), whereas only landfill materials are present in the other eastbound section (Figure 10.11). The time–settlement response at the two locations is very similar, indicating that the landfill materials responded similarly to natural organic materials. Data for the westbound roadway are represented in Figure 10.12. The surcharge load was increased between 400 and 500 days, at which time the settlement rate increased. This trend is more clearly evident starting at about 80 days where the height of the added surcharge load was considerably greater.

10.8 Settlement of mineral and industrial waste

Many of the wastes composed of fine-grained materials are disposed of behind retaining dams or dikes in the form of slurries. Land disposal of these wastes must consider the limited availability of land area and impact on the environment. To make efficient use of the land, the volume of the sediments must be reduced and its strength increased. The techniques used consist of desiccation, drainage, consolidation under self-weight, surcharging, stabilization with chemicals, lime, cement, fly ash, or the use of mechanical methods to reduce the water content before disposal.

10.8.1 Fly ash

Tests on compacted fly ash show a low value of compression index (0.02–0.25). C_c is dependent upon the type of the fly ash and the period of curing. Field data indicate that the actual settlements are usually much smaller than those computed based on the laboratory tests (Seals *et al.*, 1977). Consolidation test results for samples prepared from a virgin fly ash where load increments were applied at different time intervals are shown in Figure 10.13. Note that the sample loaded at 1 day time interval showed considerably less settlement. This is attributed to the extra time available to this sample compared with that loaded every hour. Since most laboratory

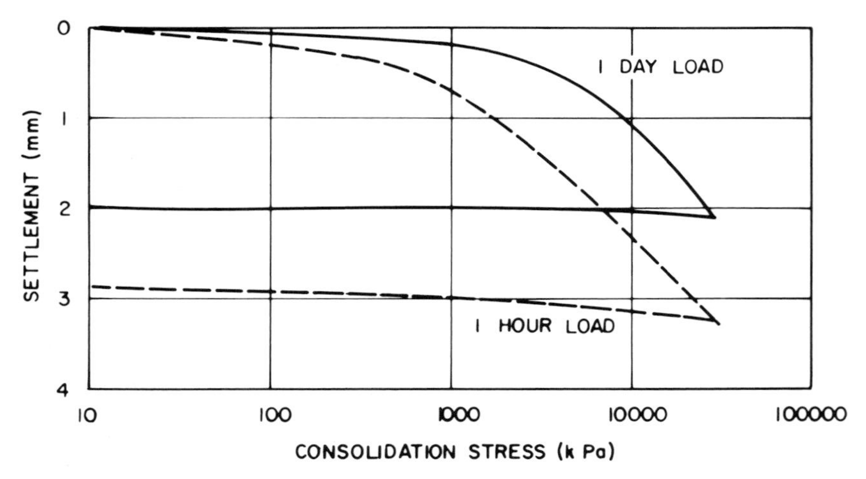

Figure 10.13 Effect of load duration on consolidation of fly ash (After Gatti and Tripiciano, 1981)

tests are carried out in a relatively short period of time when compared with loading in the field, the reported discrepancy between field and laboratory tests may be partially due to the differences in curing time.

For undisturbed specimens of a ponded fly ash from Virginia, Newman *et al.* (1987) reported a void ratio of 1.6 and C_c and C_α values of 0.65 and 0.07, respectively, with coefficient of consolidation of 2.25 cm^2/min. From a site in Illinois (Cunnigham *et al.*, 1977) for void ratios 1.26–1.40, C_c was 0.46–0.32. Considering the relative void ratios for the two cases, these value appear to be reasonable. The value of precompression stress was determined to be larger at shallower depth than at greater depth. Note that these C_c values are much larger than those for compacted fly ash, indicating a distinct advantage of disposal by compacting. Also, the dry unit weight of slurried fly ash, which is generally less than 70 lb/ft^3 (10.2 kN/m^3) is considerably lower than that of compacted fly ash (13–21 kN/m^3).

Because of the difficulty of predicting settlements from laboratory tests, field techniques such as standard penetration resistance (N values) and cone penetration resistance, q_c, assuming fly ash to be cohesionless, have been studied for settlement predictions. The computed settlements based on N values and q_c did not agree with the measured values but those based on the plate load tests did (Leonards and Bailey, 1982).

10.8.2 Flue gas desulphurization sludge

The compression index of FGD sludges at a stress level of 500 kPa is about 2–4 times greater than at a stress level of 25 kPa (Hagerty *et al.*, 1977; Ullrich and Hagerty, 1987). At higher stress levels, the compression index for sulphite sludges was considerably larger (0.2–0.8) when compared with sulphate sludges (0.02–0.07). The corresponding values reported by Krizek *et al.* (1987) are larger; 0.5–1.0 for sulphite (lime and double alkali) and 0.2–0.3 for sulphate sludges. The time for completion of primary consolidation is very short for sulphite sludge.

Fly ash has been frequently added to FGD sludges to reduce their moisture content and enhance their stability. The addition of small amounts of lime to FGD sludge–fly ash mixture may further add to its stability. However, the results are not always predictable. A sulphite sludge–fly ash–lime designed to yield maximum strength showed a higher compression index at a consolidation stress of 35 kPa after 28 days curing than after 7 days curing. A similar mixture with sulphate sludges consistently showed a lower compression index with increasing curing time. When sulphite sludge was mixed with 60–70% fly ash and was compacted to its maximum dry density its C_c value was 0.02–0.25, which is more like fly ash (Ullrich and Hagerty, 1987). Krizek *et al.* (1976) reported higher values for compression index (0.35) with similar amounts of fly ash.

10.8.3 Pulp and papermill waste

The compression index of pulp and paper mill sludges is approximately $0.4e_0$ (Wardwell *et al.*, 1977). For a given solid content, C_c increases linearly with increasing organic content. In the laboratory consolidation test pore pressure dissipated within 15 min and the coefficient of consolidation decreased from 0.11 to 0.03 cm^2/min as the stresses increased from 10 to 320 kPa. These values were about four times smaller than the field values. Similarly, the laboratory values of coefficient of secondary compression (0.015) were about four times smaller than the field values (0.056) and were also non-linear with log time (Vallee and Andersland, 1974). It is clear that the predictions based on the laboratory tests would be grossly in error. The use of horizontal drainage blanket and application of small earth surcharge can result in appreciable decrease in volume and increase in strength (Vallee and Andersland, 1974; Charlie, 1977).

10.8.4 Phosphatic clays

Settlement of phosphatic clays under self-weight may require several decades before the disposal area can be re-utilized. The use of Terzaghi theory, which assumes linear infinitesimal strain, overestimates the pore pressure dissipation at the top of a layer and underestimates it at the bottom (Gibson *et al.*, 1981). McVay *et al.* (1986) compared several large strain theories (McNabb, 1960; Gibson *et al.*, 1967, 1981; Somogyi, 1980) and concluded that they all predicted identical pore pressure, void ratio and settlement. Good agreement was reported between predicted and measured values for the permeability and void ratio relationships of:

$$k = 4.0 \times 10^{-6} \times e^{4.11} \qquad (10.28)$$

$$e = 29.43(\sigma')^{-0.29} \qquad (10.29)$$

The relationships of void ratio with permeability and effective stress usually hold for solid contents about 15% or higher (Scully *et al.*, 1984). Addition of sand to clay results in reduced settlements and improved engineering properties (Bromwell, 1978).

10.8.5 Dredged waste

Dredging is a periodic operation and it allows desiccation as a viable alternative for reducing slurry volume in many regions. To improve the effectiveness of desiccation, surface water from the disposal and rainfall runoff should be drained quickly and effectively. In the application of desiccation the slurry is deposited in thickness from about 50 cm to 1 m. As the materials dry, a few millimetre to several centimetre wide cracks are formed at the surface. They extend to depths of up to a metre or so. The resulting larger exposed surface areas aid in rapid drying. When more dredge waste is placed on partially dry material these cracks provide additional drainage paths. High capillary forces resulting from drying cause large reduction in soil volume and the drop in water table increases overburden stresses on the underlying soils.

Experiences with dredged material in the USA and Germany indicate surface drying to be an efficient means of stabilization and volume reduction (Rizkallah, 1987).

10.8.6 Surcharging dredging waste

Krizek *et al.* (1977) pointed out the difficulty of ensuring proper drainage at the bottom of a dredged fill. Bromwell (1978) recommended that, while surcharging dredge fill, single drainage should be assumed as the consolidated slurry at the lower boundary obstructs free drainage. The cost of improving bottom drainage is high, and is not justifiable unless flooding from deposition of dredging persists or occurs frequently. Vertical drains may prove successful in reducing the time needed to stabilize. Field evaluation must be made to determine their effectiveness. For details see the section on vertical drains. For undisturbed samples Salem and Krizek (1976) reported the compression index to have a range of 0.3–0.7. These values were lower than those obtained for specimens prepared in the laboratory from slurry. A relationship between liquid limit, natural water content and compression index was given as:

$$C_c = 0.04(w_l + 2w_n - 50) \quad (10.30)$$

From the standard consolidation tests on undisturbed samples the coefficient of consolidation was $6 \times 10^{-4}\,\mathrm{cm^2/s}$. The specimens consolidated from slurry in a 150 mm deep consolidation ring had a c_v value of $10^{-4}\,\mathrm{cm^2/s}$. The coefficient of secondary compression ranged between 0.002 and 0.013 for a natural water content range of 45–65%. C_α showed a linear increase as C_c increased from 0.32 to 0.55.

Surcharging may be accomplished by water load (see the section on preloading) or moving load. In the case of moving load, the surface of the dredged fill is either covered with hydraulically placed coarse fraction or allowed to desiccate to support a free draining surcharge. The surcharge placed on the surface is moved at a rate commensurate with primary consolidation. The slope in the direction of the movement of the surcharge is maintained sufficiently flat to avoid bearing capacity failure (Bishop and Vaughan, 1972).

Summary

The methods for the reduction of waste volume and improving its load-carrying capacity include compaction, dynamic compaction, and preloading. Compaction is the most commonly used method and is used for municipal waste and some industrial waste where water content of the waste can be controlled, such as natural soils, blended soils, fly ash, and coal waste. If the compaction is for a soil liner then the design and construction require extensive supervision, quality assurance and quality control. Based on coefficient of variation, the number of tests needed can be determined.

Dynamic compaction consists of dropping heavy blocks in a grid pattern. The volume of the treated materials decreases, bearing capacity increases, and post-construction settlements are reduced. It is most suited for loose coarse-grained soils, rubble fills, and non-hazardous landfills. Materials below the water table cannot be treated very efficiently.

Preloading with surcharge provides an effective means of treating very soft soils such as dredged waste and municipal waste. They require longer time than the previous techniques but can be applied to deep very soft materials. Vertical drains are used to accelerate the completion of the process.

Notation

B	drain width (L)
C_c	compression index (with respect to e)
c_h	coefficient of consolidation in the horizontal direction (L^2/T)
CR	compression ratio (with respect to ε)
C_t	secondary compression index (with respect to e)
c_v	coefficient of consolidation in the vertical direction (L^2/T)

C_α	secondary compression ratio (with respect to ε)
d_d	diameter of the disturbed zone (L)
d_e	equivalent drain diameter (L)
D_e	equivalent diameter of the soil cylinder (L)
D_{max}	maximum influence depth in dynamic compaction (L)
D_r	relative density
dt	time since the removal of surcharge load (T)
E	allowable sampling error of the expected mean
e	actual void ratio
e_{max}	void ratio of the soil in the loosest state
e_{min}	void ratio of the soil in the densest state
e_0	initial void ratio
H	drop height of weight (L)
I	influence factor
k	permeability (L/T)
k_h	radial hydraulic conductivity (L/T)
k_d	radial hydraulic conductivity, disturbed zone (L/T)
l	drain length for a typical drain closed at bottom (L)
L	drain spacing
m_v	coefficient of volume compressibility
N	standard penetration resistance
n	D_e/d_e
n_u	number of units in the sample
p_f	final or permanent load
p_s	surcharge load
q_c	cone penetration resistance
q_w	discharge capacity of the drain
RR	recompression ratio (with respect to ε)
s	d_d/d_e
t	probability factors from the t-distribution tables, time
t_1	time of surcharge load
t_d	thickness of drain
t_p	completion time for primary consolidation
t_{sec}	completion time for secondary compression under p_f
T_h	dimensionless time factor for horizontal drainage
u_e	excess pore pressure
U_{f+s}	average consolidation (final + surcharge)
U_h	average degree of consolidation for horizontal flow
u_i	initial excess pore pressure
U	degree of consolidation for three-dimensional consolidation
U_v	average degree of consolidation for vertical flow
$U_{z(f+s)}$	consolidation ratio (final + surcharge)
v'	coefficient of variation
w	weight of the dropping block
w_l	liquid limit
w_n	natural water content
z	distance from open end of the drain
γ	unit weight of the soil
γ_{max}	unit weight of the soil in the densest state
γ_{min}	unit weight of the soil in the loosest state

$\delta_{c(f+s)}$ settlement from final plus surcharge load
δ_{cf} primary consolidation settlement from final load
δ_{sf} secondary compression from permanent load
σ'_{vo} effective overburden stress

References

Baker, W. H. (1982) *Highway 71 – Springdale, Arkansas, Dynamic Deep Compaction – Sanitary Landfill*, Hayward Baker Company Report to Arkansas State Highway and Transportation Department

Bergstrom, W. R., Sweatman, M. B. and Dodt, M. E. (1987) Slurry trench construction – Collier Road Landfill, *Geotechnical Practice for Waste Disposal '87*, Proc. Spec. Conf. Geot. Special Publ. No. 13, Ed. Woods, R. D., pp. 260–274

Bishop A. W. and Vaughan, P. R. (1972) *Consolidation of Fine Grained Dredged Material after Hydraulic Deposition*, National Ports Council, London

Bjerrum, L. (1972) Embankments on soft grounds, *Proc. Specialty Conf., Performance of Earth and Earth-Supported Structures*, Vol. 2, Purdue University, June 1972, pp. 1–45

Bromwell, L. G. (1978) Properties, behavior and treatment of waste fills, *ASCE, Met. Section, Seminar – Improving Poor Soil Conditions*, New York, October 1978

Chae, Y. S. and Gurdziel, T. J. (1976) New Jersey fly ash as structural fill, *New Horizons in Construction Materials*, Vol. 1, Ed. Fang, H. Y., Envo Publishing Co. Inc., pp. 1–13

Charles, J. A., Burford, D. and Watts, K. S. (1981) Field studies of the effectiveness of dynamic consolidation, *Proc. 10th ICSMFE*, Vol. 3, Stockholm, pp. 617–622

Charlie, W. A. (1977) Pulp and papermill solid waste disposal – a review, *Geotechnical Practice for Disposal of Solid Waste Materials*, ASCE, Ann Arbor, Michigan, June 1977, pp. 71–86

Chen, C. Y., Elnaggar, H. A. and Bullen, A. G. R. (1976) Degradation and the relationship between shear strength and various index properties of coal refuse, *New Horizons in Construction Materials*, Envo Publishing Co. Inc., Vol. 1, pp. 41–52

Cunnigham, J. A., Lukas, R. G. and Anderson, T. C. (1977) Impoundment of fly ash and slag – a case study, *Proc. Geotechnical Practice for Disposal of Solid Waste Materials*, ASCE, University of Michigan, Ann Arbor, Michigan, June 1977, pp. 227–245

D'Appolonia, D. J. (1978) Foundation improvement by dynamic consolidation, American Society of Civil Engineers, New York Metropolitan Section, *Foundations and Soil Mechanics Group Seminar, Improving Poor Soil Conditions*

Dodt, M. E., Sweatman, M. B. and Bergstrom, W. R. (1987) Field measurement of landfill surface settlement, *Geotechnical Practice for Waste Disposal '87*, Geot. Spec. Tech. Publ. No. 13, Ed. Woods, R. D., pp. 406–417

EPA/530-SW-86-007 (1986) *Design, Construction, and Evaluation of Clay Liners for Waste Management Facilities*, p. 663

Gatti, G. and Tripiciano, L. (1981) *Mechanical Behaviour of Coal Fly Ashes, Proc. 10th ISCMFE*, Stockholm, Vol. 2, pp. 317–322

Gibson, R. E., England, G. L. and Hussey, M. J. (1967) The theory of one-dimensional consolidation of saturated clays. I. Finite nonlinear consolidation of thin homogeneous layers, *Geotechnique*, **17,** No. 3, 261–273

Gibson, R. E., Schiffman R. L. and Cargill, K. W. (1981) The theory of one-dimensional consolidation of saturated clays. II. Finite nonlinear consolidation of thick homogeneous layers, *Canadian Geotechnical Journal*, **18,** No. 2, 280–293

Hagerty, D. J., Ullrich, C. R. and Thacker, B. K. (1977) Engineering properties of FGD sludges, *Geotechnical Practice for Disposal of Solid Waste Materials*, ASCE, Ann Arbor, Michigan, June 1977, pp. 23–40

Hansbo, S. (1979) Consolidation of clay by band-shaped prefabricated drains, *Ground Engineering*, **12,** No. 5, 16–25.

Harrop-Williams, K. (1987) Acceptance sampling for clay liner design, *Geotechnical Practice for Waste Disposal '87*, Proc. Specialty Conference, ASCE Geot. Spec. Publ. No. 13, Ed. Woods, R. D., pp. 515–521

Hilf, J. W. (1975) Compacted fill, In *Foundation Engineering Handbook*, Eds Winterkorn, H. F. and Fang, H. Y., Van Nostrand Reinhold, New York

Holtz, R. D. and Wagner, O. (1975) Preloading by vacuum-current prospects, *Transportation Research Record*, **548,** 26–29

Holubec, I. (1976) Geotechnical aspects of coal waste embankments, *Canadian Geotechnical Journal*, **13,** No. 1, 27–39

Johnson, S. J. (1970) Precompression for improving foundation soils, *ASCE, Journal of the Soil Mechanics and Foundation Engineering Division*, **96,** No. SM1, 111–144

Keene, P. (1977) Sanitary landfill treatment, Interstate Highway 84, *Proc. Conf. Geotechnical Practice for Disposal of Solid Waste Materials*, ASCE, Ann Arbor, Michigan, June 1977, pp. 632–644

Krizek, R. J., Roderick, G. L. and Jin, J. S. (1977) Chemical stabilization of dredged materials, *Proc. Conf. Geotechnical Practice for Disposal of Solid Waste Materials*, ASCE, Ann Arbor, Michigan, June 1977, pp. 517–540

Leonards, G. A. and Bailey, B. (1982) Pulverized coal ash as structural fill, *ASCE, Journal of the Geotechnical Engineering Division*, **108,** No. GT4, 517–532

Leonards, G. A., Cutter, W. A. and Holtz, R. D. (1980) Dynamic compaction of granular soils, *ASCE, Journal of the Geotechnical Engineering Division*, **106,** No. GT1, 35–44

Lukas, R. G. (1980) Densification of loose deposits by pounding, *ASCE, Journal of the Geotechnical Engineering Division*, **106,** No. GT4, 435–446

Lumb, P. (1974) Application of statistics in soil mechanics, Chapter 3 in *Soil Mechanics – New Horizons*, Ed. Lee, I. K., Newnes-Butterworths, London, pp. 44–111

Manz, O. E. and Manz, B. D. (1985) Utilization of fly ash in road bed stabilization: some examples of western U.S. experience, *Fly Ash and Coal Conversion Byproducts: Characterization, Utilization and Disposal*, I, Materials Research Society, Vol. 43, pp. 129–144

Mayne, P. W., Jones, Jr, J. S. and Dumes, J. C. (1984) Ground response to dynamic compaction, *ASCE, Journal of the Geotechnical Engineering Division*, **110,** No. 6, 757–774

McLaren, R. J. and DiGioia, A. M. (1987) The typical engineering properties of fly ash, *Geotechnical Practice for Waste Disposal '87*, Proc. Spec. Conf. Geot. Spec. Publ. No. 13, Ed. Woods, R. D.

McNabb, A. (1960) Mathematical treatment of one-dimensional soil consolidation, *Quarterly Journal of Applied Mathematics*, **17,** No. 4, 337–247

McVay, M., Townsend, F. and Bloomquist, D. (1986) Quiescent Consolidation of Phosphatic Waste Clays, *ASCE, Journal of the Geotechnical Engineering Division*, **112,** No. 11, 1033–1052

Ménard, L. and Broise, Y. (1975) Theoretical and practical aspects of dynamic consolidation, *Geotechnique*, **15,** No. 1, 3–18

Mitchell, J. K. (1976) *Fundamentals of Soil Behavior*, John Wiley & Sons, New York

Mitchell, J. K. (1981) *Soil Improvement Methods and their Application in Civil Engineering*, 16th Annual Henry M. Shaw Lecture, North Carolina State University, Raleigh

Monahan, E. J. (1986) *Construction Off and On Compacted Fills*, John Wiley & Sons, New York

Moore, P. J. and Pedler, I. V. (1977) Some measurements of compressibility of sanitary landfill materials, Speciality Session, *Proc. 9th ICSME*, Tokyo, pp. 319–330

Morrison, A. (1982) The booming business of wick drains, *ASCE, Civil Engineering*, **53,** No. 3, 47–51

Moulton, L. K., Rao, S. K. and Seals, R. K. (1976) The use of coal associated wastes in the construction and stabilization of refuse landfills, *New Horizons in Construction Materials*, Ed. Fang, H. Y., Envo Publishing Co. Inc., Vol. 1, pp. 53–65

Oweis, I. A. and Khera, R. (1986) Criteria for geotechnical construction on sanitary landfills, *Int. Symp. on Environmental Geotechnology*, Vol. 1, Ed. Fang, H. Y., pp. 205–222

Parker, D. G., Thornton, S. I. and Cheng, C. W. (1977) Permeability of fly ash stabilized soils, *Proc. Geotechnical Practice for Disposal of Solid Waste Materials*, ASCE, Ann Arbor, Michigan, June 1977, pp. 63–70

Rainbow, A. K. M. and Nutting, M. (1986) Geotechnical properties of British minestone considered suitable for landfill projects, *Int. Symp. on Environmental Geotechnology*, Vol. 1, Ed. Fang, H. Y., pp. 531–539

Richart, F. E. (1958) Review of the theories of sand drains, *Transactions ASCE*, **124,** 709–736

Rizkallah, V. (1987) Geotechnical properties of polluted dredged material, *Geotechnical Practice for Waste Disposal '87*, Proc. Spec. Conf. Geot. Spec. Publ. No. 13, Ed. Woods, R. D., pp. 759–771

Salem, A. M. and Krizek, R. J. (1976) Stress–deformation–time behavior of dredgings, *ASCE, Journal of the Geotechnical Engineering Division*, **102,** No. GT2, 139–158

Saxena, S. K., Lourie, D. E. and Rao, J. S. (1984) Compaction criteria for eastern coal waste embankments, *ASCE, Journal of the Geotechnical Engineering Division*, **110,** No. 2, 262–284

Scully, R. W., Schiffman, F. L., Olsen, H. W. and Ko, H.-K. (1984) Validation of consolidation properties of phosphatic clay at very high void ratio, *Sedimentation/Consolidation Models*, Ed. Yong, R. N. and Townsend, F. C., pp. 1–29

Seals, R. K., Moulton, L. K. and Kinder, D. L. (1977) *In situ* testing of a compacted fly ash fill, *Proc. Geotechnical Practice for Disposal of Solid Waste Materials*, ASCE, Ann Arbor, Michigan, June 1977, pp. 493–516

Sheurs, R. E. and Khera, R. P. (1980) *Stabilization of a Sanitary Landfill to Support a Highway*, National Academy of Science, TRR 754, pp. 46–53

Shoemaker, N. B. (1972) Construction techniques for sanitary landfill, *Waste Age*, March/April

Somogyi, F. (1980) Large strain consolidation of fine grained slurries, Presented at the Canadian Society for Civil Engineers, Winnipeg, Manitoba, 20–30 May 1980

Sowers, G. F. (1973) Settlement of waste disposal fills, *8th Int. CSMFE*, Moscow

Spigolon, S. J. and Kelly, M. F. (1984) *Geotechnical Quality Assurance of Construction of Disposal Facilities*, EPA-600/2-84-040, pp. 49

Srinivasan, V., Beckwith, G. H. and Burke, H. H. (1977) Geotechnical investigations of power plant wastes, *Proc. Conf. on Geotechnical Practice for Disposal of Solid Waste Materials*, ASCE, Ann Arbor, Michigan, June 1977, pp. 169–187

Ullrich, C. R. and Hagerty, D. J. (1987) Stabilization of FGC wastes, *Geotechnical Practice for Waste Disposal '87*, Proc. Spec. Conf. Geot. Spec. Publ. No. 13, Ed. Woods, R. D.

Usmen, M. A. (1986) Properties, disposal and stabilization of combined coal and refuse, *Int. Symp. on Environmental Geotechnology*, Vol. 1, pp. 515–526

Vallee, R. P. and Andersland, O. B. (1974) Field consolidation of high ash papermill sludge, *ASCE, Journal of the Geotechnical Engineering Division*, **100,** No. GT3, 309–328

Van Zanten, R. V. (1986) *Geotextiles and Geomembranes in Civil Engineering*, John Wiley & Sons, New York

Wardwell, R. E. and Nelson, J. D. (1981) Settlement of sludge landfills with fiber decomposition, *Proc. of the Tenth ICSMFE*, Stockholm, Vol. 2, pp. 397–401

Wardwell, R. E., Walker, S. E. and Atwell, J. S. (1977) Geotechnical aspects of papermill sludge disposal, *Proceedings of the Conference on Geotechnical Practice for Disposal of Solid Waste Material*, ASCE, Ann Arbor, Michigan, pp. 756–772

Welsh, J. P. (1983) Dynamic compaction of sanitary landfill to support superhighway, *Proc. Eighth European Conference on Soil Mechanics and Foundation Engineering*, Helsinki

Winterkorn, H. F. and Fang, H. Y. (1975) *Foundation Engineering Handbook*, Van Nostrand Reinhold, New York

Wiss, J. F. (1981) Construction vibrations: state-of-the art, *ASCE, Journal of the Geotechnical Engineering Division*, **107,** No. GT2, 167–182

Yen, B. C. and Scanlon, B. (1975) Sanitary landfill settlement rates, *ASCE, Journal of the Geotechnical Engineering Division*, **101,** No. GT5, 475–490

York, D., Lesser, N., Bellatty, T., Israi, E. and Patel, A. (1977) Terminal development on a refuse fill site, *Proc. Conf. on Geotechnical Practice for Disposal of Solid Waste Material*, ASCE, Ann Arbor, Michigan, June 1977, pp. 810–830

Chapter 11

Design considerations

11.1 Introduction

There are several components that make up a land disposal facility. Geotechnical design criteria are usually required for:

1. Subgrade.
2. Liner/cut-off wall.
3. Leachate collection and detection systems.
4. Foundations for truck scales, pump stations, manholes, tanks and leachate collection, and conveyor pipes.
5. Gas venting.
6. Monitoring wells (see Chapter 6).
7. Geotechnical instrumentation to monitor settlements and assess stability.
8. Capping.

The method of waste placements and the type of waste affect the geotechnical design parameters such as its unit weight, compatibility with the liner and collection system, and foundation stability during filling, especially on sites with soft underlying soils.

11.2 Waste placement

In the past the primary consideration in the disposal of waste has been economics. Now the waste facility must be designed in accordance with the legislated and regulatory requirements with the primary factors being environmental. The elements that are common to most waste placement sites are the quality of surface and groundwater, surface and underground drainage, leachate reduction, collection and treatment system, groundwater monitoring system for pre- and post-work completion, proper maintenance and operation plan, stability of underlying soils, safe and stable slopes of the waste pile, contingencies plans if failure were to occur, and so on.

11.2.1 Sanitary landfill

In a sanitary landfill, waste is usually contained within a cell (Figure 11.1). At the end of each working day the cell is covered with a layer of soil and compacted. The dimensions of the cell depend on the volume of waste received and the availability of cover material. The cell thickness is typically 15 ft (4.6 m) but may range from 8 to

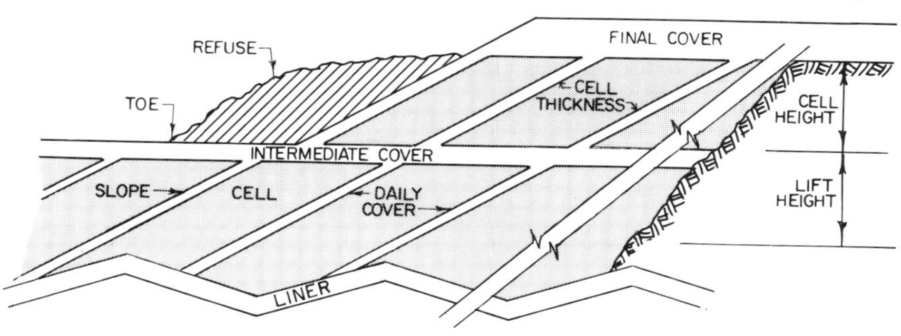

Figure 11.1 Cell construction

30 ft (2.4–9.1 m). For a landfill founded on soft clay in northern New Jersey the lift thickness was limited to 6 ft (1.8 m) to help maintain foundation stability.

The portion where waste is discharged from trucks is termed the working face. The usual slope of the working face is 3 horizontal to 1 vertical (3 : 1). This slope allows reasonably efficient compaction and easier capping and vegetative growth along the side slopes of the landfill. The width of working face is usually limited to 100–150 ft (30–45 m). For good compaction, thickness of loose waste is limited to 2 ft (0.6 m). The first lift of the waste is usually 5 ft (1.5 m) or less, with careful control to remove oversize pieces that could damage the underlying leachate collection system. The compaction equipment moves from the bottom to the top of the working face. The thickness of the daily cover is 6–12 in (150–300 mm) and the purpose is to control flying paper, minimize gas and moisture percolation, and to provide rodent control, fire control, and general aesthetics.

A series of adjoining cells constitutes a lift. If a lift surface is expected to be exposed over 30 days then an intermediate cover is mandated by many regulations. The intermediate cover is typically 1 ft thick and more resistant to erosion than the daily cover.

11.2.2 Hazardous waste

In a hazardous waste facility, cells as deep as 100 ft are lined with hydraulic barriers and wastes of various types are deposited as illustrated schematically in Figure 11.2. Wastes are brought to the facility either in drums or in bulk. Drums are placed on their ends or sides. Liquids are solidified in place with soil, ash or other materials using temporary dikes within a cell. Liquids in drums are solidified in the drums or decanted and solidified within a cell. The thickness of the waste layer depends on its type. For waste in drums, the thickness of the layer would be the length or diameter of the drum. Each waste layer is covered with an intermediate cover of typically 1.5 ft (0.46 m) thick dry soil, crushed stone or the soil excavated to develop the facility.

11.2.3 Mineral and industrial waste

Mineral and industrial waste have been typically disposed of in the form of slurry, which is generally the by-product of the process. In other instances water is added to the fine, such as fly ash, to facilitate transportation through pipes. In past practice slurry was discharged into open mine pits and covered with overburden, pumped into

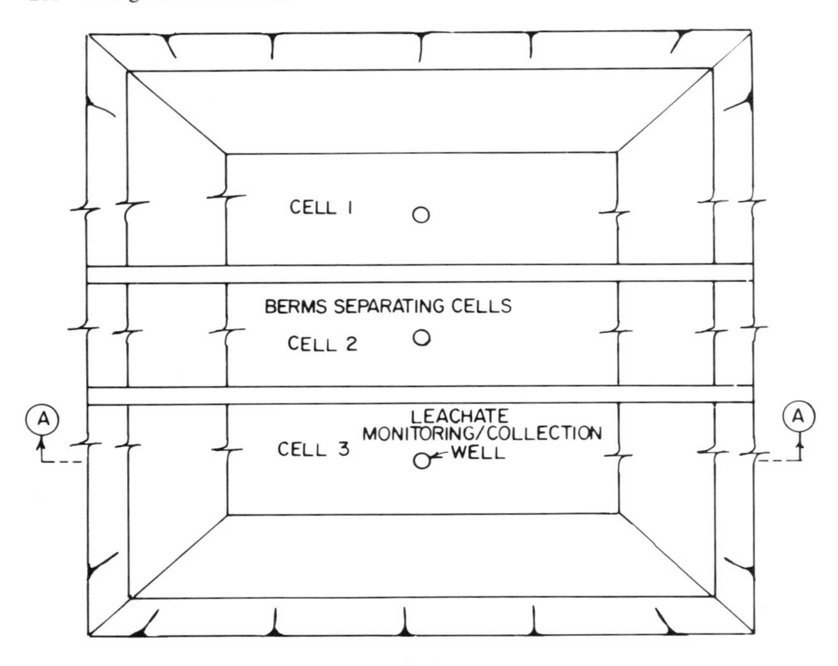

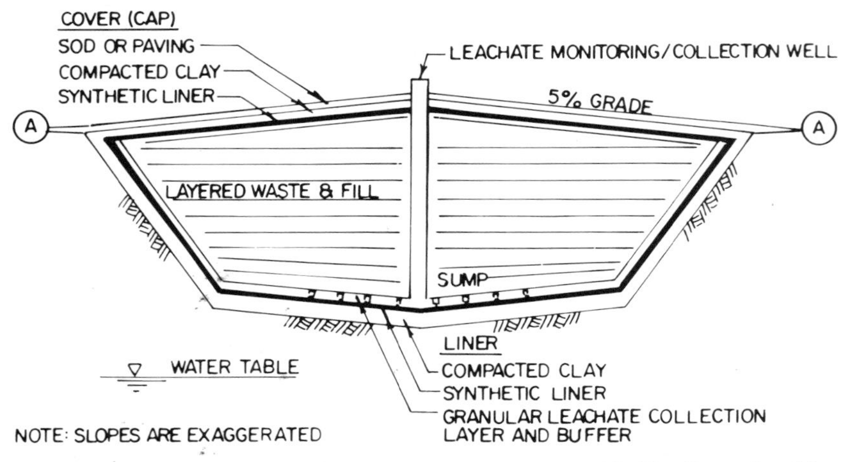

Figure 11.2 Generic RCRA controlled hazardous waste disposal facility (Reproduced from Murphy and Gilbert, 1985, by permission)

impoundments or open bodies of water, placed in stream valleys, etc., with little regard to environmental consequences. The materials so deposited are very loose, require a large disposal area, and cannot be reclaimed for extended periods of time as they lack the necessary shear strength.

Disposal in the form of compacted fill may require the removal of excess water through mechanical means, drying through the use of settlement pools before final disposal, or adding dry materials such as fly ash before compaction. Other needs include loading ports, transportation equipment, haul road, etc. The advantages are a better environmental control, larger disposal capacity, relatively short waiting period before reclaiming the land, and so on.

Typically a slurry disposal requires an earth-retaining structure. Its construction stipulates the availability of large amounts of fine-grained materials for sealing and controlling leakage, and coarse-grained materials to provide filter, drainage and structural strength. At a power plant fly ash may be used as an impermeable material and bottom ash as a substitute to coarse-grained materials. The retaining structure design is similar to that of a zoned earthen dam (*Design of Small Dams*, US Department of Interior, 1977) but it is constructed in stages as the amount of waste behind it keeps building up. There are considerable data available on the design of such retaining structures. Additionally, the environmental requirements are incorporated in the design. This may include diversion of surface water, impervious pond linings, grouted curtains, downstream seepage pools for collection and treatment of contaminated water, plans for monitoring, subsequent reclamation and restoration so that it blends with the existing topographical and geological features.

Taylor and D'Appolonia (1977) describe the design and construction of a 150 ft (45.7 m) high dam built from coarse fractions of tailings on a soft peat deposit of a thickness of up to 30 ft (9 m) which had low strength, and high compressibility and permeability. To allow the peat to gain strength and have reduction in permeability, construction of the dike was carried out at a rate so that there was enough time for the peat to gain sufficient strength. The placement of successive lifts was based on 50% average consolidation, which was selected to allow an increase in dike height that permitted the placement of slurry behind it at a rate consistent with its production. Vane shear tests were used to verify increase in shear strength of the peat, and piezometer readings provided information about the increase in effective stress. The safety of the dike was determined using slope stability analysis. The decreasing rate of settlement was an indicator of the permeability reduction, which was an essential component of the environmental requirements. The principles governing soil improvement using preloading are discussed in Chapter 10. Note that, unlike the construction of a water-retaining structure, the flexibility of stage construction at this taconite tailing disposal site was used effectively to improve the marginal site. Cross-sections at various stages of construction are shown in Figure 11.3.

Caldwell *et al.* (1986) took advantage of attenuating properties of the materials underlying an embankment constructed to retain tailings from a gold mine. Their model studies indicated that sulphate, total cyanide, arsenic, cobalt and copper would all be reduced to their detection level before the seepage reached the downstream toe of the embankment. Groundwater quality monitoring wells were installed to examine actual seepage and water quality effects.

Newman *et al.* (1987) described the conversion of a power plant ash pond disposal site to a compaction disposal facility. For construction purposes the level of the pond was lowered, which caused the surface ash to dry and allow light equipment to operate on it. In areas where drainage did not occur reasonably quickly a filter fabric covered with sand provided satisfactory trafficability. A 20 mil thick liner of high-density polyethylene (HDPE) was placed on top of the existing ponded ashes, and before its installation band drains were laid horizontally in the area at 10 ft (3 m) centres. The ashes were expected to consolidate about a foot (0.3 m) under the load of the compacted materials.

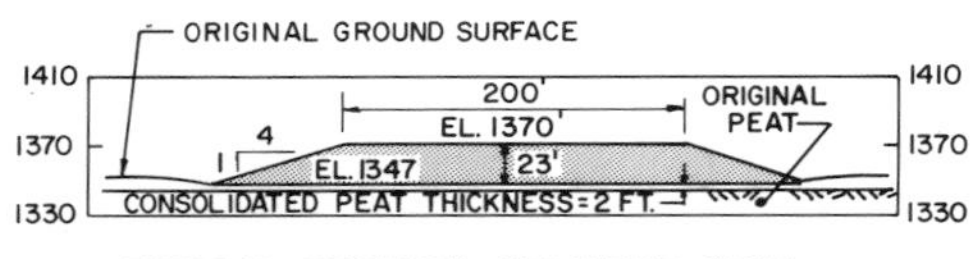

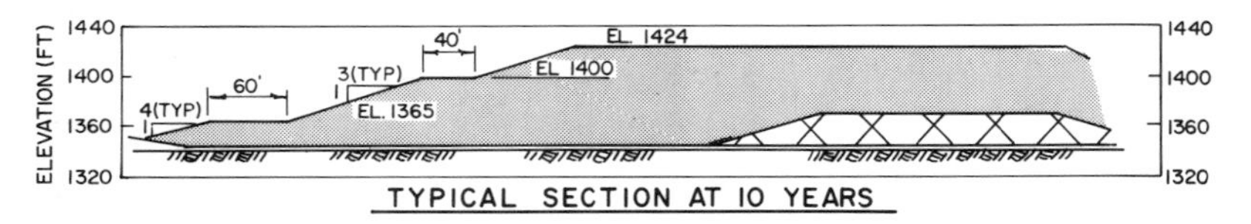

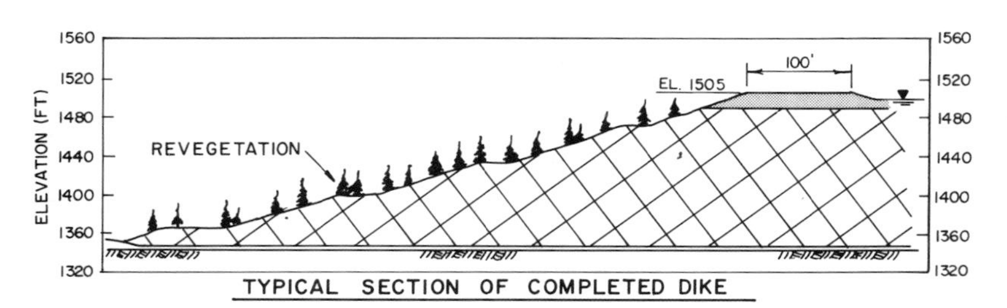

Figure 11.3 Construction stages for a tailing dam (Reproduced from Taylor and d'Appolonia, 1977, by permission)

11.3 Liners

Figure 11.4 shows different liner–leachate collection arrangements that are mandated by various regulations. Consider for example the system in Figure 11.4a. The primary or the upper liner is a flexible membrane liner (FML) with a minimum thickness of 30 mil (EPA). In the Federal Republic of Germany the required minimum thickness is 80 mil. Because of its very low hydraulic conductivity, the FML is believed to be capable of preventing the migration of hazardous waste into the underlying liner during the operation of the facility and a post-closure period of 30 years. The 30-year period is selected based on the assumption that after such period the concentrations of various contaminants are reduced to acceptable limits. The bottom liner usually consists of 3 ft (0.9 m) of clay with a hydraulic conductivity of 10^{-7} cm/s or less.

A liner is permitted to accept contaminants into it but not permit migration of contaminants through it. Because of the concern that 3 ft (0.9 m) of clay would not achieve this objective, a double liner system (Figure 11.4b) is preferred, in which another FML liner is introduced on the top of the lower earthen liner. Considering that construction defects of the upper FML may occur, a double dual liner system was developed as illustrated in Figure 11.4c. Both the primary and the secondary liners should be of the same thickness.

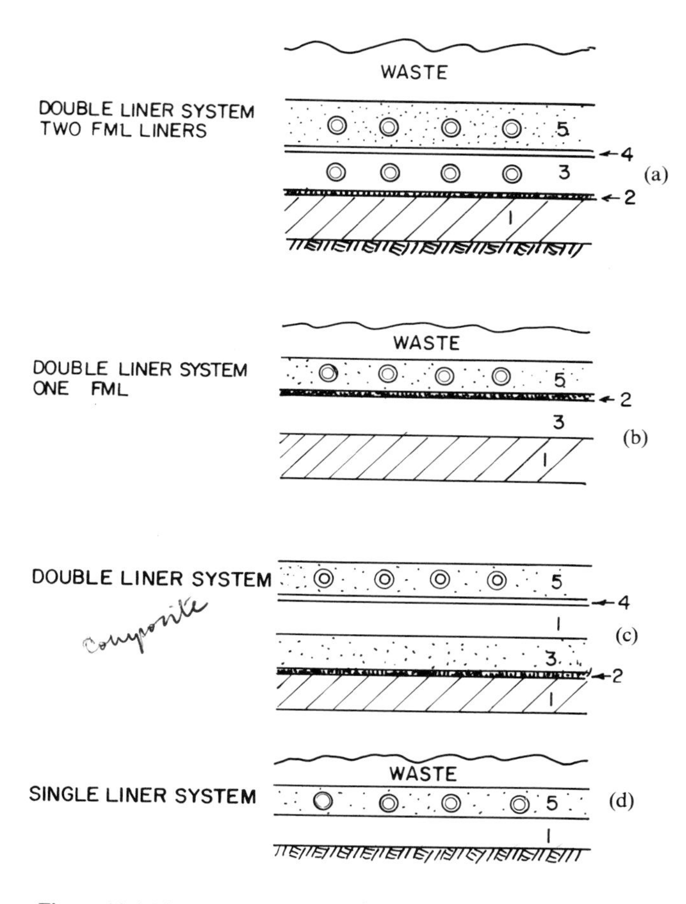

Figure 11.4 Liner systems: 1, compacted, low-permeability clay; 2, flexible membrane liner (FML); 3, leachate collection/detection system; 4, FML; 5, primary leachate collection system; ◎, collection pipes

The geomembrane liner systems in valleys at mine waste disposal sites are subjected to very high pressures. Brendel *et al.* (1987) used pressures up to 4 ton/ft^2 (400 kPa) on various types of geomembranes with thickness between 30 and 55 mil and embedded in a variety of coarse-grained materials. The embedment in a crusher run angular sand showed significant surface distress with the thickness reduced to one-quarter of the original thickness at particle contact points. When AASHTO No. 8 gravel was the bedding and cover material, the membranes with thickness up to 50 mil failed. The failure loads ranged below those expected from the embankment to 1.5 times the expected value. These tests where woven and non-woven geotextiles protected the flexible geomembrane from the coarse-grained materials showed non-woven geotextile layers to provide most protection against the point loads. However, the friction angle between the membrane and the bedding material was reduced considerably, the largest reduction being for HDPE embedded in sand. The φ value dropped from 30.5° to 6°.

The purpose of the leachate collection system (LCS) is to maintain a specified design hydraulic head, usually 1 ft (300 mm) or less, above the liner and convey the leachate to the collection pipes. Maintaining a minimum head reduces the release of pollutants in the event of a failure of the system. The secondary leachate collection system provides a medium for detecting, collecting and conveying any leachate migrating through a failed primary system.

11.3.1 Clay liners

The key elements controlling the behaviour of a clay liner are the compaction, moisture content, method of compaction, quality control during compaction, and potential increase in the hydraulic conductivity due to interaction with waste liquids. Clay liners compacted at moisture content wet of optimum are less permeable compared with those compacted dry of optimum (see Chapter 10).

Clay liners compacted by kneading (e.g. using a sheepsfoot roller) are less permeable due to the dispersion of soil particles. Clay placed in the field with higher percentages of clods and chunks could have a conductivity several orders of magnitude higher than 10^{-7} cm/s.

Plastic clays are more difficult to break and compact in the field at the optimum moisture content, are more sensitive, take relatively less time to dry when above optimum, or to moisten when below optimum. While clays with higher plasticity index are usually of lower permeability for greater ease of installation it is advisable to use lean clay with a plasticity index <20.

A demonstrated effect of waste fluids on clay liners is the increase in hydraulic conductivity (see Chapter 9). There is no established procedure other than the laboratory permeability tests to investigate the effect of waste fluids on liners. Bowders *et al.* (1984) suggested 'batch equivalent tests' to determine the absorption characteristics by measuring the number of pore volumes of flow required to achieve a breakthrough of a constituent into the effluent liquid. The time required to do the test places a limitation on utilizing this test in practice. Instead, sedimentation experiments were utilized to investigate the effect of waste fluids on the reduced concentration of clay particles and hence the sensitivity of the soil to the test liquid (Bowders *et al.*, 1984; Haxo and Dakessian, 1987). Khera *et al.* (1987) showed that expansion index was a good indicator of the effect of chemicals on soil permeability (Chapter 9).

For hazardous waste sites with high expected concentrations of organic contaminants, the use of both the permeability test and other tests such as sedimentation and expansion index tests may be necessary. Engineering properties tests and index properties tests could be conducted on contaminant-free laboratory-constituted samples and samples mixed with expected concentrations of contaminants (Chan *et al.*, 1986; Bowders *et al.*, 1984).

The effect of contaminants on the physical and engineering properties of the clay depends on the clay mineral and the nature of the waste fluid. Testing for the clay mineral is not a routine investigation requirement since the source of clay for a liner is not usually known prior to construction. Compatibility testing is usually performed after a construction contract is awarded. As a minimum, a compatibility testing program should include conducting hydraulic conductivity tests using a mixture of expected contaminants as permeant. One or perhaps more pore volumes may have to be passed through the test sample. While not regulated, testing for clay minerals is desirable as it helps to interpret the data.

The state of New Jersey requires testing of at least three separate samples from each clay source. The following tests are required on each source of material:

1. Classification (usually utilizing the unified system).
2. Grain size and Atterberg limits.
3. Compaction (moisture–density relationship).
4. Hydraulic conductivity.
5. Porosity.
6. pH.
7. Cation exchange capacity.
8. Pinhole test (required for clay liner construction over coarse-grain subgrade).
9. Mineralogy (recommended but not required).

The current state of construction is that a clay liner is accepted if it meets the regulatory requirements, which may or may not require specific compatibility testing. The leachate from municipal landfills is typically neutral and the risk of soil dissolution is insignificant.

11.3.2 Flexible membrane liners

Flexible membrane liners (FMLs) are synthetic polymers. Many common polymers are made by the chemical combination of small molecules called monomers. A monomer contains a carbon-to-carbon double bond (atoms sharing two pairs of electrons) that breaks into single bonds upon polymerization. The polymer polyethylene for example is made of ethylene monomers, and polyvinyl chloride (PVC) is made of vinyl chloride (see Figure 11.5). Other ingredients are added to PVC polymers to increase their flexibility. Vulcanizers alter the chemical structure of rubber, which becomes less plastic and more resistant to swelling by organic fluids.

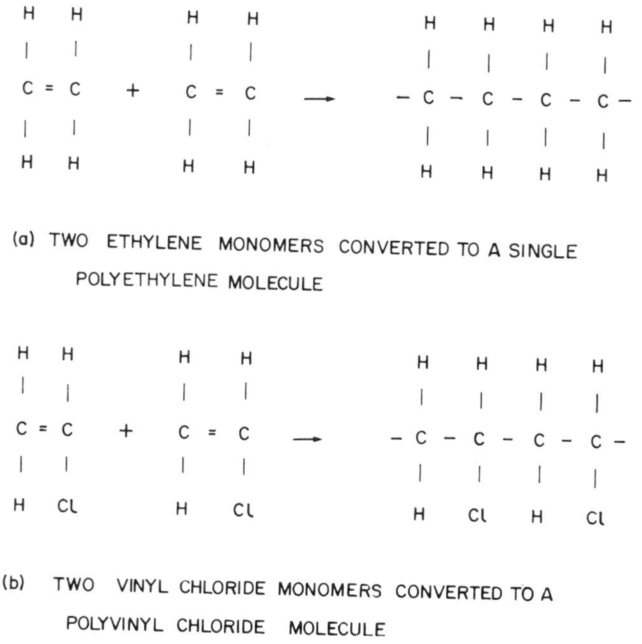

```
H   H          H   H              H   H   H   H
|   |          |   |              |   |   |   |
C = C    +     C = C    →       - C - C - C - C -
|   |          |   |              |   |   |   |
H   H          H   H              H   H   H   H
```

(a) TWO ETHYLENE MONOMERS CONVERTED TO A SINGLE POLYETHYLENE MOLECULE

```
H   H          H   H              H   H   H   H
|   |          |   |              |   |   |   |
C = C    +     C = C    →       - C - C - C - C -
|   |          |   |              |   |   |   |
H   CL         H   CL             H   CL  H   CL
```

(b) TWO VINYL CHLORIDE MONOMERS CONVERTED TO A POLYVINYL CHLORIDE MOLECULE

Figure 11.5 Molecular structures of polyethylene and polyvinyl chloride

Reinforcement is introduced into some types of FML to improve their strength and deformation properties and their resistance to cracking. In the spread-coating method woven or non-woven fabric is impregnated and coated with polymeric or asphaltic compounds on one or both sides. In calendered reinforced geomembranes the reinforcing (support) is achieved by using open fabrics such as scrims (loosely woven strong fabric) of nylon or other lightweight fibres.

11.3.2.1 Types of membranes

There are four broad classes of flexible membrane liners (National Sanitation Foundation):

1. *Thermoplastics*: Thermoplastic polymers such as PVC become soft when heated and can be moulded, extruded and shaped. Others include oil-resistant PVC (PVC-OR) and thermoplastic nitrile-PVC (TN-PVC).
2. *Thermoplastics with crystalline content*: Some of the polymeric chains tend to be arranged in an ordered crystal lattice. Examples include low-density polyethylene (LDPE), high-density polyethylene (HDPE), polypropylene and elasticized polyolefin.
3. *Elastomers* (rubbers) such as butyl rubber, ethylene propylene diene monomer (EPDM) and polychloroprene (CR), also called Neoprene.
4. *Thermoplastic elastomers* such as chlorinated polyethylene (CPE) and chlorosulphonated polyethylene (CSPE), also known as Hypalon.

Mechanical properties of unreinforced polymers are shown in Table 11.1. Of special design interest are the strength and elongation relationships. PVC could withstand larger strains compared with other polymers and a typical design value for elongation is 100%. HDPE is less flexible and, to avoid yield, a typical design value is a strain of 5%. Some polymers such as CSPE (Hypalon) are improved by reinforcement to withstand larger strains (e.g. 10%) and the controlling factor will be the breakage of the reinforcing fabric at about 20% strain.

Table 11.1 Some physical and mechanical properties of unreinforced polymers

	PVC		*HDPE*			*EPDM*			*CPE*
Nominal gauge (mil):	*30*	*45*	*30*	*40*	*60*	*30*	*45*	*60*	*30*
Relative density (minimum)	1.2	1.2	0.94	0.94	0.94	1.15	1.15	1.15	1.2
Tensile properties (minimum)									
Break tensile strength (lb/in)	69	104	90	120	180	42	63	84	43
Yield tensile strength (lb/in)	NA	NA	50	70	120	NA	NA	NA	NA
Elongation at break (%)	300	300	500	500	500	300	300	300	300
Elongation at yield (%)	NA	NA	10	10	10	NA	NA	NA	NA
Tear resistance (lb) (minimum)	8	11	15	20	30	4	6	8	4.5

NA, not applicable.

11.3.2.2 Testing of FML

Standard procedures are used by the industry to test liner products. Some of the tests are (National Sanitation Foundation, 1985 – NSF 54; Koerner, 1986):

1. *Puncture resistance* (ASTM D2582): This is used as an index to characterize the ability of the liner to withstand contacts with sharp angular stones, and the activities of man or vehicles. The laboratory test consist of clamping the specimen to a ring on which a plunger is forced until it punctures the membrane.
2. *Tear resistance or strength* (ASTM D1004, DieC, ASTM D751, ASTM 624, DieC): This is used as an index to characterize the potential for failure due to tearing stress. Because the liner is very thin compared with the subgrade, tearing stresses may be generated. The tearing stress may also result from the activities of man and the operation of equipment. In a tearing test a notch or a cut is made into the test specimen and a tensile force is applied such that tearing occurs at right angles to the direction of the load.
3. *Strength properties* (ASTM D638): These tests characterize the liner tensile strength, and stress–strain behaviour at initial yield and breakage.
4. *Resistance to soil burial* (ASTM D3083).
5. *Dimensional stability* (ASTM D1204, 212°F, 15 min).

11.3.2.3 Compatibility test

A method for testing the compatibility of waste fluid and FML has been developed by EPA under EPA Method 9090, 'Immersion Test for Membrane Liner Material, for Chemical Compatibility of Waste'. In this test, samples are immersed in the liquid waste at two different temperatures for four months. At the end of 1, 2, 3 and 4 months the samples are removed from the immersion tray and tested for strength, puncture, tear, and weight change. Any substantial variation of these indicators is used to characterize the effects of waste fluid on the liner. The nature of waste fluids alters with changes in the type of industrial waste. It is, therefore, essential that compatibility testing be carried out using the anticipated waste fluids. Compatibility tests should also be performed on pipes, filter elements, drainage and other materials that come in contact with leachate.

Another parameter that should be considered is the range of temperature that may exist in a landfill due to various chemical and biological actions and their effect on the properties of flexible membranes. It is important to realize that none of the methods currently in use for determining the suitability of these materials for use in a landfill are able to predict their behaviour on a long-term basis. In addition, there are few field data available to validate the laboratory tests because the use of synthetic materials in a landfill environment is very recent.

11.3.2.4 Liner selection

Each of the FML types has its own attributes and liabilities. Both the high density polyethylene with crystalline content and low density polyethylene with crystalline content have shown resistance to a variety of wastes. In manufacturing polyethylenes different catalytic processes and co-monomers are used, yielding materials of diverse quality with considerable cost variations. The polyethylene resin parameters that influence the properties of the final product are molecular weight and its distribution, and co-monomer type and its concentration, and type and amount of crystallinity

Table 11.2 Broad advantages and disadvantages of some polymers

	Advantages	*Disadvantages*
Polyvinyl chloride	Resistant to inorganics, good tensile strength, elongation, puncture and abrasion resistance, easy to seam	Low resistance to organic chemicals, including hydrocarbons, solvents and oil, poor resistance to exposure
High-density polyethylene	Good resistance to oils, chemicals and high temperature, available in thick sheets (20–150 mil)	Require more field seams, subject to stress cracking, punctures at low thickness, poor tear propagation
Ethylene propylene rubber	Resistant to dilute concentrations of acids, alkalis, silicates, phosphates and brine, tolerates extreme temperatures, flexible at low temperatures, excellent resistance to exposure	Not recommended for petroleum solvents or halogenated solvents, difficult to seam or repair, low seam strength
Chlorosulphonated polyethylene	Good resistance to ozone, heat, acids and alkalis, easy to seam	Poor resistance to oil, good tensile resistance if reinforced

(Cadwallader, 1986). FMLs made from HDPE and LDPE are most frequently used at hazardous waste sites.

At sanitary landfills where buried liners are exposed predominantly to inorganic chemicals, several types of liner could be used. Table 11.2 shows the broad advantages and disadvantages of various products (US Environmental Protection Agency, 1985).

11.3.2.5 Quality control

Installation of FMLs requires seaming the sheets together. Seams are the most critical stage of construction. As few seams as possible should be used and these should be factory seams wherever feasible. They must be carefully controlled, inspected and tested. The strength of a seam is measured in shear, dead loads, or peel modes. Tests are performed on both factory seams and samples cut out of field seams. Extrusion welding is considered the most reliable method of seaming. This technique involves placing a molten parent liner material between overlapping sheets or over the top of adjoining sheets.

Other types of seams are (National Sanitation Foundation, 1985):

1. *Adhesive seams*: A chemical adhesive is used to develop bond strength between the membrane surfaces. Adhesive seams are used for softened PVC and CPE with an overlap of 100 mm.
2. *Bodied solvent seams*: The parent material dissolved in a solvent is used to soften and bond the membrane materials with an overlap of 100 mm.
3. *Dielectric seams*: High-frequency dielectric equipment is used to generate heat and pressure on an overlap seam joint, resulting in a homogeneous melt of the two membrane surfaces. The seam is very strong and an overlap of 20 mm is adequate.
4. *Solvent seams*: Solvents are used to soften and bond the membrane surfaces.
5. *Tape seams*: A chemical adhesive is used to tape a seam joint. The tape provides the bond and the tensile strength of the joint.
6. *Thermal seams*: High temperature is produced between an overlap seam joint to

melt the membrane surfaces, followed by the application of pressure, which results in the homogeneous bond of the two membrane surfaces.

7. *Vulcanized seams*: Overlapped unvulcanized sheets are cured together using heat and pressure. This method is suitable for PVC, LDPE, HDPE and CPE requiring a seam of about 20 mm width.

The quality of the seams is verified by visual examination, vacuum testing, or ultrasonic testing. Visual examination is possible if adjoining sheets are welded together. The vacuum method employs a transparent suction box about 3 ft long placed over a seam portion wetted with soap solution. As suction is applied, bubbles form in the event of a leak. The ultrasonic technique verifies the thickness of the weld to ensure the pressure of the weld. It is used for non-reinforced geomembranes.

Other quality control measures are designed to ensure protection of the liner during installation. These include having a minimum cover of soil or geotextile to avoid puncturing the liner, elimination of folds and prohibition of seaming in cold weather, elimination of sharp objects in the subgrade, allowing no equipment on top of the liner, and specifying shoes for workers. Some of the other factors to be considered are handling, transporting, storage conditions, and protection during and after construction. Underliners and covers may be used to protect the geomembranes from puncture, tearing, abrasion, ultraviolet rays, etc. Where granular soils are used adjacent to the membranes the material should be founded and no more than 0.5 in in size. Other elements of considerable importance are where the membrane meets the inlet and outlet pipes, walkways, sampling sites, and so on. Guidelines for quality assurance are provided in the US Environmental Protection Agency (1985) document, 'Minimum Technology Guidance Document on Double Liner System'.

11.3.3 Admixed liners

Admixed liners (Matrecon, 1980) include a variety of materials that are admixed or mixed in place. Examples of these liners are the soil bentonite, hydraulic asphalt, and soil cement.

11.3.3.1 Soil–bentonite

Bentonite is added to the borrow soil in percentages typically varying from 3 to 6% by weight. Sands with plastic fines of 20–30% are usually preferable. The design mix (percentage of bentonite, grain size, optimum moisture content and conductivity) is determined based on laboratory tests. Compatibility with the anticipated waste fluid should be evaluated because sodium bentonite is known to be vulnerable to attacks by some organic chemicals. The design thickness is usually 1 ft (305 mm) or more.

The mixture is prepared in a central mixing plant within the site and applied wet of the optimum using a mechanical spreader in typically 8 in (200 mm) loose lifts and compacted using a vibratory compactor. The quality control procedures in the field are similar to those for the clay liner.

11.3.3.2 Hydraulic asphalt cement

Asphalt cement (6.5–9.5% in weight) and high-quality mineral aggregates are hot mixed to produce hydraulic cement concrete (HAC). This is compacted in place in a manner similar to that used for asphalt concrete paving. The proposed mixes are usually tested for:

1. Grain size of mineral aggregates.
2. Density of the mineral aggregates and asphalt mix.
3. Percentage of soil in the compacted mix.
4. Penetration test.
5. Stability – triaxial compression test.
6. Hydraulic conductivity of the mix.

11.3.3.3 Soil–cement

Soil–cement is a mixture of portland cement, water and selected borrow soil. Soil with fines content $>50\%$ is difficult to mix. A fine content (soil passing No. 200 standard US sieve) of $\leqslant 35\%$ is preferable. A minimum design compacted thickness of 1 ft (305 mm) is typically used. Soil cement liners can be degraded in environments that tend to degrade cement such as highly acidic environments. Low resistance to cracking and shrinkage and the difficulty in attaining a low permeability of 10^{-7} cm/s inhibits the use of soil cement. The trial mix testing covers the following:

1. Grain size analysis of aggregate.
2. Soil cement content.
3. Wetting and drying.
4. Freezing and thawing.
5. Compressive strength.
6. Compaction.
7. Hydraulic conductivity.

Field quality control tests include grain size distribution, cement content, and percentage compaction. The mixture is prepared in a central mixing plant and spread by mechanical spreader in maximum 9 in lifts.

11.4 Leachate collection

The best procedure to prevent waste fluids from reaching the groundwater is obviously to collect the leachate by placing a collection system above the liner. Figure 11.6 shows a simple model for a leachate collection system. The drains are spaced at a distance L apart and the liner is sloped toward the drains. Considering the distance s from the drain, the volume of flow at section s is equivalent to the percolation e from the base of the landfill times the distance from the ridge line to the point where the section is taken. Considering no flow through the liner, and no change in saturation, the input flow is equivalent to the flow passing through the vertical section. Thus,

$$e(L/2 - s\cos\alpha) = k_{\mathrm{d}} h \frac{\mathrm{d}}{\mathrm{d}s}(h + s\sin\alpha) \qquad (11.1a)$$

where k_{d} is the hydraulic conductivity of the drainage layer.

By integrating Equation 11.1a,

$$\int e(L/2)\,\mathrm{d}s - \int es\cos\alpha\,\mathrm{d}s = \int k_{\mathrm{d}} h\,\mathrm{d}h + \int k_{\mathrm{d}} h\sin\alpha\,\mathrm{d}s \qquad (11.1b)$$

Equation 11.1a is non-linear and the second integral of the right-hand side of Equation 11.1b is not directly obtainable. If h is assumed uniform over the distance s,

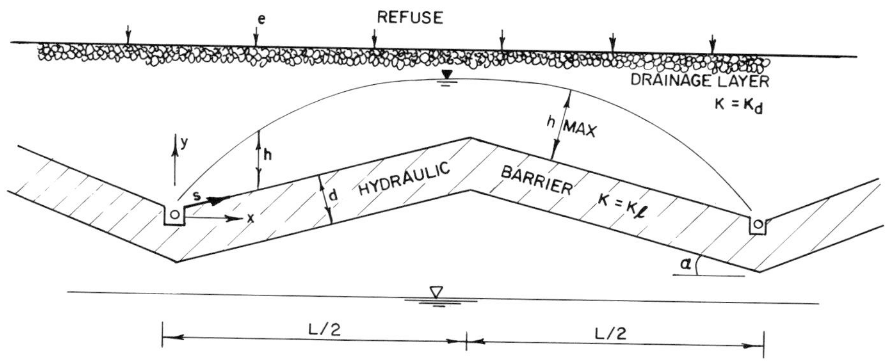

Figure 11.6 Leachate collection with a sloping base

and for small values of α, the maximum head h_{max} is conservatively computed as:

$$h_{max} = (L/2)[(\tan^2 \alpha + e/k_d)^{1/2} - \tan \alpha] \qquad (11.1c)$$

An accurate solution to Equation 11.1a could be obtained by a trial and error method (Demetracopoulos and Korfiatis, 1984). Typical values are $\tan \alpha = 0.05$, $k_d = 10^{-2}$ cm/s.

The flow to drain pipe per unit length of drain, Q_d, is computed from Equation 11.2:

$$Q_d = eL \qquad (11.2)$$

For a saturated liner of thickness d and hydraulic conductivity k_l, the maximum quantity of leachate passing through is:

$$q_l = [k_l(d + h_{max} - h_b)/d]\, A\, dt \qquad (11.3)$$

where q_l is the volume of flow passing through the liner, $(d + h_{max} - h_b)/d$ the hydraulic gradient, A the area of the flow, dt the time interval, and h_b the pressure head (negative for suction) at the bottom of the liner.

Equation 11.3 provides a conservative estimate of leakage through the liner. During the early phase of leakage, the full depth d will not be saturated. The time t required for the saturated front to reach a depth z can be estimated by solving the one-dimensional flow Equation 11.4a assuming a constant suction of capillary tension at the wetting front.

$$(w_s - w_i)\, dz = [(h_{max} + \Phi_s + z)/z]k_u\, dt \qquad (11.4a)$$

To determine the time for the saturation front to reach a depth d_s, Equation 11.4a is rearranged and integrated by parts over z from 0 to d_s. The solution is given by Equation 11.4b.

$$t = [(w_s - w_i)/k_u]\{d_s - (h_{max} + \Phi_s) \ln[1 + d_s/(h + \Phi_s)]\} \qquad (11.4b)$$

where w_s is the saturated volumetric moisture content, w_i the initial moisture content, d_s the depth of the saturated front, k_u the unsaturated hydraulic conductivity at the wetting front, and Φ_s the suction or capillary head.

k_u and Φ_s are functions of the moisture content and determined experimentally (Olson and Daniel, 1979; Elzeftway and Cartwright, 1979). The relations are non-linear. For moisture content larger than the field capacity, k_u may be expressed as

$$k_u/k_s = [(w_i - w_f)/(w_s - w_f)]^m \tag{11.5}$$

where w_f is the volumetric moisture content at field capacity, m is an experimental factor, typically 1.5–3, and k_s is the saturated hydraulic conductivity of the clay liner.

Example 11.1

Assume that in a landfill the percolation through the cover is 55 cm/year as calculated in Example 7.1. The liner slopes at 6% and the drain spacing is 30 m (Figure 11.6). Calculate the minimum thickness of the drainage layer and estimate the flow volume through 300 m drain length if $k_d = 10^{-2}$ cm/s.

Solution Use Equation 11.2:

$$e = 55/60 \times 60 \times 24 \times 365$$

$$= 1.744 \times 10^{-6} \text{ cm/s}$$

$$\text{Flow volume} = 1.744 \times 10^{-6} \times 30 \times 300$$

$$= 157 \times 10^{-6} \text{ m/s}$$

$$= 157 \text{ mm/s}$$

The maximum head, from Equation 11.1c, is:

$$h_{max} = 30 \times 100/2 \left[\left(\frac{0.06^2 + 1.744 \times 10^{-6}}{10^{-2}} \right)^{0.5} - 0.06 \right]$$

$$= 2.15 \text{ cm}$$

Minimum thickness of drainage layer = 30 cm > 2.15 cm

11.4.1 Liner performance

The liner efficiency could be evaluated by considering a slug of fluid with head h_0 instantaneously applied above the liner (see Figure 11.7). It is assumed that all the materials are at field capacity, the hydraulic conductivity of the drainage blanket is the same as that of refuse, and the leachate collection pipes are free draining. The efficiency is assessed by calculating the volume of flow into the drain and through the liner (Moore, 1983; Wong, 1977; Kmet *et al.*, 1981). The fraction of the liquid (length of saturation) moving into the drain at time t is:

$$s/s_0 = 1 - t/t_1 \tag{11.6}$$

where t_1 is the time for the trailing edge of leachate to drain to the pipe:

$$t_1 = ns_0/k_d \sin \alpha$$

and n is the porosity of the drainage layer.

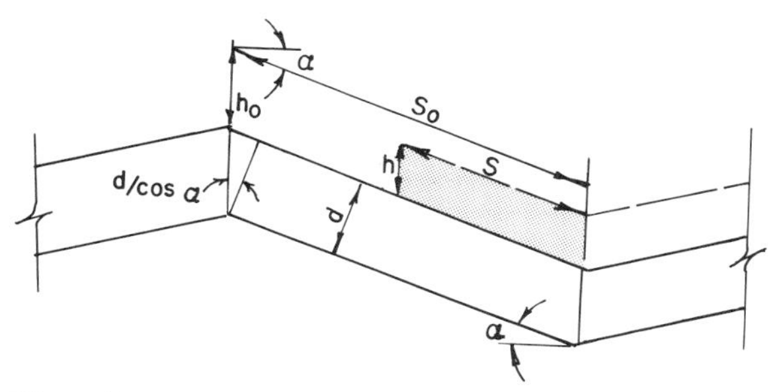

Figure 11.7 Geometry of the Wong model

The head h at the liner at time t is given by:

$$h/h_0 = (1 + d/h_0 \cos\alpha)\exp(-K't/t_1) - (d/h_0 \cos\alpha) \tag{11.7}$$

for $0 < t < t_1$, where

$$K' = s_0 \cot\alpha\, k_1/dk_d$$

$$h_0 = e/n$$

The quantity of leachate leakage q_1 through the liner per unit width of the module is given by:

$$q_1/q_0 = \left(\frac{1 + d/h_0 \cos\alpha}{K'}\right)\left[(\exp(-K't/t_1))\left(-K' + 1 + \frac{tK'}{t_1}\right) + (K' - 1)\right] \tag{11.8}$$

where $q_0 = h_0 s_0 n$.

The time t_2 for the leachate volume to leak through the liner, i.e. the time it takes for head h to become zero, is:

$$t_2 = (t_1/K') \ln[1 + h_0/(d/\cos\alpha)] \tag{11.9}$$

The amount of flow to the drain q_d is:

$$q_d = q_0 - q_1 \tag{11.10}$$

and the efficiency E (%) is:

$$E = 100\left(\frac{1 - q_1}{q_0}\right) \tag{11.11}$$

The initial value of h_0 in Equation 11.8 should be the maximum expected head during a given spring recharge event. Since t_1 in Equation 11.6 could be more than a year, h_0 must reflect residual values from previous years. The initial instantaneous head in any year, y, is calculated assuming the leftover leachate on the liner to be redistributed over s_0. The new $h_{0(y)} = h_{0(y-1)}\, s/s_0 + e/n$.

Example 11.2

For Example 11.1, determine the efficiency of the liner. Assume liner thickness = 1 m, hydraulic conductivity of the liner = 10^{-7} cm/s, hydraulic conductivity of the drainage layer = 10^{-2}cm/s and porosity = 0.4.

Solution

$d = 1$ m

$k_1 = 10^{-9}$ m/s

$\alpha = \tan^{-1} 0.06 = 3.43°$

$s_0 = 15$ m

Assume that 55 cm/year percolation occurs instantaneously.

$h_0 = 55/0.4 = 137.5$ cm/year percolation occurs instantaneously

$t_1 = (0.4 \times 15)/10^{-4} \sin 3.43$

$= 1.003 \times 10^6$ s

$= 278.6$ h

$$k^1 = \frac{15 \cot 3.43 \times 10^{-5}}{1} = 0.0025$$

$t_2 = (278.6/0.0025) \ln\{1 + [1.375/(1/\cos 3.43)]\}$

$= 96280$ h $\qquad t_1$ controls

$q_1/q_0 = [(1 + 1/1.375 \cos 3.43)/0.0025][(e^{-0.0025})(-0.0025 + 1 + 0.0025) + 0.0025 - 1]$

$= 0.0022$

$E = (1 - 0.0022)100 = 99.8\%$

11.4.2 Drainage

The function of the drainage layer is to facilitate the flow of leachate to the drain pipe and hence avoid a leachate mound build-up. To minimize clogging, some designers specify a filter between the solid waste and the drainage layer.

11.4.2.1 Soil filter

The purpose of the filter is to allow leachate to pass through but prevent the migration of fines. Filters are usually selected based on the grain size distribution of the base materials. Since it is usually not possible or practical to characterize the grain sizes of solid waste, the selection of a filter becomes a matter of experience and judgement. EPA guidelines (US Environmental Protection Agency, 1985) recommend a minimum filter thickness of 6 in (150 mm). Considering the waste material as a base, the following criteria are applicable for soil filters (NAVFAC, 1982).

To avoid movement of particles from the base:

$$\left.\begin{aligned} D_{15F}/D_{85B} &< 5 \\ D_{50F}/D_{50B} &< 25 \\ D_{15F}/D_{15B} &< 20 \end{aligned}\right\} \qquad (11.12)$$

To avoid head loss in a filter, D_{15F}/D_{15B} must be >4, and the permeability of the filter must be large enough to suffice for the particular drainage system, where D_{15}, D_{50} and D_{85} are the particle sizes at which 15, 50 and 85% of the materials (filter, F; base, B) by weight are finer.

For a very uniform base material ($C_u < 1.5$), D_{15F}/D_{85B} may be increased to 6.

For a broadly graded base material ($C_u > 4$), D_{15F}/D_{15B} may be increased to 40.

To avoid segregation, the filter should contain no size larger than 3 in and, to avoid internal movement of fines, it may have less than 3% passing a No. 200 sieve.

For typical municipal refuse, the grain size could vary over a wide range. Grain size distributions of shredded refuse, Fungaroli and Steiner (1979) suggest that D_{15} varies from 0.1 to 6 mm, D_{85} from 2 to 100 mm, and D_{50} from 0.8 to 100 mm. It is clear that, based on the above criteria, the filter size could vary over a wide range. In the absence of grain size data a very porous filter consisting of $\frac{3}{4}$ to $\frac{3}{8}$ in gravel could be used on the first layer of refuse. The suitability of the selected filter with respect to clogging could be assessed in the laboratory by utilizing representative leachate as a permanent factor in a hydraulic conductivity test, and by determining the steady-state conductivity value.

11.4.2.2 Geotextile filter

Geotextiles in geotechnical application are pervious textile materials made from synthetic polymers such as polypropylene, polyester and polyethylene. They are either woven or non-woven fabrics. Woven fabrics are made using conventional textile machinery with fibres systematically woven. The non-woven fabrics are made of fibres arranged at random. Geotextiles are available in thickness ranging from 0.3 to over 5 mm.

The hydraulic characteristics of a geotextile are expressed in terms of its opening size, gradient ratio, area ratio, and hydraulic conductivity. The equivalent opening size (EOS) is expressed in terms of the US standard sieve of opening size equal to the size of the particles of which the fabric will retain 95%. For most geotextile fabrics the EOS values are between US standard sieve No. 40 and No. 100. The apparent opening size (AOS) of a fabric is that which allows 5% of the particles of uniform size to pass through it (ASTM D 4751-87).

The gradient ratio test was developed by the US Army Corps of Engineers to test for clogging. The test result is expressed in terms of the gradient ratio (GR) defined as the hydraulic gradient through the geotextile and 1 in of soil immediately above it divided by the hydraulic gradient across the next 2 in (51 mm) of soil above it. A maximum value of 3 for the gradient ratio was established as acceptance standard (US Army Corps of Engineers, 1977a and b). For important jobs, filter clogging should be evaluated by performing long-term tests using the proposed fabric and the site soils. Flow rate can decrease significantly over time in some cases.

Permittivity is the volume flow rate of water per unit area per unit head in the normal direction. It may be determined using a falling head or constant head (ASTM D 4491) method. The permittivity, Φ, the coefficient of normal hydraulic conductivity, k_n, and the fabric thickness, f_t, are related by:

$$\Phi = k_n/f_t \tag{11.13}$$

Permittivity ranges from 0.02 to over 30 s^{-1}.

The transmissivity, θ, the coefficient of hydraulic conductivity in the plane of the fabric, k_p, and thickness, f_t, are related by:

$$\theta = k_p f_t \tag{11.14}$$

For a method of determining θ see ASTM D 4716.

The mechanical properties are based on the grab tensile strength, puncture tests, and stress–strain behaviour. For further details see Koerner (1986).

Table 11.3 FHA filtration and clogging criteria

Function	*Criteria*	*Soil type*
Filtration	0.149mm < EOS < D_{85}	Soils with ≤ 50% by weight particles passing No. 200 US standard seive
	0.211mm > EOS > 0.149mm	Soils with > 50% by weight passing No. 200 sieve
Clogging (severe conditions)	Non-woven fabric Gradient ratio < 3 Woven fabric open area > 4%	
Clogging (other applications)	Fabric hydraulic conductivity $K_{fabric} > 10 \times K_{soil}$	

Several criteria have been proposed in conjunction with the use of geotextile for filtration when the fabric is in contact with soil. Essentially, the criteria are based on EOS as an index. For a granular base (less than 50% by weight passing a No. 200 sieve) several published criteria were summarized by Rollin and Denis (1987). The FHA criteria require the AOS to be less than D_{85} but larger than 0.149 mm. Other requirements are given in Table 11.3. The severe conditions shown in Table 11.3 may be generated by a combination of high hydraulic gradient, gap-graded particle size distribution, and cohesionless soils (Koerner, 1986). Experimental evidence suggests that woven monofilament fabrics are more resistant to clogging under severe test conditions (Koerner, 1986). The Giroud criterion (Giroud, 1982) allows consideration of soil relative density and gradation.

Geotextiles used as filters in contact with municipal waste are prone to very high risk of clogging (Rollin and Denis, 1987). Limited laboratory experimental evidence suggests a substantial decrease in permittivity under the combined effect of surface deposition of suspended solids and internal clogging (Cancelli and Cazzuffi, 1987). If this laboratory evidence depicts field behaviour, the result could be an undesired build-up of leachate. Careful testing and evaluation should therefore be conducted.

11.4.2.3 Soil drainage layer

The drainage layer should have the characteristics to limit the leachate build-up to less than the specified value of 1 ft (US Environmental Protection Agency, 1985). The hydraulic conductivity is usually chosen to be at least one order of magnitude higher than that of the waste (typically 10^{-2} cm/s) and a minimum thickness of 12 in (305 mm). In the event that a filter is not installed, the drainage layer should be designed to avoid clogging as described in the previous section.

11.4.2.4 Synthetic drainage layers

Synthetic drainage layers are made of polymers with a hydraulic conductivity, k_p, in their plane large enough to transmit fluid or gas. The available products (referred to as geonets or geocomposites) take the form of a plastic net composed of plastic channels, plastic mat with geotextile filler, or corrugated plastic plates with a geotextile filler.

Commercial products of synthetic drainage layers are available in thickness from 4 to over 40 mm. Several layers may be used to achieve the drainage capacity. The two controlling factors in performance are the conductivity under the expected stress from the landfill and long-term reduction in permeability due to creep. Creep effects have been shown to reduce the in-place flow capacity by several orders of magnitude (Smith and Kraemer, 1987).

11.4.2.5 Drainage pipes

Drainage pipes are usually surrounded by a drainage envelope which acts as a filter drain and increases the hydraulic efficiency. The drainage blanket in this case would be the base material. A minimum thickness of 12 in (305 mm) is usually recommended. In order to avoid movement of the filter into the drain pipe, the following criteria may be used (NAVFAC, 1982):

$$\frac{D_{85F}}{\text{slot width}} > 1.2\text{–}1.4 \qquad (11.15)$$

$$\frac{D_{85F}}{\text{hole diameter}} > 1.0\text{–}1.2 \qquad (11.16)$$

Filter systems (filter layers, fabrics, pipes) can become clogged by ferruginous (iron) and carbonate depositions and incrustations. In modern landfill designs, provisions are made for clean-out of drainage pipes are illustrated in Figure 11.8. This is accomplished by the use of high-pressure water jets, the introduction of weak solutions of hydrochloric acid, or other options.

If the drain system cannot be restored to its full efficiency and a leachate mound develops, pumping of leachate could be considered for leachate removal. In this case, the design of the pumping system would require field assessment of the hydraulic properties of refuse utilizing a pumping test (Oweis *et al.*, 1990).

11.5 Foundation stability and settlement

The stability of landfills could be assessed based on the usual limit equilibrium methods commonly used in geotechnical engineering (Oweis *et al.*, 1985). The waste could be characterized by pseudo friction angle φ and cohesion c. The cohesion could be interpreted as apparent cohesion of a reinforced granular mass. Typically, a friction angle φ of 20° to 25° and a cohesion of 200–400 lb/ft^2 (10–20 kPa) are appropriate values for municipal landfills. The procedure for a wedge-type analysis is illustrated in Figure 11.9.

Figure 11.10 illustrates an example of stability analysis. Two potential failure planes are shown. One of the failure planes is taken along the geomembrane–clay liner interface. In this case,

$$\alpha_1 = (45 + \varphi)$$

$$\alpha_2 = (45 + \varphi/2)$$

α_3 is the slope of the liner along the base, β_1 is the slope of the liner as it meets the ground surface, and the friction angle along the geomembrane–soil contact is assumed to be 12°. Analyses of this type are sensitive to the slope of the liner. Several

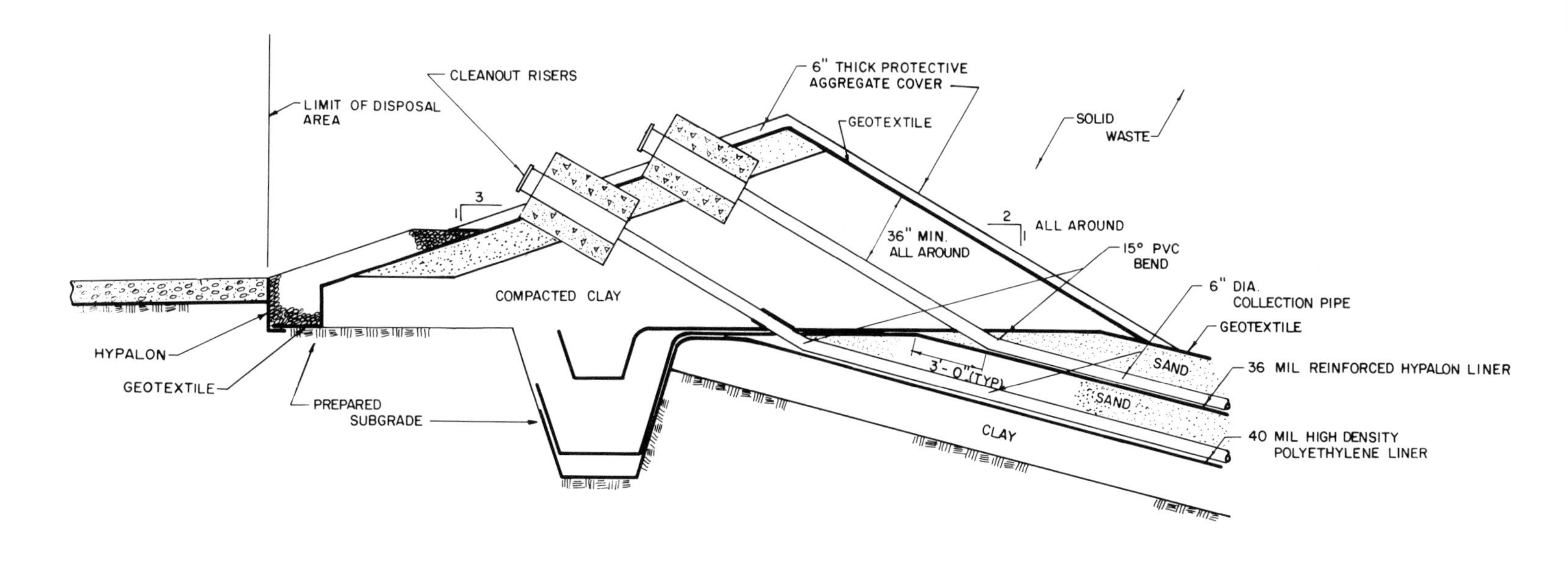

Figure 11.8 Landfill toe schematic: cross-section through cleanout risers

wedges should be tried, and the design (side slope, liner slope, and other geometric parameters) is determined based on the critical wedge producing the minimum factor of safety. In order to limit foundation movements, and potential liner failure, a safety factor of 1.5 or higher is necessary utilizing conservative strength parameters. The second failure plane is taken to be within the soft clay where wedges make an angle of 45° with the horizontal.

Data on the frictional characteristics of the geomembrane (or geotextile–soil interface) are limited. Based on the available data and in the absence of specific testing, the following interface friction angle of values could be used:

Geomembrane on sand	= 15–20°
Geomembrane on compacted clay	= 10–15°
Geonets on geomembrane	= 6–10°
Geomembranes on geotextile	= 12–18°
Geotextiles on geonets	= 10–15°
Geotextiles on sand	= 22–28°

The above values could be exceeded based on field tests. The conditions of the surface of the polymer (cleanliness, roughness, etc.) influence the interface frictional resistance. It is suggested that a factor of safety of at least 1.5 be used in conjunction with the above parameters.

Excessive differential settlements could damage the liner. The methods of Chapter 5 are used to asses the foundation settlement under the waste loading. Differential settlement produces tensile stresses that tend to cause the liner to develop cracks.

An assessment of the tolerable differential settlement of a liner could be made by considering the maximum deflection of a simply supported long beam of unit width with uniform load. The combined deflection due to moment and shear is expressed as (Timoshenko and MacCullough, 1949):

$$\delta_{max} = \left(\frac{5M_{max} \times l^2}{48 \times EI}\right)\left(1 + \frac{9.6EQ_{max}}{Gl^2}\right) \tag{11.17}$$

where δ_{max} is the maximum deflection at the centre, l is the span length, E is Young's modulus, G is the shear modulus, I is the moment of inertia about the neutral axis, Q_{max} is the static moment of cross-sectional area from neutral axis to extreme fibre ($3I/2t$), and t is the depth of the beam.

The maximum moment, M_{max}, is:

$$M_{max} = wl^2/8 = \varepsilon_{max}EI/C \tag{11.18}$$

where ε_{max} is the maximum strain, C the distance from the neutral axis to extreme fibre, and w the load per unit length of beam.

Substituting Equation 11.18 in Equation 11.17,

$$\delta_{max}/l = \left(\frac{5\varepsilon_{max}l}{48C}\right)\left(1 + \frac{14.4IE}{tGl^2}\right) \tag{11.19}$$

Tension tests on compacted clay suggest that E in tension is comparable with or higher than E in compression (Leonards and Narain, 1963; Ajaz and Parry, 1975). The tensile strain ε_{max} at failure (ε_f) is small and published data suggest a range from about 0.03% to over 0.2% depending on the moisture content, plasticity index, unit weight, and clay minerals.

DEFINITION OF TERMS

P_α = RESULTANT HORIZONTAL FORCE FOR AN ACTIVE OR CENTRAL WEDGE ALONG POTENTIAL SLIDING SURFACE a b c d e.

P_β = RESULTANT HORIZONTAL FORCE FOR A PASSIVE WEDGE ALONG POTENTIAL SLIDING SURFACE e f g.

W = TOTAL WEIGHT OF SOIL AND WATER IN WEDGE ABOVE POTENTIAL SLIDING SURFACE.

R = RESULT OF NORMAL AND TANGENTIAL FORCES ON POTENTIAL SLIDING SURFACE CONSIDERING FRICTION ANGLE OF MATERIAL.

P_w = RESULTANT FORCE DUE TO PORE WATER PRESSURE ON POTENTIAL SLIDING SURFACE CALCULATED AS:

$$P_w = \frac{hw_i + hw_{ii}}{2}(L)(\gamma_w)$$

ϕ = FRICTION ANGLE OF LAYER ALONG POTENTIAL SLIDING SURFACE.

C = COHESION OF LAYER ALONG POTENTIAL SLIDING SURFACE.

L = LENGTH OF POTENTIAL SLIDING SURFACE ACROSS WEDGE.

h_w = DEPTH BELOW PHREATIC SURFACE AT BOUNDARY OF WEDGE.

γ_w = UNIT WEIGHT OF WATER.

PROCEDURES

1. EXCEPT FOR CENTRAL WEDGE WHERE α IS DICTATED BY STRATIGRAPHY USE $\alpha = 45° + \frac{\phi}{2}$, $\beta = 45° - \frac{\phi}{2}$ FOR ESTIMATING FAILURE SURFACE.

2. SOLVE FOR P_α AND P_β FOR EACH WEDGE IN TERMS OF THE SAFETY FACTOR (F_s) USING THE EQUATIONS SHOWN BELOW. THE SAFETY FACTOR IS APPLIED TO SOIL STRENGTH VALUES (TAN ϕ AND C).
MOBILIZED STRENGTH PARAMETERS ARE THEREFORE CONSIDERED AS $\phi_m = \text{TAN}^{-1}\left(\frac{\text{TAN}\,\phi}{F_s}\right)$ AND $C_m = \frac{C}{F_s}$.

$$P_\alpha = \left[W - C_m L \text{ SIN } \alpha - P_w \text{ COS } \alpha\right] \text{ TAN}\left[\alpha - \phi_m\right] - \left[C_m L \text{ COS } \alpha - P_w \text{ SIN } \alpha\right]$$

$$P_\beta = \left[W + C_m L \text{ SIN } \beta - P_w \text{ COS } \beta\right]\left[\text{TAN}(\beta + \phi_m)\right] + \left[C_m L \text{ COS } \beta + P_w \text{ SIN } \beta\right]$$

IN WHICH THE FOLLOWING EXPANSIONS ARE TO BE USED:

$$\text{TAN}(\alpha - \phi_m) = \frac{\text{TAN}\,\alpha - \frac{\text{TAN}\,\phi}{F_s}}{1 + \text{TAN}\,\alpha\,\frac{\text{TAN}\,\phi}{F_s}} \qquad \text{TAN}(\beta\ \phi_m) = \frac{\text{TAN}\,\beta + \frac{\text{TAN}\,\phi}{F_s}}{1 - \text{TAN}\,\beta\,\frac{\text{TAN}\,\phi}{F_s}}$$

3. FOR EQUILIBRIUM $\Sigma P_\alpha = \Sigma P_\beta$. SUM P_α AND P_β FORCES IN TERMS OF F_s, SELECT TRIAL F_s, CALCULATE ΣP_α AND ΣP_β. IF $\Sigma P_\alpha \neq \Sigma P_\beta$, REPEAT. PLOT P_α AND P_β VS. F_s WITH SUFFICIENT TRIALS TO ESTABLISH THE POINT OF INTERSECTION (i.e., $\Sigma P_\alpha = \Sigma P_\beta$), WHICH IS THE CORRECT SAFETY FACTOR.

4. DEPENDING ON STRATIGRAPHY AND SOIL STRENGTH, THE CENTRE WEDGE MAY ACT TO MAINTAIN OR UPSET EQUILIBRIUM.

5. NOTE THAT FOR $\phi = 0$, ABOVE EQUATIONS REDUCE TO:

$$P_\alpha = W \text{ TAN } \alpha - \frac{C_m L}{\text{COS}\,\alpha}, \quad P_\beta = W \text{ TAN } \beta + \frac{C_m L}{\text{COS}\,\beta}$$

6. THE SAFETY FACTOR FOR SEVERAL POTENTIAL SLIDING SURFACES MAY HAVE TO BE COMPUTED IN ORDER TO FIND THE MINIMUM SAFETY FACTOR FOR THE GIVEN STRATIGRAPHY.

Figure 11.9 Stability analysis: wedge method

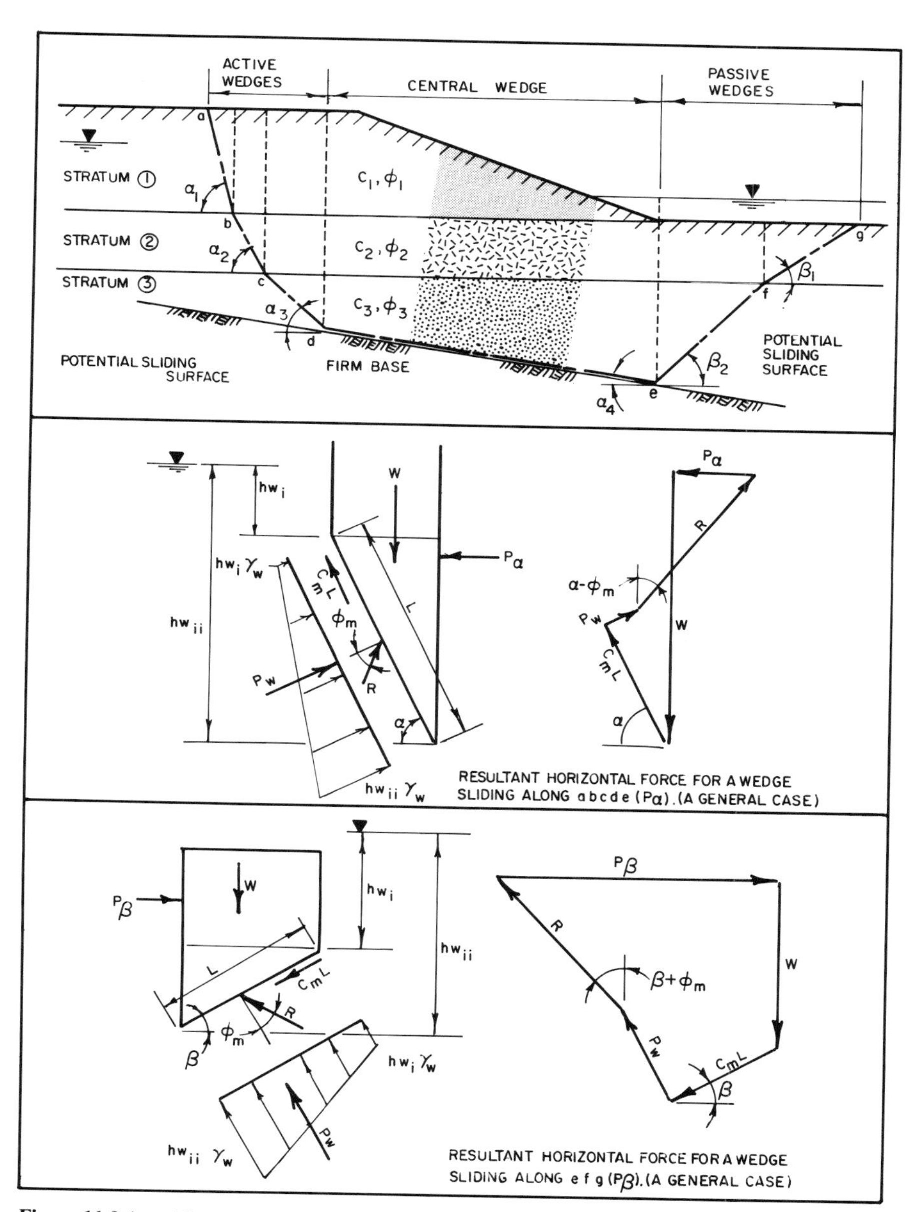

Figure 11.9 (*cont'd*)

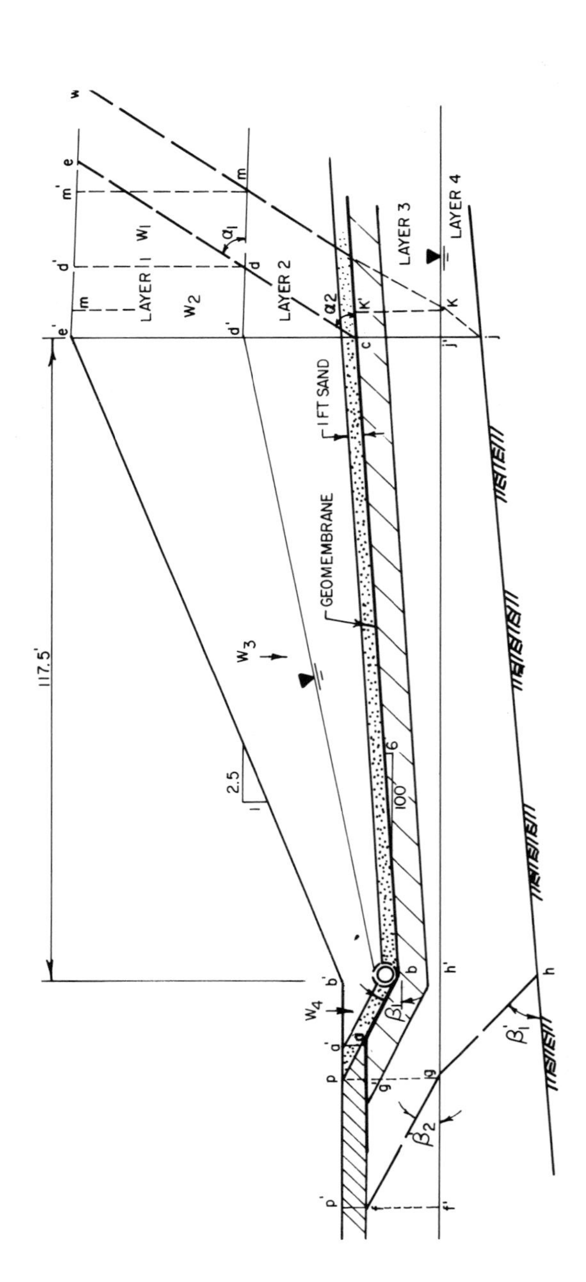

117.5'
LAYER 1
LAYER 2
LAYER 3
LAYER 4
GEOMEMBRANE
1 FT SAND
2.5
1
100
6

SHALLOW POTENTIAL FAILURE SURFACE

REFUSE AND SOIL PROPERTIES

LAYER 1, REFUSE $\phi = 26^\circ$, $C = 100\ lb/ft^2$, $\gamma = 35\ lb/ft^3$

2, REFUSE $\phi = 26^\circ$, $C = 100\ lb/ft^2$, $\gamma = 70\ lb/ft^3$

3, SAND $\phi = 30^\circ$, $C = 0$, $\gamma = 120\ lb/ft^3$

4, SOFT CLAY, $\phi = 0$, $C = 500\ lb/ft^2$, $\gamma = 100\ lb/ft^3$

GEOMEMBRANE : INTERFACE FRICTION ANGLE WITH CLAY LINER = 12°

CLAY LINER: $\phi' = 30^\circ$, $C' = 0$, $\gamma = 120\ lb/ft^3$

CONSIDER THE ONE FOOT SAND LAYER ABOVE THE MEMBRANE AS PART OF LAYER 2.

CONSIDER POTENTIAL FAILURE ALONG GEOMEMBRANE WEDGE abcde

d'e' = 30', d'c = 20' bb' = 10', a'a' = 5', a'b' = 10'

$\alpha_1 = 45 + \phi/2 = 58^\circ$

$W_1 = (30)^2/2\ \text{TAN}\ 32\ (.035) = 9.8K,\ P_{W_1} = 0$

$\alpha_2 = 58^\circ$

$W_2 = (20\ \text{TAN}\ 32)(30)(.035) + \frac{20^2}{2} \times 0.07 \times \text{TAN}\ 32$

$= 13.1K + 8.8K = 21.9K$

$P_{W_2} = \frac{20}{\text{SIN}\ 58} \times \frac{20 \times 0.0625}{2} = 14.7K$

$W_3 = (\frac{30+10}{2})(117.5)(.035) + (\frac{20+0}{2})(117.5)(.07)$

$= 82.3 + 82.25 = 164.55K$

$\alpha_3 = \text{TAN}^{-1}(0.06) = 3.43$

$P_{W_3} = 0$ NO LEAKAGE THROUGH MEMBRANE

$= \frac{20(.0625)(117.5)}{2\ \text{COS}\ 3.43} = 73K$ WITH LEAKAGE

$\beta_1 = \text{TAN}^{-1}\ 1/2 = 26.56$

$W_4 = \frac{5+10}{2}(10)(0.035) = 2.6^K,\ P_W = 0$

$P_{\alpha_1} = [9.8 - \frac{0.1}{F_s}(30)]\left[\frac{\text{TAN}58 - \frac{\text{TAN}\ 26}{F_s}}{1 + \frac{\text{TAN}\ 58\ \text{TAN}\ 26}{F_s}}\right]$

$- [\frac{0.1}{F_s}\ \frac{30}{\text{SIN}\ 58}\ \text{COS}\ 58]$

$= (9.8 - \frac{3}{F_s})(\frac{1.6F_s - 0.49}{F_s + 0.78}) - \frac{1.88}{F_s}$

$P_{\alpha_2} = [21.9 - (\frac{0.1}{F_s})20 - 14.8\ \text{COS}\ 58][\frac{1.6F_s - 0.49}{F_s + .78}]$

$- [\frac{0.1(20)}{F_s \cdot \text{TAN}\ 58} - 14.8\ \text{SIN}\ 58]$

$= (21.9 - \frac{2}{F_s} - 7.8)(\frac{1.6F_s - 0.49}{F_s + 0.78})$

$- \frac{1.25}{F_s} + 12.6$

P_{α_3} (IMPERVIOUS LINER)

$= [164.55]\left[\frac{\text{TAN}3.43 - \frac{\text{TAN}\ 12}{F_s}}{1 + \frac{\text{TAN}\ 3.43\ \text{TAN}\ 12}{F_s}}\right]$

$= 164.55\ [\frac{0.06F_s - 0.212}{F_s + 0.013}]$

$P_{\beta_1} = 2.6\left[\frac{\text{TAN}\ 26.56 + \frac{\text{TAN}\ 12}{F_s}}{1 - \frac{\text{TAN}\ 26.56\ \text{TAN}\ 12}{F_s}}\right]$

$P_{\beta_1} = 2.6\left[\frac{0.5F_s + 0.212}{F_s - 0.106}\right]$

F_s	$P\alpha_1$	$P\alpha_2$	$P\alpha_3$	$\Sigma P\alpha$	$\Sigma P\beta$
1.1	3.02	19.68	-21.5	-1.2	1.99
1.2	3.73	20.54	-19.0	-5.27	1.93

$F_s \simeq 1.12$

Figure 11.10 Stability analysis: Sussex wedge method

For a homogeneous material with $E/G = 2.8$ and assuming compression = tension, a factor of safety of 1.0, and maximum strain equal to failure strain ε_f, the limiting δ_{max}/l for crack development is:

$$\delta_{max}/l = (0.21\varepsilon_f l/t)\,[1 + 3(t/l)^2] \tag{11.20a}$$

If the neutral axis is at the top or bottom, Equation 11.20a becomes

$$\delta_{max}/l = (0.1\varepsilon_f l/t)\,[1 + 12(t/l)^2] \tag{11.20b}$$

Laboratory soil beam bending tests have been suggested for assessing the flexibility of compacted soils (Ajaz and Parry, 1975).

The assessment of tolerable settlement of a geomembrane could be made by considering the stresses and strains under a given ratio of (δ_{max}/l). Considering a thin square membrane with a side l, the strain ε and stress σ along the neutral axis are expressed as (Timoshenko and MacCullough, 1940):

$$\varepsilon = 1.848(\delta_{max}/l)^2 \tag{11.21}$$

$$\sigma = 2.464E(\delta_{max}/l)^2 \tag{11.22}$$

where ε is the tensile strain and δ_{max} is the maximum deflection at the centre of the membrane.

If the membrane is subjected to load over a distance l with simple end supports, the maximum deflection at the centre of the membrane with zero bending stiffness (Timoshenko and MacCullough, 1940) is:

$$\delta_{max}/l = 0.4(ql/2Et)^{1/3} \tag{11.23}$$

where q is the stress generated by materials above the membrane, and t the thickness of the membrane.

11.6 Health and risk assessment

For successful liner installation, the leakage through an effective liner system is usually insignificant. In the event of a failure of the system, the leachate with a variety of chemical constituents enters the natural geological materials under the waste and could contaminate the groundwater resources. An event like this could occur as a result of the cracking of a clay liner and/or imperfect installation of a geomembrane. Assuming saturated conditions beneath the failed liner, a plume develops in the underlying aquifer. The vertical and lateral development of such a plume and the rate at which it spreads depend on the hydraulic properties of the geological materials and the hydraulic gradients.

Some regulations require modelling to assess the risk of contamination to water resources (wells, lakes, streams, etc.) in the event of a liner failure. The liner design is regulated with the intent that no release of contaminant is to be permitted through the liner of a disposal facility. For a sanitary landfill, a period of 30 years after closure may be selected. The assumption is that after such period the concentrations of various contaminants in leachate are reduced to acceptable limits.

The processes by which leachate flows through a porous medium are complex. The best understood are the physical processes involving groundwater flow, dispersion and diffusion. The geochemical and biological processes are significant in attenuating the contaminants or changing their mobility.

11.7 The geochemical process

This process involves primarily adsorption/desorption and precipitation/dissolution (Kincaid *et al.*, 1984). Adsorption refers to the mechanism of ion exchange, or chemical elements in an aqueous solution binding strongly with the solid surface of soil. In the first case, for example, a potassium ion in solution may be exchanged with a hydrogen ion on the solid soil surface and potassium in this case is retained. In the second case, an element in solution (e.g. copper) reacts with a surface hydroxyl group (OH) and forms a complex leading to retention of some metals. The hydrogen ion concentration characterized by the pH value affects the adsorption process.

The total capacity of soils to exchange ions characterizes their ability selectively to retain certain elements in the leachate. Fine-grained soils have higher CEC because of larger availability of exchange surfaces. Major cations in leachate (e.g. magnesium, potassium, sodium) attenuate primarily through cation exchange (Chapter 8).

Precipitation occurs when the soluble concentration of a contaminant in leachate is higher than that of an equilibrium state (Kincaid *et al.*, 1984). The form of attenuation is significant in attenuating metals at higher than neutral pH (Bagchi, 1986).

11.8 Physical processes

Contaminants dissolved in water (solutes) could migrate through the liner and underlying soils in three identifiable ways:

1. *Transport by advection*: Under a given hydraulic gradient, the groundwater and solute move with a certain advective velocity depending on the hydraulic conductivity and variations of hydraulic head (Lu *et al.*, 1985).
2. *Diffusion*: Diffusion is caused by the molecular motion of a solute. Because of a difference of molecular concentrations, the solute is transported toward regions of lesser concentrations. This results in the spread of solute into an increasingly larger volume of the aquifer.
3. *Hydrodynamic dispersion*: Dispersion of a solute is caused by the variation of fluid velocity within the pore space. These processes are considered in a solute mass transport model (Kincaid *et al.*, 1984; Lu *et al.*, 1985).

Several parameters must be determined before attempting to solve the mass transport equations. Considering the difficulty of acquiring credible input data, experience coupled with the simple Darcy equation could provide answers perhaps as reliable as the complex numerical models necessary to solve the flow equations.

The relative importance of various modes of transport depends on the advective velocity and the hydraulic conductivity. In an analytical study of flow through a liner (Rowe, 1987), mechanical dispersion appeared to be insignificant for an advective velocity less than 0.1 m/year. For advective velocities greater than 0.02 m/year advection dominates over diffusion. For advective velocities less than 0.0001 m/year, diffusion dominates over advection. For advective velocities between 0.02 and 0.0001 m/year, both advection and diffusion are important.

11.9 Landfill covers

11.9.1 Function and performance

The major function of a cover is to minimize percolation. Other functions are vector control, gas control, future site use, and aesthetics. The relative suitability of soils for various functions are illustrated in Table 11.4 (Lutton *et al.*, 1979). The key geotechnical elements for the design are the stability of the cover against sliding and its resistance to cracking. A cover with a multi-component structure is shown in Figure 11.11. The vegetative cover helps to reduce percolation by drawing moisture for transpiration and to mitigate erosion. The drainage layer beneath the top soil has the function of diverting the infiltration to a conveyance system and, therefore, reduces percolation. Thus, the drainage layer for the cover has a similar function to that of a liner. In addition, the drainage layer helps future utility of the site by enhancing mobility.

The hydraulic barrier has the same function as the liner. In this case, however, maintaining the integrity of the barrier is difficult on refuse because of the large differential settlement, which could cause cracking of a clay cover or failure of synthetic polymer (Oweis, 1988). Other causes of cracking are desiccation and the build-up of gas pressure if a venting system is not functional. The criteria for installing a clay cover are similar to those for liners. A hydraulic conductivity of 10^{-7}cm/s is usually specified but perhaps never attained because of cracking. Synthetic polymers have been utilized in covers, but have the following disadvantages (Hatheway and McAneny, 1987):

1. May be vapour or gas degradable.
2. Cannot be exposed to the elements.
3. Uncertain life span under various in-place conditions.
4. Low tensile strength.

Table 11.4 Ranking of soil types according to performance of some cover function

Soil type USCS symbol	*Impedence of water percolation*	*Hydraulic conductivity (approx.)* (cm/s)	*Support vegetation*	*Impedance of gas migration*	*Resistance to water erosion*	*Frost resistance*	*Crack resistance*
GW	X	10^{-2}	X	X	I	I	I
GP	XII	10^{-1}	X	IX	I	I	I
GM	VII	5×10^{-4}	VI	VII	IV	IV	III
GC	V	10^{-4}	V	IV	III	VII	V
SW	IX	10^{-3}	IX	VIII	II	II	I
SP	XI	5×10^{-2}	IX	VII	II	II	I
SM	VIII	10^{-3}	II	VI	IV	V	II
SC	VI	2×10^{-4}	I	V	VI	VI	IV
ML	IV	10^{-5}	III	III	VII	X	VI
CL	II	3×10^{-8}	VII	II	VIII	VIII	VIII
OL			IV		VII	VIII	VII
MH	III	10^{-7}	IV		IX	IX	IX
CH	I	10^{-9}	VIII	I	X	III	X
OH			VIII				IX
PT			III				

Ranking: I (best) to XIII (poorest).

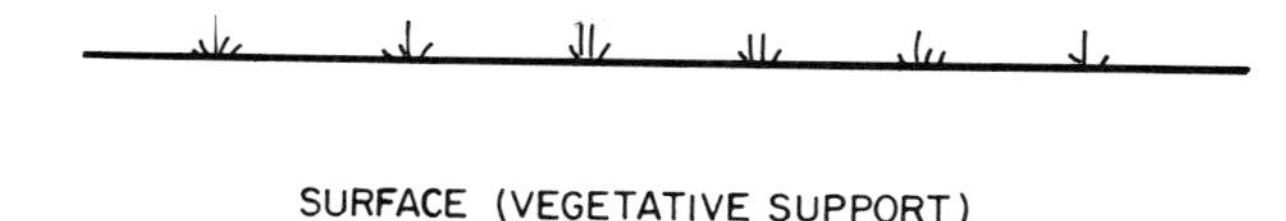

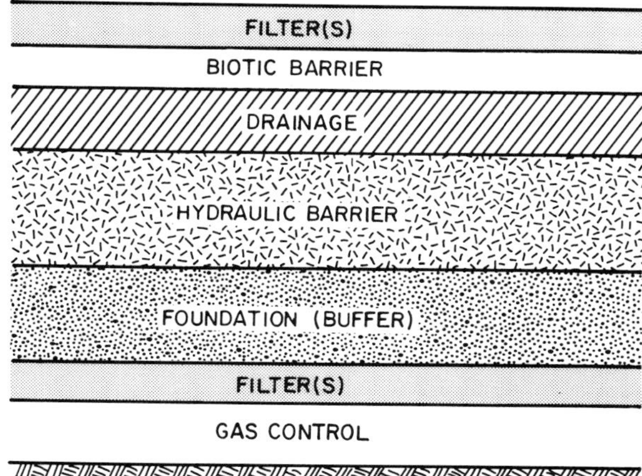

Figure 11.11 Structure of a multi-component cover system

5. Require careful installation.
6. May break because of severe differential settlements of the landfills or loss of support beneath.

Considering the problem of cracking of a clay hydraulic barrier, a highly efficient drainage layer above the barrier and/or thick covers are necessary if limiting the percolation is an important design factor. Controlling percolation may not be critical if the treatment of excess generated leachate is considered economically more feasible than an expensive cap. The water balance method discussed in Chapter 7 does not allow for inclusion of lateral drainage or variation of conductivity of various components of the cap. Computer-oriented solutions such as those used in the HELP model (Schroeder *et al.*, 1983) are useful for this purpose. Figures 11.12 and 11.13 show cover efficiency as a fraction of hydraulic conductivity using the HELP model with and without a drainage layer, respectively. The cover efficiency, E_c, is defined as:

$$E_c = 1 - e/p$$

where e is the percolation through cover, and p the precipitation.

The foundation or separation layer provides the platform for installing the hydraulic barrier and transition to the underlying gas control medium. The separation layer is usually composed of granular soil with high mechanical strength when compacted to help reduce the potential for barrier cracking. The gas control medium is composed of coarse to medium gravel.

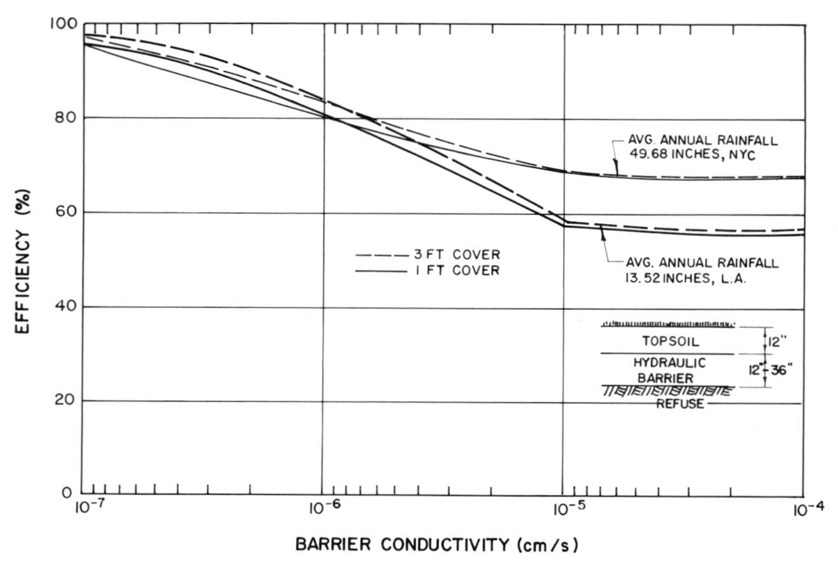

Figure 11.12 Cover efficiency *versus* barrier hydraulic conductivity, with drainage layer

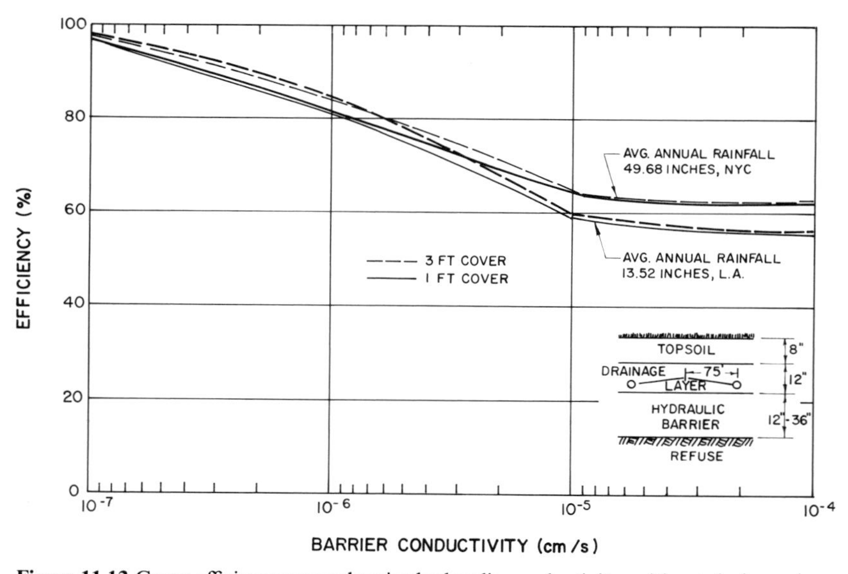

Figure 11.13 Cover efficiency *versus* barrier hydraulic conductivity, without drainage layer

Construction of a clay cover resistant to cracking is difficult to accomplish, so regular and often expensive repairs may be needed. There is some evidence to suggest that reinforcement with high-strength geotextile on the top and bottom improves the tensile resistance and hence reduces the cracking potential. This type of construction is not as yet a design practice. Stabilizing the refuse by dynamic compaction or surface compaction helps to reduce differential settlements.

11.9.2 Leakage from liners and covers

Modern designs for landfills require the construction of a secondary drainage layer to act as a leak detection and control system. A secondary liner underlies the leak detection system. Synthetic polymers usually fail along the seams or by puncture. Leakage through the imperfections could be approximated by flow through a pinhole percolating directly into the subgrade (Figure 11.14). This is a case of hemispherical flow (Harr, 1962):

$$Q_l = \frac{2\pi k_l(h_0 - h_R)}{1/r_0 - 1/R} \tag{11.24}$$

where Q_l is the leakage volume (L^3/T), k_l the hydraulic conductivity of the subgrade, r_0 the radius of the pinhole, h_0 the hydraulic head at $r = r_0$, and h_R the head at a spherical radius R.

For $R \gg r_0$ and $h_R \approx 0$,

$$Q_l = 2\pi k_l h_0 r_0 \tag{11.25}$$

For $r_0 = 2.5$ mm, $h_0 = 305$ mm and $k = 10^{-7}$ cm/s,

$$q = 2.9 \times 10^{-7}\ \text{l/min}$$

If the imperfection is approximated by a slot width of $2r$, the volume of flow per unit length at a radial distance r is:

$$q_l = \pi r k_l \,\mathrm{d}h/\mathrm{d}r \tag{11.26}$$

and

$$h = -(q_l \ln r/\pi k_l) + C$$

The constant C is determined by the specification that $h = h_0$ at $r = r_0$, hence

$$C = h_0 - (q/\pi k_l)\ln(r/r_0)$$

Considering that at $r = R$ and $h = h_R$,

$$q_l = \frac{\pi k_l(h_R - h_0)}{\ln r/r_0} \tag{11.27}$$

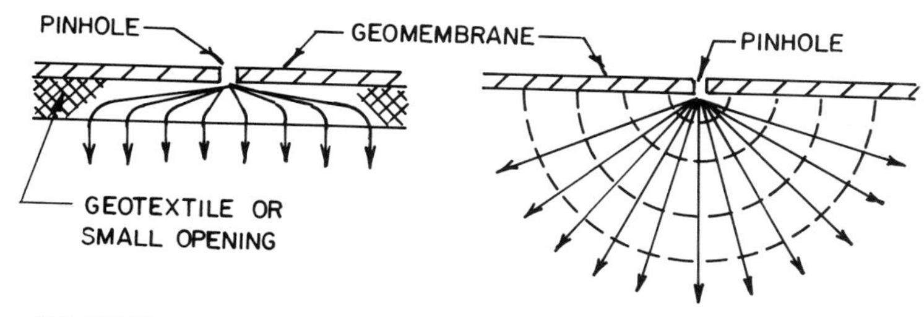

Figure 11.14 Two types of percolation (Reproduced from Fukuoka, 1986, by permission)

Clay liners underlying synthetic liners are usually 2 ft (610 mm) thick and have a pervious sand layer underneath them. To solve Equation 11.27 it is assumed that the head varies within the radial distance. As evident from the equation, q is not very sensitive to the ratio r/r_0. For example, at $r/r_0 = 50$, $\ln r/r_0 = 3.9$, whereas at $r/r_0 = 100$, $\ln r/r_0 = 4.6$ and the error would be about 15%, which is insignificant, considering the uncertainty in the k value.

The other leakage mechanism presumes a gap between the membrane and soil (Figure 11.14). The gap could be occupied by a geotextile. Assuming a large hole with a radius r_0 and free drainage (gap), the approximate limiting Q_1 would be

$$Q_1 = \pi r_0^2 (2gh_0)^{0.5} \qquad (11.28)$$

where g is a gravity constant. For a long slot with width $2r_0$, the flow per unit length is:

$$q = 2r_0 (2gh_0)^{0.5} \qquad (11.29)$$

Considering a gap or a geotextile underlying the synthetic liner and assuming the flow pattern shown in Figure 11.14b, the flow through a pinhole with a radius r_0 is (Fukuoka, 1986):

$$Q_1 = \frac{2\pi T k_p (h_0 - h_R)}{\ln(R/r_0)} \qquad (11.30)$$

where T is the thickness of the geotextile, and k_p the plane conductivity of the geotextile (or Tk_p the apparent permeability index); h_0 and h_R are defined at distances r_0 and R from the pinhole.

Cracking is the major reason for an apparent increase in hydraulic conductivity of clay liners compared with laboratory values. For a crack width of b extended through the liner leakage could be modelled as a flow through parallel plates. The flow per unit length could be expressed as (Albertson and Simons, 1964):

$$q = \gamma_w b^3 i / 12u \qquad (11.31)$$

where b is the crack width, q the flow per unit length of crack, γ_w the unit weight of water, u the dynamic viscosity of water, and i the average hydraulic gradient.

Assuming u to be 30 $(\text{ft s})^{-1}$,

$$q = 2.5 b^3 i \ (\text{ft}^3/\text{s per ft}) \qquad (11.32)$$

If the head loss across the depth of the crack is assumed to be the same for the intact liner, then

$$k_e \approx k_1 + 2.5 b^2 r_c \qquad (11.33)$$

where k_e is the average effective conductivity of a cracked liner, and r_c is the ratio of crack area to total area.

If the crack is filled with sand of permeability k_s, then the effective permeability k_e will be expressed as:

$$k_e/k_1 = 1 + r_c(k_s - k_1)/k_1 \qquad (11.34)$$

11.9.3 Side slope and stability

The stability of covers along the side of a landfill could be assessed using the wedge method. Figure 11.15 shows two cases where seepage occurs either parallel to the

$$F_s = \frac{\tan\delta}{\tan i}\left(1 - \frac{X}{T}\frac{\gamma_w}{\gamma}\right) + \frac{C}{\gamma T}\,\frac{1}{\sin i \cos i}$$

δ = INTERFACE FRICTION
i = SLOPE ANGLE
T = COVER THICKNESS
γ = UNIT WEIGHT OF COVER
γ_w = UNIT WEIGHT OF WATER
X = DEPTH OF SEEPAGE SURFACE
C = COHESION ALONG INTERFACE
F_s = FACTOR OF SAFETY

(A) SEEPAGE LINE PARALLEL TO SLOPE

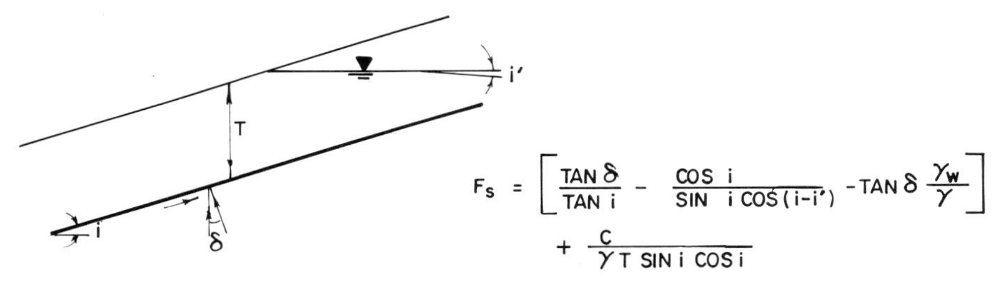

(B) SEEPAGE INTERSECTING THE SLOPE

Figure 11.15 Side slope stability in two cases in which seepage occurs

slope or intersecting the slope. The strength parameters Φ' and c' should be taken along the interface, where these parameters provide the lowest factor of safety.

The preferred side slope in various regulations is 3 horizontal to 1 vertical or flatter. This slope allows a reasonable operation of track-mounted construction equipment and easier growth of vegetation on the slope. In a municipal landfill, near vertical cuts were made in excess of 25 ft with no apparent instability. For temporary construction in municipal waste a slope of 1 vertical to 1 horizontal is conservative provided the potential sliding surface is within the refuse material.

The effect of cover reinforcement could be included by considering the equilibrium of a wedge restrained by a tensile force F_t parallel to the slope. This force would be provided by anchoring the reinforcement at the top of the slope and running the reinforcement along the full length of the slope. If the passive resistance is ignored at the toe, the resulting expression for the factor of safety with reinforcement F_{sr} is:

$$F_{sr} = (1/(1-t))F_s \quad (11.35)$$

where $t = F_t/\mathrm{T}\ L_c \cos i \sin i$

$= F_t/W_c \sin i.$

Where F_s = factor of safety without reinforcement (Figure 11.15), L_c = length of cover measured along the slope, W_c = weight of cover and F_{sr} = factor of safety for reinforced cover.

11.10 Gas management

The potential for explosion of methane is the key health and hazard problem generated by gases produced by the refuse decomposition process. Other concerns are odours and the toxic constituents of gas. For these reasons, a proper closure of a landfill requires gas management that would avoid health hazards. Control of methane gas is accomplished by one or more of the following:

1. Passive venting.
2. Impermeable barrier.
3. Forced (power operated) venting.

11.10.1 Passive venting

The objective of passive venting is to control venting of gases to places where it is either vented to the atmosphere or burned. Figure 11.16 shows arrangements for cap and an atmospheric venting systems. A series of vents could be connected to a blowe that induces suction to force vent. The objective is to limit the build-up of gas pressure in order to minimize gas discharge through the cap. The vents should be placed at locations where gas concentration and/or pressures are high. This usually occurs toward the lower sections of the landfill. A vent fully penetrating the refuse to the

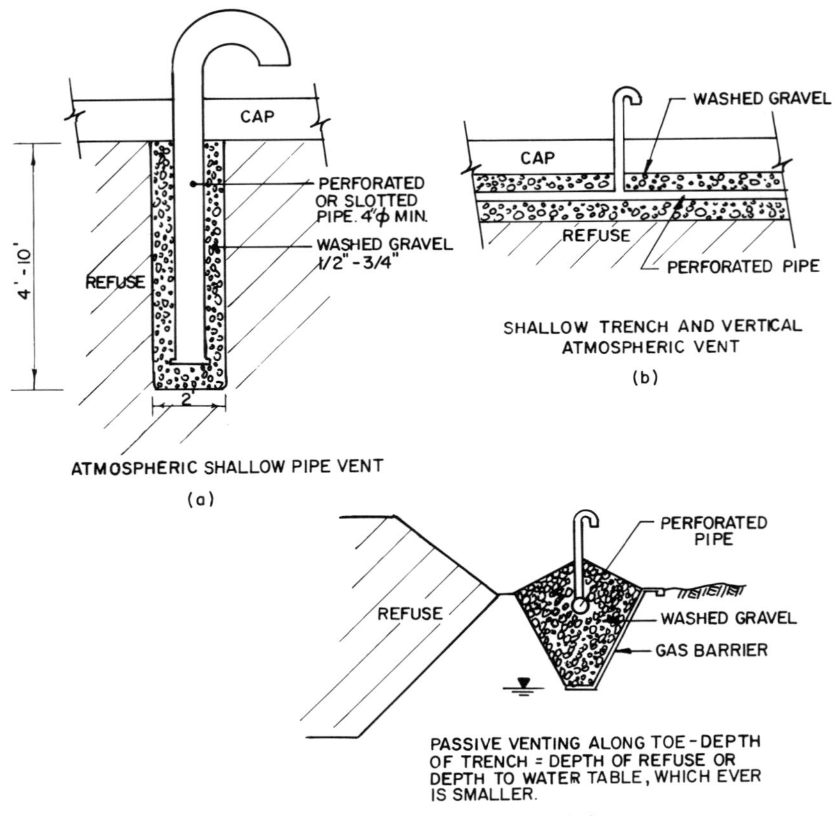

Figure 11.16 Shallow passive venting systems

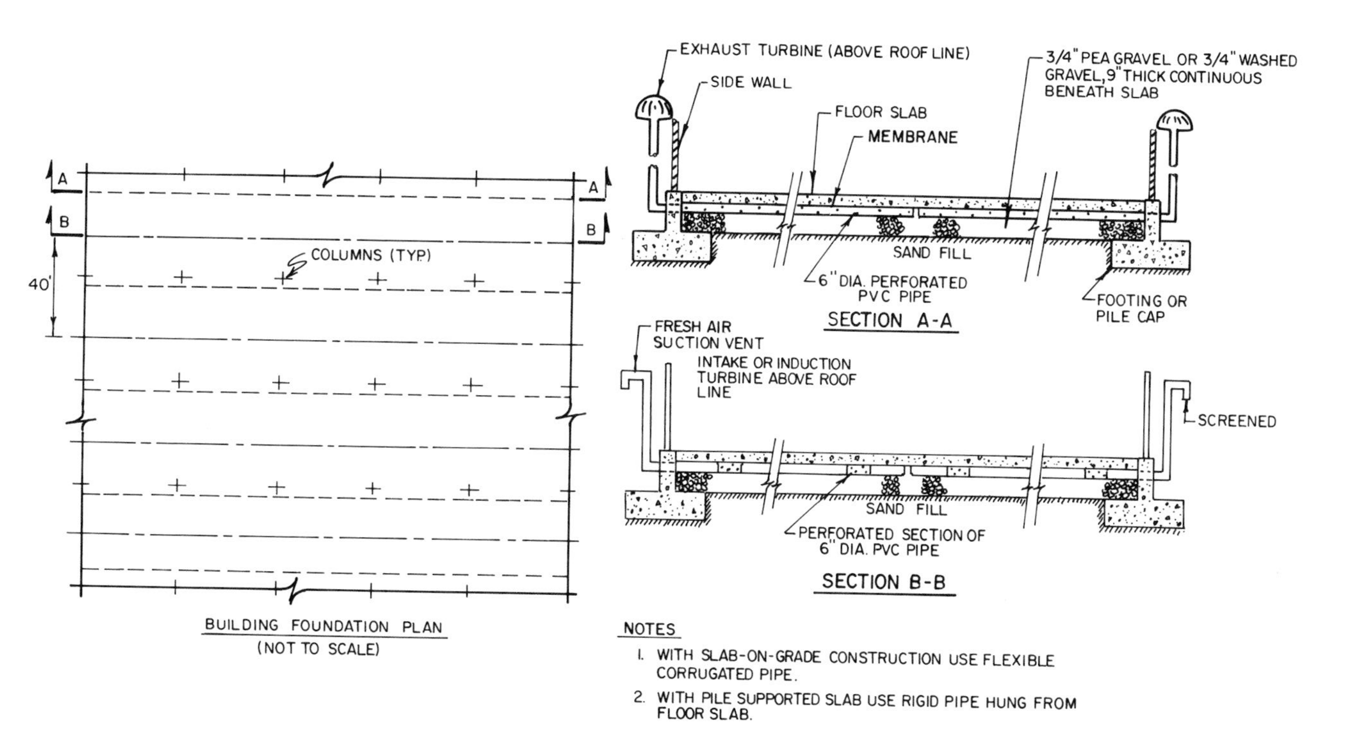

Figure 11.17 Passive gas venting below the floor slab

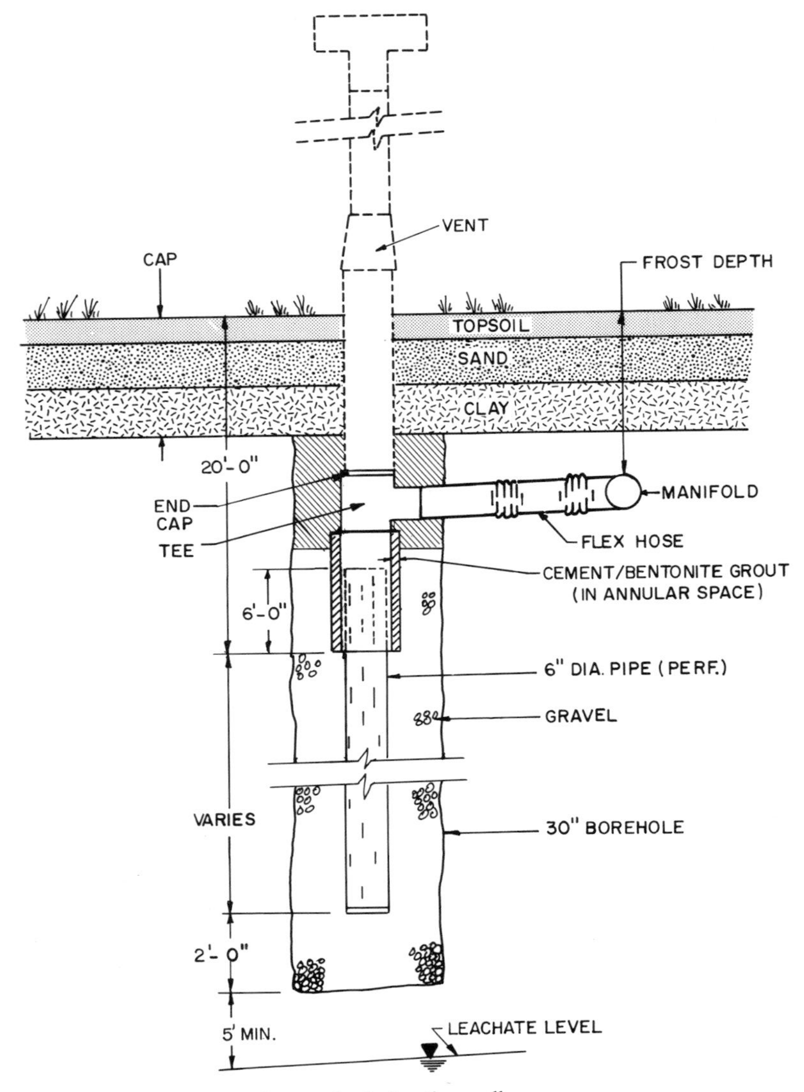

Figure 11.18 (a) Gas passive venting/extraction well

leachate level is most effective. The atmospheric vents are typically spaced at 100–150 ft (30–45 m). Such vents are not effective in controlling lateral gas migration. Where there are no geological constraints preventing lateral gas migration perimeter vents are used (Figure 11.16c). The gas migrating laterally is intercepted by a trench of much higher conductivity to gas. This system, if properly installed, could be effective in reducing the methane concentration in areas away from the toe of the landfill. The perimeter vents are typically spaced at 50 ft (15 m) intervals.

Figure 11.17 shows a typical passive venting system for buildings that has been found to be effective. Fresh air is introduced through suction pipes while gases are vented to the atmosphere above the roof. The air turbines would be operated naturally by wind or by power.

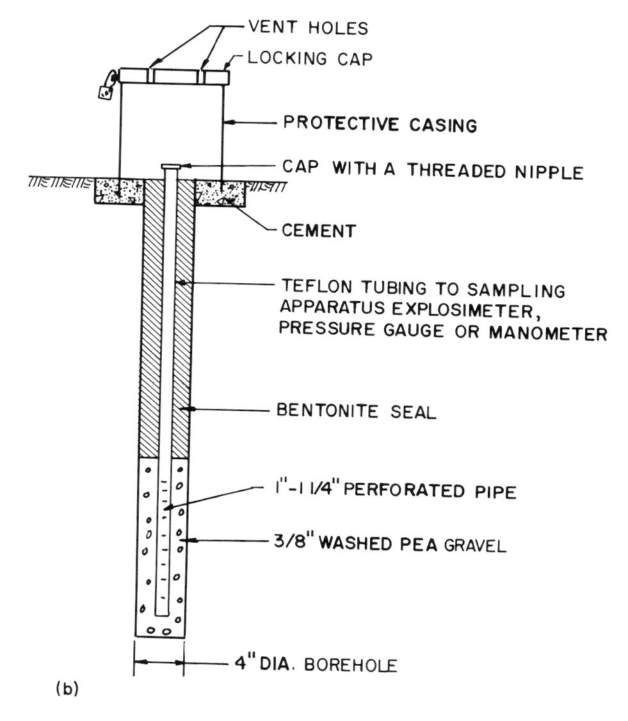

Figure 11.18 (b) Gas probe

11.10.2 Active venting

Where atmospheric venting is considered insufficient for controlling gas discharge, an active or forced venting system is considered. Forced ventilation is more effective in controlling the vertical or lateral migration of gases. It is accomplished by connecting a vacuum pump or blower at the discharge end of a vent pipe. When gas wells are used, negative pressures are induced at the well head and through a radius of influence around the well. The flow rate for the gas pump is at least equivalent to the gas generation rate. The radius of influence depends primarily on the flow rate and the depth of the well screen. Carlson (1977) reported a radius of 150 ft for a pumping rate of 50 ft^3/min. Norton Construction (1983) reported radii of influence of 20 ft at 30 ft^3/min flow rate to 75 ft at a flow rate of 50 ft^3/min depending on the depth and location of the well. Where the gas is used for commercial purposes the flow rate is limited so that the amount of air from the atmosphere introduced into the system is not excessive. The flow rate as well as the depth of well screen below the surface affects the amount of air drawn into the system. The selected rate should be based on site-specific pumping tests (Carlson, 1977; Norton Construction, 1983). The reported limiting quantities are in the neighbourhood of 40 ft^3/min.

Condensate from gas vapour tends to obstruct gas flow through the pipes leading from gas wells. The pipes are sloped to condensate traps, where condensate is collected or recirculated through the landfill. Figure 11.18 shows a schematic for a gas extraction well. The performance of the well is assessed by monitoring gas quality and pressure through probes at various locations. A schematic of a probe is shown in

Figure 11.18b. A periodic maintenance of the system is necessary to accommodate the large total and differential settlements of the landfill.

11.11 Remedial action

Existing disposal facilities with deficient leachate and gas control management measures are required to implement remedial measures to lessen the impact of the pollutants on the environment. The remedial technologies include many of the contruction items required in the design and construction of a new landfill. These may include capping, gas venting, and installing shallow trenches for leachate collection. For landfills with no liners, remedial action may require the installation of a vertical hydraulic barrier, a leachate collection system around the landfill, pumping from around the landfill, or pumping the leachate from inside the landfill.

11.11.1 Vertical barrier walls

The principal qualities affecting the selection of a vertical barrier wall are its hydraulic conductivity, resistance to pollutants, strength, and cost. Slurry trench walls are installed to control leachate in areas where the trench needed to build the wall could be maintained stable at a reasonable cost. Conventional excavating equipment can be used where depth to an impervious stratum is limited to about 50 ft (15.25 m). At greater depths the wall may be advanced to the depth limit of the conventional equipment and another type of equipment, such as dragline or clamshell, may be used at greater depths (Figure 11.19).

The two most common types of wall used in the USA are soil–bentonite (SB) walls and cement–bentonite (CB) walls. Both of these are constructed using a vertical trench which is excavated under a slurry that keeps its sides from caving in. For the SB wall the slurry is a water–bentonite mixture and for the CB wall it is a water–cement–bentonite mixture that acts as a stabilizing fluid. The desirable range of pH is 6.5–10. If the pH exceeds 10.5 the slurry will have a tendency to flocculate. The density of the slurry is adjusted to suit the ground conditions and ranges from 65 to 90 lb/ft^3 (10.2–12.5 kN/m^3).

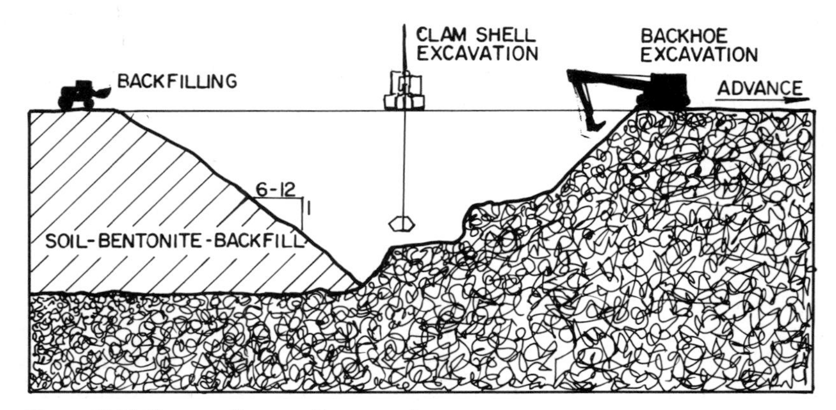

Figure 11.19 Slurry wall excavation operation

The slurry pressure should be sufficient to resist at least the active earth pressure plus the hydrostatic fluid pressure. The fluid pressure should be based on piezometric heads along the trench alignment. The slurry level in the trench is at least 3–5 ft (0.9–1.5 m) above the highest groundwater level. In cases where structures or utilities are within a distance equal to the trench depth, it is prudent to design the slurry to resist the at-rest earth pressure conditions. Soil movements on both sides of the trench of the order of 0.5% of trench depth are common. This may cause damage to various structures along the trench alignment.

11.11.2 Soil–bentonite walls

The slurry stabilized trench is backfilled with a mixture of soil, sodium bentonite and water. To facilitate placement, the mixture usually has a slump of 6–8 in. The amount of bentonite by dry weight is 3–5%, fines are 10–40%, and water content is typically 25–35%. Unit weight ranges from 105 to 120 lb/ft^3 (16.5–18.9 kN/m^3).

Quality control requires specifications on the methods of mixing and requirements of soil, water, and bentonite. Verification of the trench during construction is a key function.

11.11.2.1 Materials

The hydraulic conductivity of a soil–bentonite wall depends on the soil gradation and quantity of bentonite. Soils with significant amounts of clay are preferable since clay fines would be more resistant to organic chemicals of the leachate. The amount and plasticity of fines as well as bentonite content affect the hydraulic conductivity of the SB material as illustrated in Figures 11.20 and 11.21 (D'Appolonia, 1980). For hazardous waste sites (Ryan, 1987), the percentage of clayey fines may have to be greater than 35%. For ordinary municipal waste landfills, sands with lesser fines but with a larger percentage of bentonite could be utilized. An assessment of leachate quality and its potential impact on bentonite must be made before the backfill gradation is specified.

The fines should be inorganic (CL or CH). The gradation and density of the backfill material are usually tested for every 200–1000 yard^3 (50–760 m^3), depending on the project conditions (size, uniformity of borrow sources, etc.). Potable water is used to hydrate the bentonite. A manufacturer's certificate of compliance with the American Petroleum Institute (API) standard 13A on bentonite quality is usually sufficient. API specification 13A covers oil well drilling-fluid material and 13B covers field testing of drilling fluids such as its density, viscosity and gel strength, filtration, etc. Typically, the slurry minimum viscosity is specified at 40–50 s using the marsh funnel. The slurry tests are performed daily (once to twice per day).

The design backfill slump is checked 3–5 times per shift. A laboratory hydraulic conductivity test on the backfill material is usually performed for each 1000–2000 yard^3 (750–150 m^3) of material.

11.11.2.2 Mixing and placement

A typical plan for SB installation consists of a slurry preparation area consisting of a water source (in tanks or a pond), a pond or tank for storage and hydration of the slurry pumped into the trench, and another pond or tank for storage and hydration of slurry used in backfill mixture (see Figure 11.22). Mixing slurry in the trench must be avoided.

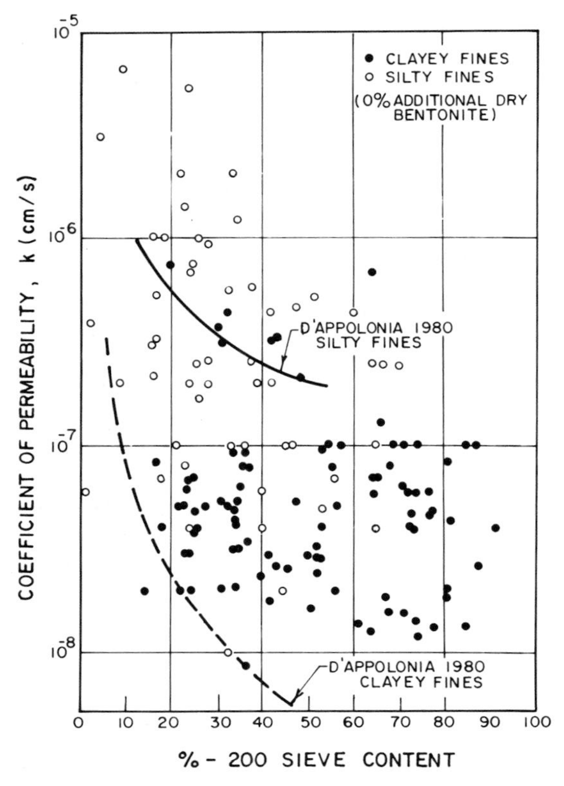

Figure 11.20 Permeability of backfill *versus* fines content (Reproduced from Ryan, 1987, by permission)

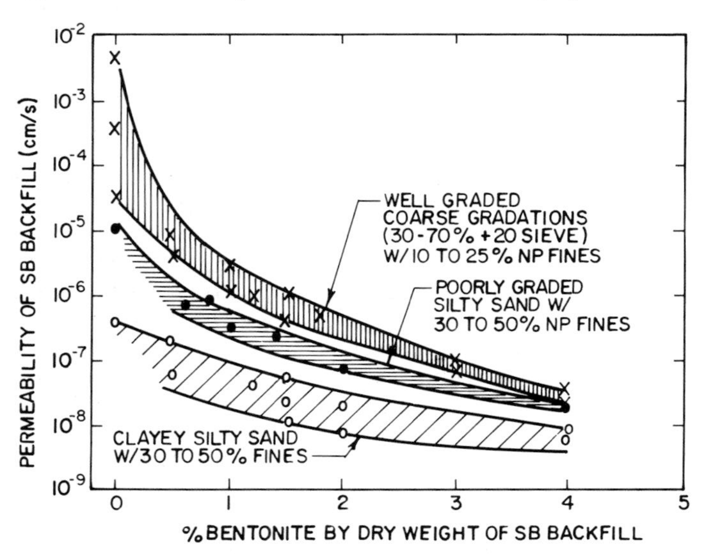

Figure 11.21 Permeability of SB backfill *versus* bentonite content (Reproduced from d'Appolonia, 1980, by permission)

An experienced backhoe operator is a key element for a successful operation. An approach section is dug first. The length of trench open at any one time should not exceed the reach of the equipment. The soil backfill is placed alongside the trench where it is bladed and tracked by a bulldozer for proper consistency and density, and then pushed into the trench. This method does not allow continuous control over the quality of mixture. A large number of tests would be required for quality assurance. Because of the proximity of the operations to the trench, free dropping of mixture into the trench is unavoidable. This could cause segregation of the mixture. The preferred technique is to have the selected soil backfill, dry bentonite and bentonite slurry mixed inside regular concrete mixer trucks (Figure 11.23). An excellent control of the mix can be maintained with this technique. The toe of the backfill slope (Figure 11.23) at the end of each shift should be at or near the excavation point. This slope is 1 vertical to 6–12 horizontal. When resuming excavation, the top of the backfill slope is cleaned with the excavating equipment. Free dropping of the SB mixture should be avoided. A preferred procedure is to place the SB mixture 20–30 ft (6–9 m) along the completed wall and away from the slurry–backfill interface as illustrated in Figure 11.24.

The depth of the trench should be measured and recorded after cleaning every 50 ft (15.25 m) or more often if necessary. If a section of the trench is left unfilled, the bottom of the trench must be measured before the start of the second shift. Deviation of bottom elevation is an indication of trench collapse or instability.

11.11.2.3 Wall verification

The most critical area where SB wall failures could occur is around the key into the impervious stratum. Quality control regarding proper cleaning of the trench is essential. It is difficult to repair the wall after construction. The current procedure for verification is to recover backfill samples at various depths and test them for hydraulic conductivity. Thin-walled samplers used for soft clay are employed. Installing closely placed piezometers on the upgrade and downstream site of the wall is another technique of verification. If such piezometers are installed at various depths, any significant leakage through the wall could be detected. The 'variable head permeability' test with rising head has been suggested for measuring the in-place hydraulic conductivity (Teeter and Clemence, 1986). For a low-permeability wall (10^{-7} cm/s or where the well screen depth is limited) the accuracy of the measurement may not be reliable. Such tests, however, would be useful in detecting major deficiencies in the walls at the locations tested.

Utilization of conventional borings is useful in verifying the quality of the backfill at frequent intervals. At this stage, there is no reliable technique for verification of a wall other than drilling a cased hole and sampling. Water or cleaning fluid should be avoided for cleaning the bore hole as they may cause hydraulic fracture. An electrical cone may be a better choice as it gives a continuous log of the soil strength. In cases of an originally clean site, the lowering electrical resistivity of the downstream formations could be interpreted as an indicator of wall leakage. This technique would not be applicable to existing facilities when the surrounding soils are already polluted, but it could be useful for monitoring the performance of a leachate management system for new facilities.

Figure 11.22 SB mixing plant

Figure 11.23 Introduction of the SB mix to the trench

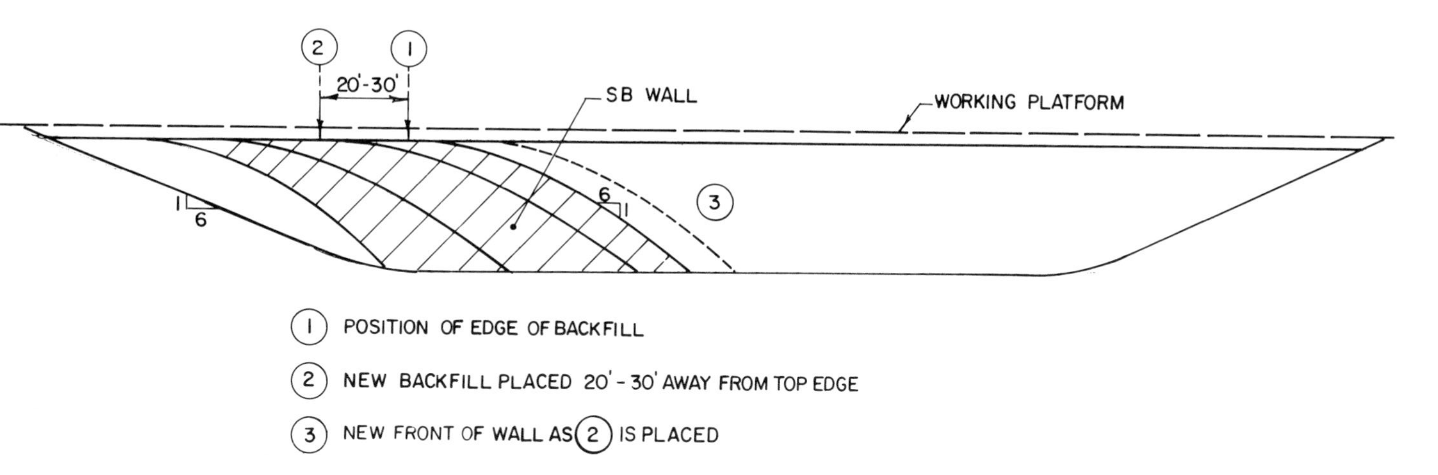

(1) POSITION OF EDGE OF BACKFILL

(2) NEW BACKFILL PLACED 20' - 30' AWAY FROM TOP EDGE

(3) NEW FRONT OF WALL AS (2) IS PLACED

Figure 11.24 Construction of an SB wall

11.11.3 Cement–bentonite walls

Cement–bentonite walls are considered in areas where suitable soil backfill is not available, working area is restricted, and some structural strength is desired. The soil–bentonite unconfined compressive strength is very low and is unlikely to exceed 50–100 lb/ft^2 (2.5–5 kPa) even under long-term conditions. By contrast, cement–bentonite strength could be controlled with 20–40 lb/in^2 (138–276 kPa) ultimate strength being typical. These walls are more pervious and brittle than SB walls. In an installation where significant lateral movements are expected, CB walls should not be used. An example of this would be a landfill undergoing lateral movements because of low factors of safety. A CB wall is typically 1.5 ft wide and contains on a weight basis 10–20% cement, 70–85% water, and 4–6% bentonite. A higher percentage of cement increases the structural strength but the wall becomes more brittle, is more prone to chemical attack, and has higher hydraulic conductivity. High bentonite contents reduce permeability and structural strength. A hydraulic conductivity less than 10^{-6} cm/s is usually difficult to achieve immediately after construction but 10^{-7} cm/s may be achieved in 6–12 months after construction.

11.11.4 Applications

Two applications for vertical barrier wall are shown in Figure 11.25. In cases where there is a leachate mound inside the landfill, the barrier wall must surround the disposal facility, and the collection drain is installed at an elevation lower than the lowest expected elevation of groundwater outside the wall. For an intact wall, the drain will function to collect leachate from inside the landfill with insignificant leakage from outside through the wall. It is clear that the collection drain could be installed as deep as the impervious layer, thus eliminating the need for the wall. This type of installation could prove expensive because of the greater trench cost and the cost of treatment of larger volumes of leachate.

Ryan (1987) reported a case where leachate containing hydrocarbons, phenol, acetone, and other organic compounds with individual concentrations less than 75 ppm had contaminated the groundwater. These concentrations are within the range where no detrimental effect has been reported on sodium bentonite (Chapter 9). However, studies on an ordinary sodium bentonite and a 'contaminant-resistant' bentonite indicated that both types of material were incompatible with the leachate. The permeability to leachate of the ordinary bentonite was three times larger and of the contaminant-resistant bentonite over five and a half times larger than that to water. This corroborates the findings of Khera *et al.* (1987) where an ordinary bentonite was found to be more resistant to the pollutants than a contaminant-resistant bentonite. Ryan (1987) tried attapulgite as a substitute material and found its permeability to be unaffected by the leachate. The project was completed using attapulgite slurry and backfill prepared with attapulgite instead of bentonite. Attapulgite is a clay mineral with needle-like particles and has a much lower swelling potential than sodium bentonite. Other instances where sodium bentonite was found unsatisfactory and attapulgite was used successfully have been reported by Xanthakos (1979). The disadvantage of attapulgite is that it is more costly, requires a larger quantity because of its low swelling potential, and is more difficult to mix into a slurry and backfill.

In an effort to find substitute materials which are comparatively more stable in a polluted environment than sodium bentonite, test sections consisting of sodium

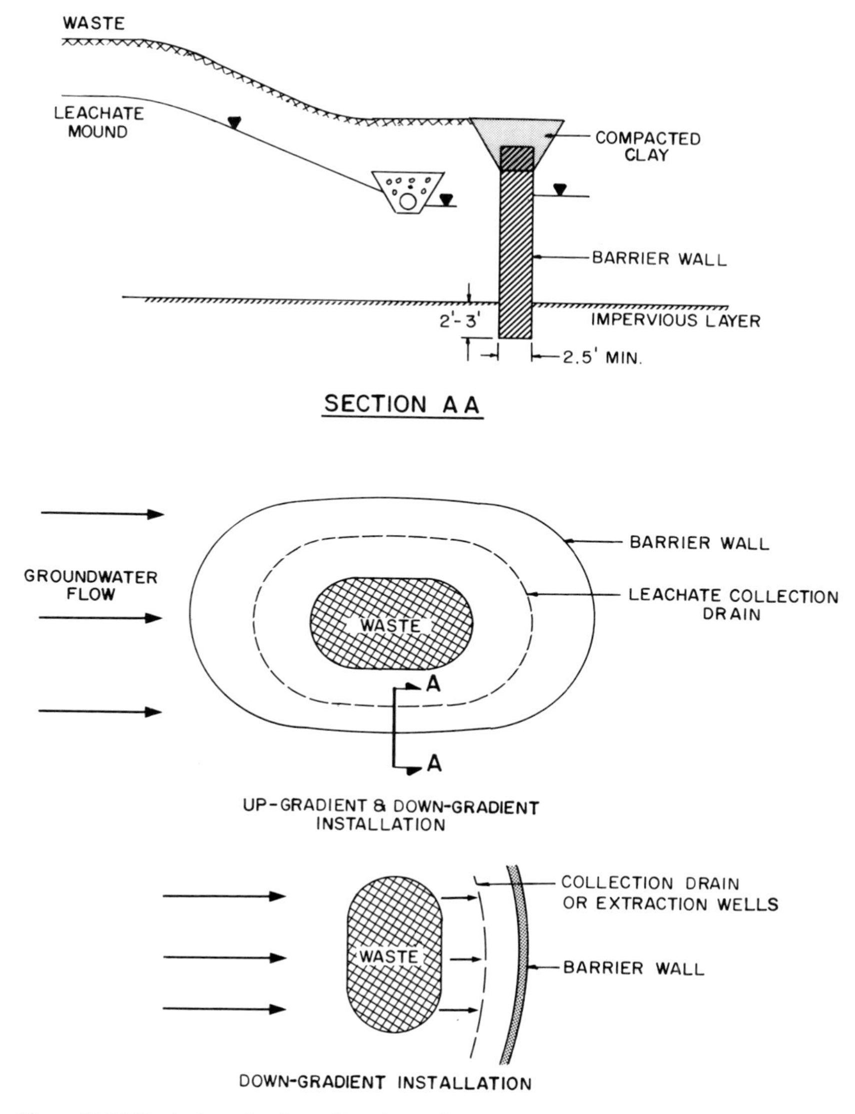

Figure 11.25 Typical application of barrier walls

bentonite–cement backfill and calcium bentonite–cement backfill were installed around an industrial waste disposal site near Heidelberg, in the Federal Republic of Germany (Nussbaumer, 1987). The proportions by weight of bentonite:cement: water were 0.030:0.194:0.776 for sodium bentonite, and 0.146:0.146:0.708 for calcium bentonite. The laboratory permeability values were 10^{-7} and 10^{-8} cm/s, respectively. The leachate in the vicinity of the test sections was highly contaminated. After one and a half years undisturbed samples were obtained from the wall for each of these sections. The specimens from the calcium bentonite section were intact and showed no changes in hydraulic conductivity. The specimens from the sodium bentonite section showed partial disintegration and an increase of two orders of magnitude in the hydraulic conductivity. The greater resistance of calcium bentonite to chemicals may be partly due to the fact that a calcium ion is more difficult to replace than a sodium ion. Since the swelling potential of calcium bentonite is considerably less than that of sodium bentonite, 3–4 times more calcium bentonite must be used for a comparable adjustment in permeability (Xanthakos, 1979). These cases show that the lower concentration of the pollutants is not a guarantee for the satisfactory performance of barriers consisting of sodium bentonite.

Because of groundwater fluctuations upper reaches of the wall are subjected to drying and wetting cycles. No data are available for the effect of desiccation on the backfill. Recent data (Khera *et al.*, 1988) with water and other chemicals indicate that the desiccation cracks develop parallel to the dry–moist interface. The size and the number of these cracks increase with increasing drying and wetting cycles (Figure 11.26). Since drying essentially proceeds from the ground surface down, horizontal cracks will develop in an SB wall. Depending on the size and distribution of these cracks considerable flow could occur through them. The flow will have a tendency to increase the size of the cracks because of soil–chemicals interaction and washing away initially of the finer soil particles, and with the increasing velocity due to widening channels coarser particles will also be carried away.

To determine the competency of a completed soil–bentonite cutoff wall Engemoen and Hensley (1986) conducted penetration tests with an electrical cone. There was no significant difference in the tip resistance at varying depths or among backfills 2 months and 18 months old. Also, little additional settlement of the trench was recorded after the first month. The lack of strength increase with either depth or age of the backfill was attributed to the low confining pressure due to arching effect. Note that the combination of desiccation cracks and arching will produce much higher hydraulic conductivity as the cracks will not close as efficiently upon subsequent saturation because of the lack or reduced values of confining pressure. The cone was also capable of measuring inclination of the tip at any depth. The horizontal deviation of up to 17 ft (5.2 m) was calculated for a penetration depth of less than 100 ft (31 m).

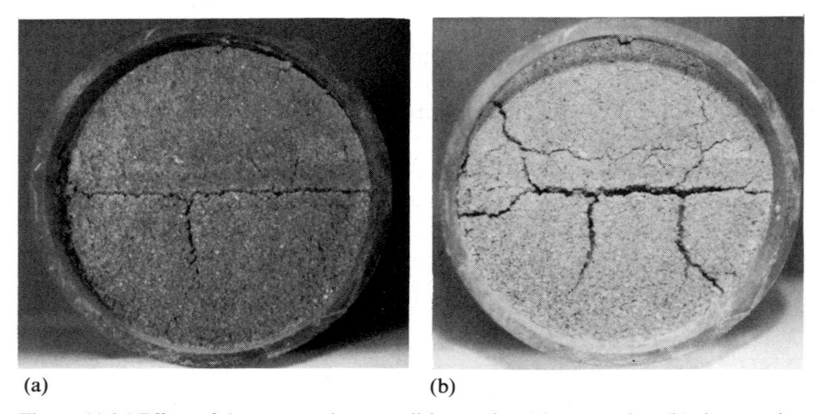

(a) (b)

Figure 11.26 Effect of dry–wet cycles on soil bentonite: (a) two cycles, (b) three cycles

11.12 Construction on landfills

Construction on or in refuse poses several problems not usually encountered on other sites. Arguments have been presented over the years as to why one should avoid construction on refuse landfills (Eliassen, 1947). The major problems are settlement, odour control, and gas control. Other problems relate to the durability of construction materials in the contaminated soil environment. Eliassen (1947) cites methods for housing construction on landfills where the objectives were to control gas and odour. He also cites an allowable bearing pressure for footings of 2000 lb/ft^2 (100 kPa) that sanitary landfills may carry without appreciable localized settlements. The early methods for gas control are similar in concept to the current passive techniques. The gas problem was handled by leaving a 12–24 in (305–610 mm) air space between the floor and the top of a 12 in (305 mm) cinder layer overlying the landfill cap (Eliassen, 1947). The air space was enclosed with wood siding but vented through screened openings on all four sides. A gastight paper membrane covered with aluminium foil was secured under the wood flooring.

Without stabilization, large total and differential settlements may preclude the support of buildings and roadways. Even with stabilization, only light structures (1–2 storeys) may be supported on treated refuse. The ongoing long-term settlement due to decomposition cannot be totally eliminated by stabilization. Restriction or prevention of moisture entering into the landfill could substantially reduce the rate of settlement from decomposition. This can be accomplished by paving or proper placement of a geomembrane over the landfill.

With the increasing cost of prime sites in an urban environment the utilization of refuse sites for building and roadway construction becomes cost effective. Examples of the behaviour of refuse under load have been presented in Chapter 10. For light structures, support of refuse may be impractical for refuse thickness in excess of about 20 ft (6 m) because of the difficulty in stabilization and subsequent settlement. For heavier structures, deep foundations would be necessary. Removal or excavation of refuse is usually not an attractive alternative because of the inherent environmental and health problems.

11.12.1 Stabilization

The usual method for stabilizing refuse is by preloading using the same concepts as for stabilization of compressible soils. With preloading, the primary (short-term) settlements could be cut by about half, compared with untreated refuse. The secondary settlement could be reduced depending on the length of time the preload is maintained. Preloading would not produce significant reduction in future settlement due to decomposition unless access to moisture was prevented. The dynamic compaction technique has been used to stabilize refuse (Chapter 10).

Moulton *et al.* (1976) described a large-scale test where a 40 × 40 × 18 ft (12 × 12 × 5.5 m) deep landfill was grouted under gravity with fly ash at 100% moisture content. Because of the lack of positive confinement the grouting did not result in any compaction. Initially the rate of settlement for the grouted section was high but in two years it dropped to half that of the ungrouted section indicating some beneficial effects.

Blacklock (1987) reported stabilization of a municipal landfill with hydrated lime–fly ash mixture containing 30–60% solids. The waste site which was closed for several years consisted of 16.5 ft (5 m) refuse and 6.5 ft (2 m) cover of expansive clay. Pressure grouting was carried out through vertical holes in a grid pattern to a depth of 5 m. An initial set of holes was followed by an additional set spaced equally between the previous locations. The grouted area was 8600 m^2 with the buildings covering 3000 m^2. About 1600 ton of dry lime–fly ash material were required. All building loads were supported on drilled piers and the grouted area provided parking facilities. Various vents were incorporated in the buildings to mitigate possible methane gas problems. It was recommended that lime slurry be pressure injected to control the pH, which has a significant influence on methane production. It was suggested that in landfills where there is a possibility of displacing leachate, well points be established prior to grouting and extracted leachate be recycled into the lime–fly ash slurry. Where excessive acidity occurs or where primary treatment is for methane gas, a 48 h curing period was recommended before successive injections. However, there are no data available to support the view that the use of lime–fly ash pressure grouting does in fact reduce decomposition, gas or leachate generation, or long-term settlement of waste.

11.12.2 Foundations

When designing foundations, in addition to the conventional soil parameters, other information such as the nature of the contaminants, soil resistivity, pH, ion contents, presence of bacteria, etc., should also be determined. Though little is known about the effect of contaminated soils on construction materials such information may have some value in selecting appropriate materials. For example sulphate, chloride and calcium are some of the ions that affect the durability of steel. The structural elements of foundation such as shallow footings, piles, piers, and underground building components such as vapour barriers, pipes, and drains are vulnerable in a waste environment. The rate of deterioration is difficult to predict as our current knowledge is limited to uncontaminated soils environments only.

Where shallow foundations are deemed not feasible because of expected large settlements, impossibility of removal of refuse or other factors, then pile foundations become necessary. The factors controlling pile design are driveability, downdrag, and corrosion potential.

Driving of piles through refuse is usually difficult. Techniques such as pre-augering or predrilling may be attempted to facilitate drilling. The use of heavy pile sections coupled with protective tips and a large hammer with more than 32 000 ft lb of delivered energy may be effective. The use of vibrating hammers is unlikely to be effective because of obstructions. Despite all this, relocating the piles may be necessary to avoid the obstructions. In this sense, the foundation will have to be designed to accommodate pile deviations of 5 ft or more from planned locations.

The corrosion potential may dictate the selection of pile types. All pile types (steel, concrete and timber) are subject to deterioration. The use of concrete-filled steel pipe piles may be considered. The heavy steel pipe allows hard driving and provides protection of the concrete. The steel contribution to the structural support is ignored. Dense concrete with acid resistance and low water–cement ratio is usually recommended. If steel piles are used, corrosion should be allowed for by using a much thicker section than required for structural support. Creosote-treated timber piles have been used in refuse. Protection of the creosote or the protective chemical during driving, and the need to pre-auger for each timber pile, are major cost factors. A summary of the existing technology of pile protection in a corrosive environment is presented by Gauffreau (1987).

Very few data are available for estimating the downdrag on piles due to settlement of refuse. The limited observations suggest that the downdrag force may be of the order of 10% of the weight of overlying refuse. For example, if the depth of refuse is 50 ft and the unit weight is 40 lb/ft^3 (6.3 kN/m^3), then the downdrag stress is 200 lb/ft^2 (31.5 kN/m^3).

Data are also lacking on the lateral pressure from refuse. It is logical to consider that the refuse is somewhat reinforced by various objects as evidenced by the near vertical slopes in refuse over 25 ft. The active lateral earth pressure coefficient, K_a, is expected to be less than that of loose sand which is about one-third. A value of 0.2 appears reasonable for unsoaked refuse.

11.13 Gas venting

Methods of venting have been discussed in Section 11.10. Regardless of initial measurements or the presence or absence of methane, some form of passive venting should be used for facilities near landfills. Methane was detected as far as 2000 ft (610 m) from a landfill. It appears that structures within 0.5 mile (800 m) may have to be vented depending on the particular site conditions.

Notation

A	area of the flow
C	distance from the neutral axis to extreme fibre
C_u	coefficient of uniformity
d	liner thickness
D_{15}	particle size at which 15% of the materials by weight are fines
D_{50}	particle size at which 50% of the materials by weight are fines
D_{85}	particle size at which 85% of the materials by weight are fines
d_s	depth of saturated front
dt	time interval
E	efficiency of the liner
e	percolation
f_t	fabric thickness
g	gravity constant
G	shear modulus
h	leachate head in the landfill
h_b	pressure head (negative for suction) at the bottom of the liner
h_{max}	maximum leachate head in landfill
h_0	hydraulic head at $r = r_0$
h_0	initial leachate head in landfill
h_R	head at a spherical radius R
I	moment of inertia about neutral axis
k_p	hydraulic conductivity in the plane of the fabric
k_d	hydraulic conductivity of the drainage layer
k_l	hydraulic conductivity of the saturated liner
k_n	hydraulic conductivity in a direction normal to the fabric
k_p	hydraulic conductivity in the plane of the fabric
k_s	saturated hydraulic conductivity of the clay liner
k_u	unsaturated hydraulic conductivity at the wetting front

L distance between drains
l span length
m an experimental factor, typically 1.5–3
M_{max} maximum moment
n porosity of the drainage
p precipitation
q loading generated by materials above the membrane
Q_d flow to drain pipe per unit length of drain
q_d volume of flow to the drain
Q_l leakage volume
q_l volume of flow passing through the liner
Q_{max} static moment of cross-sectional area from neutral axis
r radius
r_0 radius of the pinhole
s perpendicular distance from the drain along the slope
s_0 distance from the drain to the ridge
T thickness of the geotextile
t thickness of the membrane
t_1 time for leachate to drain to the pipe
t_2 time for the leachate volume to leak through the liner
w load per unit length of beam
w_f volumetric moisture content at field capacity
w_i initial volumetric moisture content
w_s saturated volumetric moisture content
α slope angle of drain
α_i failure plane inclination of wedge i – active side
β_i failure plane inclination of wedge i – passive side
δ_{max} maximum deflection at the centre
θ transmissivity
φ friction angle
Φ permittivity
Φ_s suction of capillary head
ε tensile strain
ε_{max} maximum tensile strain

References

Ajaz, A. and Parry, R. H. G. (1975) Stress–strain behavior of two compacted clays in tension and compression, *Geotechnique*, **25,** No. 3, 495–512

Albertson, M. I. and Simons, D. B. (1964) Fluid mechanics, Sec. 7. In *Handbook of Applied Hydrology*, Ed. Chow, V. T. McGraw-Hill, pp. 7–13

Bagchi, A. (1986) Landfill geostructure construction using blended soil, *International Symposium of Environmental Geotechnology* (ed. Fand, H. Y.), **1**, 43–52

Blacklock, J. R. (1987) Landfill stabilization for structural purposes, *Geotechnical Practice for Waste Disposal '87*, Proc. Spec. Conf. Geot. Spec. Publ. No. 13, Ed. Woods, R. D., pp. 275–294

Bowders, J. J., Daniel, D. E., Broderick, G. P and Liljestrand, H. M. (1984) Methods of testing the compatibility of clay liners with landfill leachate, *Hazardous and Industrial Solid Waste Testing, Fourth Symposium*, ASTM STP 886, p. 247

Brendel, G. F., Glogowski, P. E., Greco, J. L. and Smith, L. C. (1987) Use of geomembranes in deep valley landfills, *Geotechnical Practice for Waste Disposal '87*, Ed. Woods, R. D., Geot. Spec. Publ. No. 13, University of Michigan, Ann Arbor, Michigan, pp. 334–346

Cadwallader, M. W. (1986) Selection and specification criteria for flexible membrane liners of the high density polyethylene variety, *Int. Symp. on Environmental Geotechnology*, Vol. 1, Ed. Fang, H. Y., pp. 323–333

Caldwell, J. A., Thatcher, J. and Kiel, J. (1986) Environmental geotechnical considerations in the design of the cannon mine tailing impoundment, *Int. Symp. on Environmental Geotechnology*, Vol. 1, pp. 312–322.

Cancelli, A. and Cazzuffi, D. (1987) Permittivity of geotextiles in presence of water and pollutant fluid, *Conference Proceedings Geosynthetics '87*, Vol. 2, Industrial Fabrics Association International, 24–25 February 1987, pp. 471–481

Carlson, J. (1977) *Recovery of Landfill Gas at Mountain View*, Engineering Site Study, USEPA, EPA/530/SW-587, Washington, DC

Chan, P. C., Selvakumar, G. and Shih, C. Y. (1986) The effect of liquid organic contaminants on geotechnical properties of clay soils. *Proc. of the Eighteenth Mid-Atlantic Industrial Waste Conference*, Technomic Publishing Co. Inc., pp. 409–420

D'Appolonia, D. J. (1980) Soil–bentonite slurry trench cutoffs, *ASCE, Journal of the Geotechnical Engineering Division*, **106,** 64–65

Demetracopoulus, A. C. and Korfiatis, G. P. (1984) Design considerations for landfill bottom collection system, *Civil Engineering for Practicing and Design Engineers*, **3**, 967–984

Design of Small Dams (1977) 2nd edn, United States Department of the Interior, Bureau of Reclamation, p. 816

Eliassen, R. (1947) Why you should avoid housing construction on refuse landfills, *Engineering News Record*, 1 May, 90–94

Elzeftway, A. and Cartwright, K. (1979) Evaluating the saturated and unsaturated hydraulic conductivity of soil, *ASTM, Permeability and Groundwater Contaminant Transport*, **STP 746,** 168–181

Engemoen, W. O. and Hensley, P. J. (1986) ECPT investigation of slurry trench cutoff wall, *Proc. on In-Situ '86, Use of in situ Tests in Geot. Eng.*, Geot. Spec. Publ. No. 6, Ed. Clemence, S. P., pp. 514–528

Fukuoka, M. (1986) Large permeability test for geomembrane, subgrade system, *Third International Conference on Geotextiles*, Vol. III, pp. 917–922

Fungaroli, A. A. and Steiner, R. L. (1979) *Investigation of Sanitary Landfill Behavior*, Vol. 1, Final Report, USEPA 600/2-79-053a, p. 63

Gauffreau, P. E. (1987) A review of pile protection methods in a corrosive environment, *Int. Symp. on Environmental Geotechnology*, Vol. 2, Envo Publishing Co. Inc., pp. 372

Giroud, J. P. (1982) Filteration criteria for geotextile, *Proc. 2nd Int. Conf. on Geotextiles*, Vol. 1, Las Vegas, pp. 103–108

Harr, M. E. (1962) *Groundwater and Seepage*, McGraw-Hill, New York, p. 260

Hatheway, A. W. and McAneny, C. C. (1987) An in depth look at landfill covers, *Waste Age*, August, 135–156

Haxo, H. E. and Dakessian, S. (1987) Assessment of the potential for incompatibility of soil liner materials with specific organic compounds, *Proc. Hazmacon '87*, April 21–23, Associated of Bay Area Governments, ABAG, pp. 496–511

Khera, R. P., Wu, Y. H. and Umer, M. K. (1987) Durability of slurry cut-off walls around the hazardous waste sites, *Proc. 2nd Int. Conf. on New Frontiers for Hazardous Waste Management*, EPA/600/9-87/018F, pp. 433–440

Khera, R. P., Thilliyar, M. and Moradia, H. (1988) *Durability of Slurry Cut-Off Walls Around the Hazardous Waste Sites – SITE 13*, Report, NSF Industry/University Cooperative Research Center, NJIT, Newark, New Jersey

Kincaid, C. T., Morrey, J. R. and Rogers, J. E. (1984) *Geohydrochemical Models for Solute Migration*, Electric Power Research Institute, EPRI EA-3417, Vol. 1

Kmet, P., Quinn, K. J. and Slavik, C. (1981) Analysis and design parameters affecting the collection efficiency of clay lined landfills, *4th Annual Madison Waste Conference*, 28–30 September 1981, pp. 204–227

Koerner, R. M. (1986) *Design With Geosynthetics*, Prentice-Hall, Englewood Cliffs, New Jersey

Leonards, G. A. and Narain, J. (1963) Flexibility of clay and cracking of earth dams, *ASCE, Journal of the Soil Mechanics and Foundation Division*, **89,** SM 2

Lu, C. S. J., Eichenberger, B. and Stearns, R. J. (1985) *Leachate from Municipal Landfill*, Noyer Publication

Lutton, R. J., Regan, G. L. and Jones, L. W. (1979) *Design and Construction of Covers for Solid Waste Landfills*, EPA 600/2-79-165

Matrecon Inc. (1980) *Lining of Waste Impoundment and Disposal Facilities*, EPA 530/SW-870, p. 11

Moore, C. A. (1983) *Landfill and Surface Impoundment Performance Evaluation Manual*, USEPA, SW-869

Moulton, L. K., Rao, S. K. and Seals, R. K. (1976) The use of coal associated wastes in the construction and stabilization of refuse landfills, *New Horizons in Construction Materials*, Ed. Fang, H. Y., Envo Publishing Co. Inc., Vol. 1, pp. 53–65

Murphy, W. L. and Gilbert, W. L. (1985) *Settlement and Cover Subsidence of Hazardous Waste Landfills*, EPA/600/2-85-035, p. 6

National Sanitation Foundation (1985) *Standard Number 54, Flexible Membrane Liners*, National Sanitation Foundation, Ann Arbor, Michigan

NAVFAC DM 7.1 (1982) *Soil Mechanics Design Manual*, Naval Facilities Engineering Command

Newman, F. B., McGee, J. and Burns, D. (1987) Embankment over fly ash pond at Portsmouth power station, *Geotechnical Practice for Waste Disposal* '87, Proc. Spec. Conf. Geot. Spec. Publ. No. 13, Ed. Woods, R. D., pp. 713–727

Norton Construction Company (1983) *Feasibility Study for Utilization of Landfill Gas at the Royaltor Road Landfill, Broadview Height, Ohio,* US Department of Energy, DOE/RA/50363-1, Washington, DC

Nussbaumer, M. (1987) Beispiele für die Herstellung von Dichtwänden im Schlitzwandverfahren, *Mitteilung des Instituts für Grundbau und Bodernmechanik*, TU Braunschweig, Vol. 23. Ed. Meseck, H., Dichtwände und Dichtsohlen, Braunschweig, Federal Republic of Germany, pp. 21–34

Olson, R. E. and Daniel, D. E. (1981) Measurement of the hydraulic conductivity of fine-grained soils, *ASTM Permeability and Groundwater Contaminant Transport*, **STP 746,** 18–64

Oweis, I. S., Smith, D. and Ellwood, B. (1990) Hydraulic properties of refuse, Submitted for possible publication

Oweis, I. S. (1988) Sanitary landfill clay caps: do they inhibit leachate generation? *4th Int. Conf. on Urban Solid Waste Management and Secondary Materials*, Philadelphia, December 1988

Oweis, I. S., Mills, W. T., Leung, A. and Scarino, J. (1985) Stability of sanitary landfills, *Geotechnical Aspects of Waste Management*, Seminar, Met. Section, ASCE, December 1985

Rollin, A. L. and Denis, R. (1987) Geosynthetic filtration in landfill design, *Conference Proceedings Geosynthetics '87*, Industrial Fabrics Association International, 24–25 February, Vol. 2, pp. 456–470

Rowe, R. K. (1987) Pollutant transport through barriers, *Geotechnical Practice for Waste Disposal '87*, Proc. Specialty Conference, ASCE, 15–17 June 1987, pp. 159–181

Ryan, C. R. (1987) Vertical barriers in soil for pollution containment, *Geotechnical Practice for Waste Disposal '87*, Proc. Spec. Conf. Geot. Spec. Publ. No. 13, Ed. Woods, R. D., pp. 182–204.

Schroeder, P. R., Morgan, J. M., Walski, T. M. and Gibson, A. C. (1983) *Draft Manual, The Hydrologic Evaluation of Landfill Performance (HELP) Model*

Smith, A. D. and Kraemer, S. R. (1987) Creep by geocomposite drains, *Conf. Proc., Geosynthetics '87*, Industrial Fabrics Association International, 24–25 February 1987, Vol. 2

Taylor, M. J. and D'Appolonia, E. (1977) Integrated solution to tailings disposal, *Proc. Geotechnical Practice for Disposal of Solid Waste Materials*, ASCE, University of Michigan, Ann Arbor, Michigan, June 1977, pp. 301–326

Teeter, R. M. and Clemence, S. P. (1986) In place permeability measurements of slurry trench cutoff walls, *Proceedings of Institute '86*, a speciality conference sponsored by the Geotechnical Engineering Division, ASCE, pp. 1049–1069

Timoshenko, S. and MacCullough, G. H. (1949) *Elements of Strength of Materials*, Van Nostrand, p. 197

US Army Corps of Engineers (1977) *Plastic Filter Cloth*, Civil Works Construction Guide Presentation No. CW-02215, Chief of Engineers, Washington, DC

US Environmental Protection Agency (1985) *Draft minimum technology guidance on double-liner systems for landfills and surface impounds . . . Design, construction, and operation*, EPA/530-SW-85-012

Wong, J. (1977) The design of a system for collecting leachate from a lined landfill site, *Water Resources Research*, **13,** No. 2

Xanthakos, P. P. (1979) *Slurry Wall*, McGraw-Hill, New York

Index